Encyclopedia of Insecticides: Theory and Applications

Volume I

Encyclopedia of Insecticides: Theory and Applications

Volume I

Edited by **Nancy Cahoy**

New York

Published by Callisto Reference,
106 Park Avenue, Suite 200,
New York, NY 10016, USA
www.callistoreference.com

Encyclopedia of Insecticides: Theory and Applications
Volume I
Edited by Nancy Cahoy

International Standard Book Number: 978-1-63239-262-6 (Hardback)

Contents

Preface

The aim of this book is to educate the readers regarding theories and applications in the field of insecticides. This book has been compiled with the purpose of providing benefit to researchers, food-producing factories, government, leaders and consumers also. If the book proves to be satisfactory to these readers, then we will consider that the ambition with which this book and its content were written, has been fulfilled.

The researches compiled throughout the book are authentic and of high quality, combining several disciplines and from very diverse regions from around the world. Drawing on the contributions of many researchers from diverse countries, the book's objective is to provide the readers with the latest achievements in the area of research. This book will surely be a source of knowledge to all interested and researching the field.

In the end, I would like to express my deep sense of gratitude to all the authors for meeting the set deadlines in completing and submitting their research chapters. I would also like to thank the publisher for the support offered to us throughout the course of the book. Finally, I extend my sincere thanks to my family for being a constant source of inspiration and encouragement.

Editor

Part 1

Basic and Alternative Control of Insect Pests

Insecticide Thiamethoxam: A Bioactive Action on Carrot Seeds (*Daucus carota* L.)

Andréia da Silva Almeida[1], Francisco Amaral Villela[2],
Maria Ângela André Tillmann[2] and Geri Eduardo Meneghello[2]
[1]Ciência e Tecnologia de Sementes;
[2]Ciência e Tecnologia de Sementes;
University Federal de Pelotas
Brazil

1. Introduction

Explorations of the substances potential that can act as promoters of growth-inducing activities are in substantial contribution of research to agriculture. Many of these activities are related to the activation capacity of the plants resistance mechanisms, thus enabling seek control through integrated management.

With the modernization of agriculture, marked advances in farming techniques have been obtained, allowing mitigate the limiting factors with weather conditions such diseases, pests, among others. The plant physiology has promoted great advances in recent years with the advent of modern techniques such as the production of plants by tissue culture, genetic engineering and biotechnology. Among these modern techniques, the use of bioactive, capable of increasing the productive potential of plants, is an increasing use in the practice of modern agriculture and widespread in countries highly technical.

In Brazil, the use of bioactive beginning to be explored and the results of several studies have shown that these substances provide significant increases in productivity and, in quality, as observed, for example, significant increases in the amount of larger fruits. Bioactivators are natural substances of plant origin that have actions similar to the main plant growth regulators, aimed at growth and development of the plant. Provide better physiological balance, favoring closer ties to the genetic potential of culture.

Moreover, they are complex organic substances, not bioregulators, growth modifiers, capable of working in the plant transcription factors and gene expression in membrane proteins by altering the ion transport. They also act in metabolic enzymes could affect the secondary metabolism and may alter the mineral nutrition, induce the production of precursors of plant hormones, leading to hormone synthesis and more intense response to nutrients and plant hormones.

Applied to plants, bioactivators cause modification or alteration of specific metabolic and physiological processes, such as increasing the division and cell elongation, stimulation of chlorophyll synthesis and photosynthesis, flower bud differentiation, increasing the life of plants, softening the effects of adverse weather conditions and increasing the absorption of nutrients and setting their roots.

The bioactivator acts in the expression of genes responsible for synthesis and activation of metabolic enzymes related to plant growth by altering the production of amino acid precursors of plant hormones. With the increased production of hormones, plant expressed greater vigor, germination and root development. With a greater number of roots, increases the absorption and resistance of plant stomata to water loss, which benefits the metabolism and increases resistance to stress.

Thiamethoxam, 3-(2 – chloro – thiazole - 5 - ylmethyl) - 5 - methyl (1,3,5) oxadiazinan - 4 - ylidene-N- nitroamine, whose chemical structure is shown in Figure 1, is a systemic insecticide neonicotinoid group, family nitroguanidine, which acts on acetylcholine nicotinic receptor in membrane of insects, damaging the nervous system and causing them to death. It is used successfully in pest control in beginning cycle from different cultures. Due to numerous reports of field observations describing increases in vigor development and productivity, even in the absence of pests, has been considered a product that has phytotonic effect.

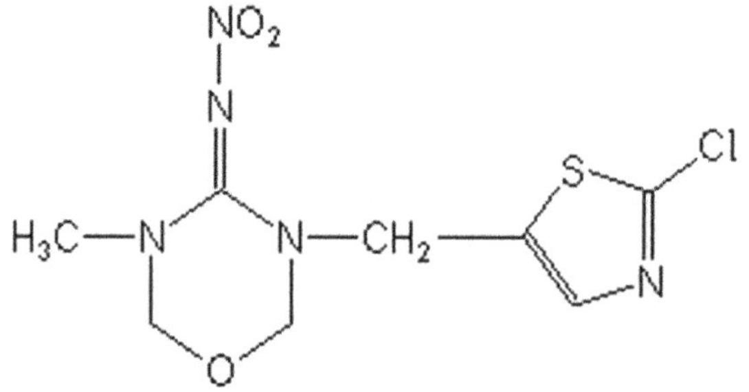

Fig. 1. Chemical structure of thiamethoxam. Source: Oliveira, et al., (2009).

With reference to the mechanism of action of thiamethoxam, the molecule has the ability to induce physiological changes in plants. Due to results obtained, it is concluded that this insecticide acts as bioactive two ways. The first, activating carrier proteins of cell membranes allowing greater ion transport by enhancing plant mineral nutrition. This increase in the availability of minerals promotes positive responses in the development and plant productivity. The second is related to increased enzyme activation caused by thiamethoxam at both the seed and the plant, thereby increasing both primary and secondary metabolism. It also increases the synthesis of amino acids, precursors of new proteins and endogenous synthesis of plant hormones. The responses of plants to these proteins and hormone biosynthesis may be related to significant increases in production.

Results of research related with soybean (*Glycine max*) (Castro et al., 2008; Cataneo, 2008), rice (*Oryza sativa*) (Clavijo, 2008; Almeida et al.; 2010), cotton (*Gossypium hirsutum*) (Lauxen et al., 2010), bean (*Phaseolus vulgaris*) (Almeida et al., 2010) e vegetables like, lettuce (*Lactuca sativa*), tomato (*Lycopersicon esculentum*), pumpkin (*Cucurbita pepo* L.) and carrot (*Daucus carota*) indicate benefit effect of thiamethoxam uses.

2. Bioactivator in physiological performance of carrot seeds

Carrot (*Daucus carota* L.) is the most economical expression of vegetables among those whose edible portion is the root, and to highlight the nutritional value, as a major vegetable sources of provitamin A (Spinola et al, 1998). Careful selection of cultivars allows sowing of carrots over the years in many regions (Filgueira, 2000).

The success of horticulture generally depends on the establishment of suitable stand for each crop; otherwise reductions may occur in the quantity and quality of final product (Silva and Vieira, 2006).

Under field conditions, carrot seed germination may have low, slow and irregular, resulting in uneven emergence and a heterogeneous population of plants (Corbineau et al., 1994).

With increasing mechanization in vegetable production, establish rapid and uniform culture becomes increasingly important, and it is desirable that the evaluation of seed quality to provide information on their performance on the field.

Establishment of appropriate stand depends on the use of seeds with high physiological potential, able to germinate rapidly and uniformly under a large variation of the environment. Speed and timing are very important because they allow reducing the degree of exposure of seeds and plants to adverse factors (Marcos Filho, 2005).

Reduced or uneven emergence can lead to developmental delays, problems with weed control, non-uniformity of culture in different phenological stages, interference on product quality and characteristics related to the efficiency of harvesting (Marcos Filho, 2005).

In vegetables, backwardness and uneven development may be reflected in product quality and reducing the commercial value, such as lettuce, cabbage, carrots, cauliflower, eggplant and onion (Kikuta and Marcos Filho, 2007).

A survey on physiological quality of seeds of different kinds of vegetables such as carrots, peas, beets, tomato and watermelon industry, marketed and / or used by farmers, it was observed that germination of these seeds do not always fit the minimum standard of marketing required for each species (Nascimento, 1994).

Therefore, failure to stand and seedling vigor at low field level are frequent, with the need for appropriate and sensitive methods to detect these differences in seed quality.

In this context, considering the lack of information concerning the effect of thiamethoxam and the potential benefits that treatment can provide, the present study was to evaluate the influence of thiamethoxam in physiological performance of carrot seeds.

3. Methodology

This work was conducted at the Laboratory of Seed Analysis Textbook and greenhouse, Faculty of Agronomy Eliseu Maciel, Universidade Federal de Pelotas (UFPel), Pelotas / Brazil.

Seed lots of carrot cultivar Brasilia represented by four lots.

To establishment concentrations of thiamethoxam, it was used the following concentrations: 0.0, 0.05, 0.1, 0.2, 0.4, 0.8 mL / L, based on germination test three were selected. Tests conducted to evaluate the quality of seeds were performed with and without water stress.

The seeds were treated in a plastic bag containing distilled water at 0.1 mL of water to 0.05 mL and 0.4 mL of product thiamethoxam to 3g of carrot seeds. The product was applied directly to the bottom of the plastic bag before putting the seeds. Then the seeds were placed in plastic bag and mixed until uniform coating of the seeds. It was used a volume of mixture (product + water) sufficient to promote a more even distribution of product on the seeds. For measurement of product and distilled water were used micropipettes.

Water stress was achieved by the water potential of -0.4 MPa, using aqueous solutions of polyethylene glycol (PEG 6000). The calculation of the solute quantities was performed according to Villela et al. (1991). Polyethylene glycol solutions, thus obtained were applied to the paper substrate, in an amount equivalent to 2.5 times its dry weight in all parameters evaluated in the laboratory involving the germination test.

To evaluate the physiological quality of seeds were conducted the following tests:

Germination: it was used four replications of 50 seeds of each batch distributed in transparent plastic boxes (gerbox) on two sheets of white blotter paper, moistened 2.5 times the weight of paper, placed in a germination chamber set to maintain the temperature constant 25 ° C. Counts were made in the seventh and fourteenth days after sowing, and assessments, carried out according to ISTA (2010) by computing the percentages of normal seedlings.

Accelerated aging: conducted with 4.0 g of seeds distributed in wire screen suspended and placed inside plastic boxes, type gerbox (mini-camera). Inside the germination boxes were placed 40 mL of water and then the boxes were taken to an incubator set at a constant temperature of 41° C for 48 hours and subsequently subjected to germination tests, as described above. The evaluation was performed seven days after sowing, by computing the percentage of normal seedlings.

Root length: four replicates of 50 seeds were sown on a line drawn in the upper third part of paper substrate. The rolls containing the seeds remained at 25 ° C for seven days, after being evaluated, the root length of normal seedlings, with the aid of a millimeter ruler. The root length was obtained by adding the measurements of each replicate and dividing by the number of seedlings, with results expressed in centimeters.

Speed of germination: performed according to the methodology of the germination test, determined by daily counts to stabilize the number of seedlings in the test and the speed calculation made according to Maguire (1962).

Emergence of seedlings in the greenhouse: four replications of 50 seeds were distributed in individual cells of polystyrene trays (Styrofoam), containing commercial substrate Plantimax ®. The trays were kept in the greenhouse and evaluations were performed at 16 days after sowing, counting seedlings in length and more than 1.0 cm. The results were expressed as a percentage of emergence.

Statistical procedure: completely randomized factorial 4x3 (four lots and three concentrations of the product) separately in the evaluation with and without water stress, with three replications. For comparison of means between control and concentrations, it was used Dunnet test, probability level of 5%.

4. Results establishment of concentrations

The concentrations selected based on the result of germination test of treated seeds with different concentrations of the product thiamethoxam, beyond control were 0.05 and 0.4 mL.

Germination of treated seeds in accordance with product concentrations were 70% (zero), 75% (0.05 mL / L), 72% (0.1 mL / L), 72% (0.2 mL / L), 75% (0.4 mL / L) and 70% (0.8 mL / L) without stress. The choice of the concentration of 0.05 mL / L was based on the fact that seeds showed germination similar to other concentrations and spend less of the product. On the other hand, the concentration of 0.4 mL / L was selected because germination test were seedlings showed well developed, open cotyledons and normal roots. At the concentration 0.8 mL / L was found that the seedlings were developed, but their roots had necrosis.

Statistical analysis performed by Dunnet's test showed significant results for comparison of means between control and concentrations of all parameters.

It is noted in Figure 1 that germination of four seed lots, without (Figure 1A) and with (Figure 1B) water stress, treated with thiamethoxam showed significant difference compared to control. Increases in germination were marked and varied according to the lots from 5 to 23 percentage points if the seeds have not been subjected to water stress and 4 to 15 to be subjected to stress.

In Figure 1B it appears that water stress reduced the percentage germination of seed lots. Lots 1 and 3 not treated, after the water stress reached below the standard of marketing, however, treatment stimulated germination of seeds and lots reached the minimum germination (70%) of the standard marketing. In soybean seeds was also observed that thiamethoxam accelerates germination, induces more growth of the embryonic axis and minimize the negative effects in situations of presence of aluminum, salinity and water deficiency (Cataneo et al., 2006).

There is a trend of germination of treated lots with different concentrations of the product showed similar results, with the exception of Lot 3, in which the concentration of 0.4 mL / L was more efficient.

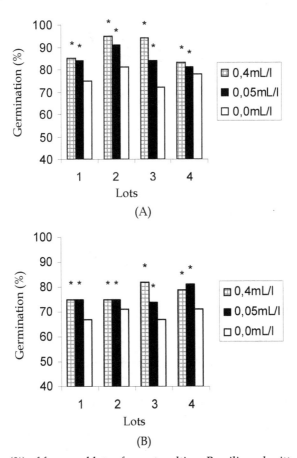

Fig. 1. Germination (%) of four seed lots of carrot, cultivar Brasilia, submitted (A) or not (B) to water stress. * Different from the control by Dunnet test at probability level of 5%.

According to Figure 2, germination after accelerated aging of treated seeds without (Figure 2A) and with (Figure 2B) water stress showed significant differences related to control. Positive difference varied according of lots, 2 to 11 percentage points in seed not submitted to stress and 2 to 9 in submitted to water stress. This superiority resistance occurs because thiamethoxam move through plant cells and actives several physiological reactions, such as functional protein expression related with plant defense mechanism avoid stress factors like drought, high temperatures, toxic effects, among others, improving productivity, leaf and radicular area, as found in soybean seed (Tavares e Castro, 2005).

Concentrations showed positive results in situations with and without water stress, but the concentration of 0.4 mL / L performed better for lots 2 and 3, without stress and 2, 3 and 4 with stress.

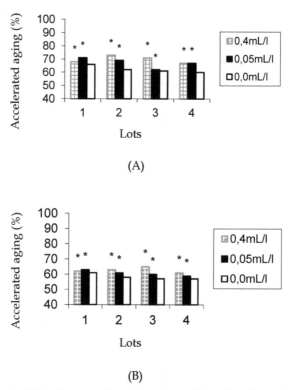

(A)

(B)

Fig. 2. Accelerated aging (%) of four seed lots of carrot, cultivar Brasilia, without (A) and with (B) water stress.* It differs from the control by Dunnet test at probability level of 5%.

According to data presented in Figures 3 and 4, treated seeds showed marked differences in root length compared to untreated, on average 4 cm, in both cases without (Figure 3 and 4 A) and with (Figures 3 and 4 B) water stress. This effect of thiamethoxam of supporting the growth of the root system, confirming the effect of rooting observed by Pereira et al. (2007) in sugar cane and potatoes; and also by Tavares et al. (2007) in soybean. It is believed that the thiamethoxam increase water uptake and stomatal resistance, improving the water balance of the plant, tolerating water deficit better (Castro, 2006). As observed in soybean root development increases the absorption of nutrients, increases the expression of leaf area and plant vigor (Tavares and Castro, 2005).

The data speed of germination, without (Figure 5A) and with (Figure 5B) stress show that the treated seeds had a higher rate compared to control. The concentrations used had similar results. Treated seeds germinated on average one day soon if they have not been subjected to water stress and two days are subject to stress. This effect is very promising because carrot seeds in field conditions have poor germination, slow and irregular resulting in uneven emergence (Corbineau et al., 1994). This increased speed of germination is caused by physiological changes that occur in the plant indirectly stimulating the production of hormones, resulting in increased vigor, root growth, water absorption and primary and secondary metabolism, as observed in the sugarcane crop (Castro, 2007).

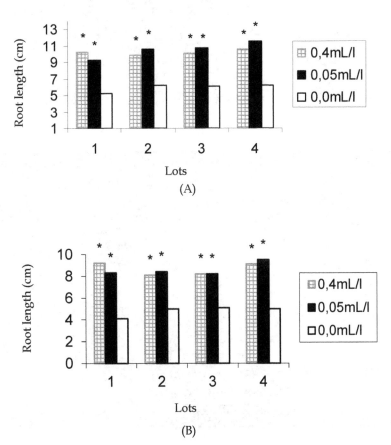

Fig. 3. Root length (cm) of seedlings of four seed lots of carrot, cultivar Brasilia, without (A) and with (B) water stress. * Different from the control by Dunnet test at probability level of 5%.

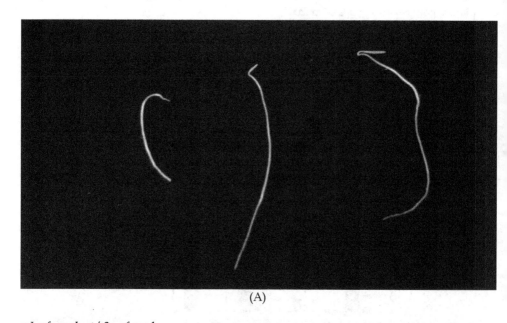

(A)

mL of product/ 3g of seed

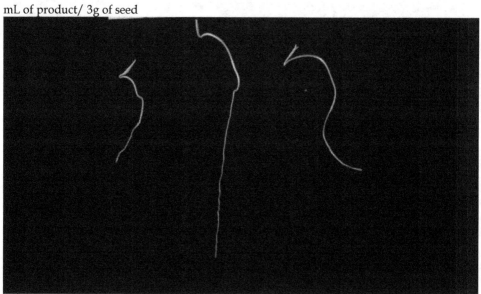

Fig. 4. Root length (cm) of seedlings of four seed lots of carrot, cultivar Brasilia, without (A) and with (B) water stress.

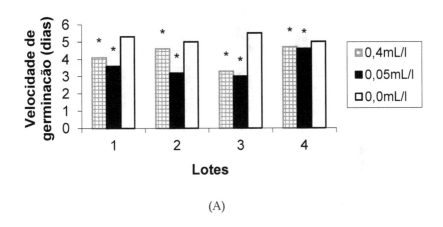

(A)

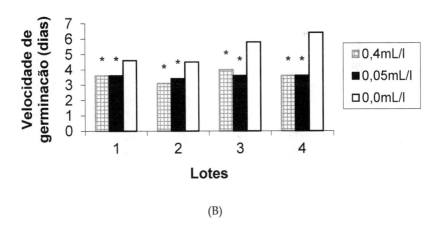

(B)

Fig. 5. Speed of germination (days) of four seed lots of carrot cultivar Brasilia, without (A) and with (B) water stress.* It differs from the control by Dunnet test at probability level of 5%.

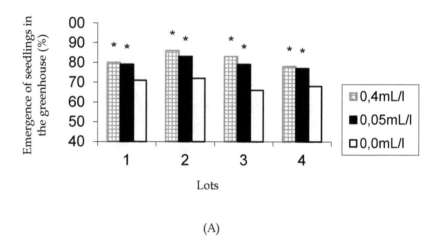

(A)

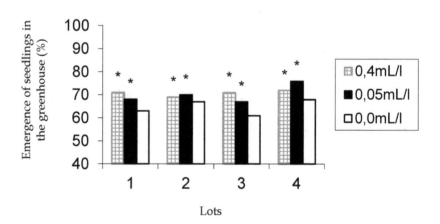

Fig. 6. Emergence of seedlings in the greenhouse for four seed lots of carrot, cultivar Brasilia without (A) and with (B) water stress. * Different from the control by Dunnet test at probability level of 5%.

In Figure 6, without (Figure 6A) and with (Figure 6B) water stress, it was observed that the emergence of seedlings in the greenhouse was stimulated, and the seeds treated with thiamethoxam showed significant differences compared to control. The positive differences compared to control vary according to lots, 9 to 17 percentage points if the seeds have not been subjected to water stress and 20 to 10 percentage points when subjected to stress. The two concentrations showed similar responses. These results confirm those found in soybean, to be seen increase in the root system and the percentage of seedling emergence also in water deficit conditions (Castro et al., 2006). According to the literature, soybean seeds treated with thiamethoxam have higher levels of amino acids, enzyme activity and synthesis of plant hormones that increase the plant responses to these proteins and these events provide significant increases in production and reducing the time of establishment of culture in the field, making it more tolerant to stress factors (Castro, 2006).

The results obtained can be described that the product stimulated the performance of carrot seeds in all parameters evaluated, both in seeds subjected to water stress or not. Carrot seeds treated with the product thiamethoxam showed significant increases in germination and vigor for all lots. Among the aspects of vigor, the product stimulated the growth of the root length, which is of great importance to the culture of carrots and this result was obtained in the laboratory confirmed in the greenhouse.

The product was more effective in stimulating the quality of seeds not subjected to water stress, with the exception of root length which positive change was similar for seeds subjected to stress or not. In all parameters evaluated, increases in the quality varied according to the lot. Concentrations of the product for most tests evaluated did not differ, however there was a trend of higher concentration to the higher values.

The application of thiamethoxam has strong interest for the culture of carrot, whose edible portion is the root and, moreover, by presenting, in field conditions, poor germination, slow, irregular with uneven emergence, the product acts as an enhancer, by allowing the expression of seed germination potential, accelerate the growth of roots and increase the absorption of nutrients by the plant. These features of thiamethoxam combined with the use of genetics and physiological high-quality seed powers the productive capacity of the culture.

5. Conclusions

Thiamethoxam product stimulates the physiological performance of carrot seeds subjected to water stress or not, with variable intensity according to lot.

Concentrations of 0.05 and 0.4 mL of the product is effective, however there is a tendency of higher concentration to the higher increases in quality.

6. References

ALMEIDA, A.S.; TILLMANN, M.A.A.; VILLELA, F. A.; PINHO, M.S. Bioativador no desempenho fisiológico de sementes de cenoura. Revista Brasileira de Sementes, Brasília,v.31, n. 3, p. 87-95, 2009.

ANANIA, F.R.; TEIXEIRA, N.T.; CALAFIORI, M.H.; ZAMBON,S. Influência de inseticidas granulados sistêmicos nos teores de N-P-K nas folhas de amendoim (*Arachis hypogaea* L.) Ecossistema, Espírito Santo do Pinhal, v. 13, p. 121-124, 1988a.

ANANIA, P,F.R.; TEIXEIRA, N.T.; CALAFIORI, M.H.; ZAMBON,S. Influência de inseticidas granulados sistêmicos nos teores de N-P-K nas folhas de limoeiro Taiti (*Citrus aurantifolia*.) cv. Peruano. Ecossistema, Espírito Santo do Pinhal, v. 13, p. 121-124, 1988b.

CALAFIORI, M.H; TEIXEIRA, N.T; SCHMIDT, H A P.; ANANIA, P.F.R.; GRANDO, F.I.; PALAZZINI, R.; MARTINS, R.C.; OLIVEIRA, C.L.; ZAMBON, S. Efeitos nutricionais de inseticidas sistêmicos granulados sobre cafeeiros. Ecossistema. Espírito Santo do Pinhal, v.14.p. 132-14, 1989.

CASTRO, P.R.C.; PITELLI, A M.C.M.; PERES, L.E.P.; ARAMAKI, P.H. Análise da atividade hormonal de thiametoxam através de biotestes. Publicatio, UEPG, 2007

CASTRO, P.R.C. Agroquimicos de controle hormonal na agricultura tropical. Boletim, n.32, Série Produtor Rural, USP/ ESALQ/ DIBD, Piracicaba, 46p., 2006.

CASTRO, P.R.C.; PITELLI, AM.C.M.; PERES, L.E.P. Avaliação do crescimento da raiz e parte aérea de plântulas de tomateiro MT, DGT E BRT germinadas em diferentes concentrações do inseticida thiametoxan. In ESCOLA SUPERIOR DE AGRICULTURA "LUIZ DE QUEIROZ". Relatório técnico ESALQ/Syngenta. Piracicaba, p.14-25, 2005.

CASTRO, P.R.C.; SOARES, F.C.; ZAMBON, S.; MARTINS, A N.; Efeito do aldicarb no desenvolvimento do feijoeiro cultivar Carioca. Ecossistema. Espírito Santo do Pinhal, v.20, p. 63-68, 1995.

CATANEO, A C.; ANDRÉO, Y.; SEIFFERT, M.; BÚFALO,J.; FERREIRA,L.C. Ação do inseticida Cruiser sobre a germinação do soja em condições de estresse. In: IVCONGRESSO BRASILEIRO DE SOJA, Resumos, Londrina, p.90, 2006.

CORBINEAU, F.; PICARDE, M.A.; CÔME, D. Effects of temperature, oxigen and osmotic pressure on germination of carrot seeds: evaluation of seed quality. Acta Horticulturae, The Hague, v.354, p.9-15, 1994.

De GRANDE, P.E. Influência de aldicarb e carbofuran na soja (*Glycine max* L.) Merrill. 137f. Dissertação (Mestrado em Entomologia) - Escola Superior de Agricultura "Luiz de Queiroz", Universidade de São Paulo, Piracicaba, 1992.

DENARDIN, N.D. Ação do thiametoxan sobre a fixação biológica do nitrogênio e na promoção de ativadores de crescimento vegetal. In: Universidade de Passo Fundo. Relatório técnico, Passo Fundo, 2005.

HORII, A; McCUE, P.; SHETTY, K. Enhancement of seed vigour following and phenolic elicitor treatment. Bioresource Technology, United States, v.98, n.3, p.623-632, 2007.

JUNQUEIRA, F.M.A; FORNER, M.A; CALAFIORI, M.H.; TEIXEIRA, N.T.; ZAMBON, S.; Aplicação de aldicarb em diferentes dosagens e tipos de adubação influenciando a produtividade na cultura da batata (*Solarium tuberosum* L.). Ecossistema, Espírito Santo do Pinhal, v. 13, p. 101-107, 1988.

LAUXEN, L.R.; VILLELA, F. A.; SOARES, R. C. Desempenho fisiológico de sementes de algodão tratadas com tiametoxam. Revista Brasileira de Sementes. Brasília, v. 32, n. 3, p. 61-68 , 2010.

LUBUS, C.A.F.; FERRAZ, J.A.D.P.; CALAFIORI, M.H.; ZAMBON, S.; BUENO, B.F. Ensaio com diferentes dosagens de aldicard e de adubo visando a produtividade na cultura da batata (Solarium tuberosum L.), Ecossistema, Espírito Santo do Pinhal, v. 10, p. 64-66, 1985.

MAGUIRE, J.D Speed of germination and in selection and evaluation for seedling emergence and vigor. Crop Science, Madison, v.2, n.2, p.176-177, 1962.

NUNES, J.C. Bioativador de plantas: uma utilidade adicional para um produto desenvolvido originalmente como inseticida. Revista SEEDNews, Pelotas, v.10, n.5, p.30-31, 2006.

OLIVEIRA, V.S.; LIMA, J.M.; CARVALHO, R.F.; RIGITANO, R.L.O. Absorção do inseticida tiametoxam em latossolos sob efeito de fosfato e vinhaça. Revista Química Nova, Lavras, v. 32, n. 6, p. 1432-1435, 2009.

PEREIRA, M.A.; CASTRO, P.R.C.; GARCIA, E.O; REIS, A. R. Efeitos fisiológicos de Thiametoxan em plantas de feijoeiro. In: XI CONGRESSO BRASILEIRO DE FISIOLOGIA VEGETAL, Resumos, Gramado: Sociedade Brasileira de Fisiologia Vegetal, 2007.

REDDY, K.R.; REDDY, V.R.; BAKER, D.N.; McKINION, J.M. Effects of aldicarb on photosynthesis, root growth and flowering of cotton. In: PLANT GROWTH REGULATION SOCIETY OF AMERICAN ANNUAL MEETING, 16., Arlington. Proceedings... Arligton: Plant Regulation Society of American, p.168-169, 1989.

REDDY, K.R.; REDDY, V.R.; BAKER, D.N.; McKINION, J.M. Is aldicarb a plant growth regulator. In PLANT GROWTH REGULATION SOCIETY OF AMERICAN ANNUAL MEETING, 17., Proceedings... Saint Paul: Plant Regulation Society of American, p.79-80, 1990.

TAVARES, S.; CASTRO, P.R.C.; RIBEIRO, R.V.; ARAMAKI, P.H. Avaliação dos efeitos fisiológicos do tiametoxam no tratamento de sementes de soja. Revista da Agricultura, Piracicaba, 2007.

TAVARES, S.; CASTRO, P.R.C. Avaliação dos efeitos fisiológicos de Cruiser 35FS após tratamento de sementes de soja. In: ESCOLA SUPERIOR DE AGRICULTURA "LUIZ DE QUEIROZ". Relatório técnico ESALQ/Syngenta Piracicaba, p. 1-13, 2005.

TEIXEIRA, N.T.; ZAMBON, S.; BOLLELA, E.R,; NAKANO; OLIVEIRA, D.A; CALAFIORI, M.H. Adubação e aldicarb influenciando os teores de N, P e K, nas folhas da cultura da batata (Solarium tuberosum L). Ecossistema, Espírito Santo do Pinhal, v.16, p.120-125, 1991.

VILLELA, F.A; DONI-FILHO,L,; SEQUEIRA,E.L. Tabela de potencial osmótico em função da concentração de polietileno glicol 6000 e da temperatura. Pesquisa Agropecuária Brasileira, Brasília, v.26,n.11/12,p.1957-1968, 1991.

WHEATON, T. A; CHILDERS, C.C.; TIMMER, L.W.; DUNCAN, L.W.; NIKDEL, S. Effects of aldicarb on the production, quality of fruits and situation of citrus plants in Florida. Proceedings of the Florida State for Horticultural Society, Tallahasse, v. 98, p. 6-10, 1985.

2

Tree Injection as an Alternative Method of Insecticide Application

Joseph J. Doccola and Peter M. Wild
Arborjet, Inc. Woburn, MA
USA

1. Introduction

Injection directly into the conductive tissues of trees was a method first investigated systematically by Leonardo da Vinci, but some of the most early tree injection experiments were not recorded until early in the 20th century (Roach, 1939, May, 1941, Costonis, 1981). Dutch elm disease, a destructive vascular wilt disease of elm renewed interest in tree injection in the 1970s (Jones and Gregory, 1971; McWain and Gregory, 1971; Jones et al., 1973; Gregory et al., 1973; Gregory and Jones, 1975; Shigo and Campana, 1979; Kielbaso et al. 1979; Shigo et al., 1980), when more common fungicide applications proved ineffective. During this time, several injection methods, including trunk infusion (Schreiber 1969), and pressurized trunk injections (Filer 1973; Helburg et al. 1973; Reil and Beutel 1976, Sachs et al., 1977; Kondo, 1978, Darvas et al., 1984, Navarro et al., 1992), were developed. Tree injection was also used for treatment of other tree pathogens (Guest et al., 1994; Fernández-Escobar et al.1994, 1999), insects, and physiological disorders (i.e., interveinal chlorosis) in the EU (Fernández-Escobar et al. 1993). Interest in tree injection technologies (McClure, 1992, Doccola et al., 2007; Smitley et al., 2010) in the US has also increased, with the introduction of several tree killing insects such as hemlock woolly adelgid (*Adelges tsugae*), Asian longhorned beetle (*Anoplophora glabripennis*) and emerald ash borer (*Agrilus planipennis*). In addition to new injection technology, formulations are being designed for injecting into trees that improve plant safety and reduce application time. Examples of the new technologies are the TREE I.V. micro-infusion system and Air/Hydraulic micro-injector (Arborjet, Inc. Woburn, MA, USA) and the Eco-ject® Microinjection System (Bioforest Technologies, Canada). Today, tree injection is an alternative method of chemical application with certain advantages: (1) efficient use of chemicals, (2) reduced potential environmental exposure, and (3) useful when soil and foliar applications are either ineffective or difficult to apply (Stipes, 1988; Sanchez-Zamora and Fernandez-Escobar, 2004). Tree injection into roots, trunks or limbs requires wounding of the tree, which has implications to the tree's health. The question often asked is, does the benefit gained by tree injection outweigh the risk of the wound caused by treatment? This question of cost-benefit is certainly valid. However, this concern must also be weighed against environmental (and off target) exposures when trees are sprayed or insecticides are applied to the soil. An underlying assumption is that the value of the tree and its treatment is greater than sustaining tree loss. Key factors weigh in to wound responses in trees that likewise demand consideration. These include (1) the tree species, (2) tree health, (3) the attributes of the chemistry applied and (4) the frequency that applications are made. Such issues present a broader and more complex paradigm and carry over into tree injection practices. In order to apply tree injections effectively, one needs a basic understanding of the (1) method of application, (2) the chemistry applied, and (3) tree condition. The aim of this paper is to recommend tree injection as an alternative application method for systemic insecticides to (1) protect trees against destructive insects, (2) to minimize potential environmental exposures, and (3) to manage tree wound responses.

2. Tree anatomy and physiology

The introduction and movement of liquid insecticides by injection is dependent upon tree vasculature. Anatomically, trees are highly connected systems (Shigo, 1989, 1991). Fibrous, non-woody roots absorb water and solutes (i.e., minerals in dissolved form) from the rhizosphere (root-soil environment). Hydraulic movement upward in the xylem is dependent upon transpiration from stomates, driven by the moisture lost from leaf surface to the ambient atmosphere (Greulach, 1973). Upward translocation of systemic insecticides also depends upon the rise of sap in trees.

Although movement of sap in the stem is generally upward (i.e., straight sectorial ascent), there is considerable variation in the path of water movement across species (Zanne et al., 2006). The ascent of water in trees follows two basic patterns, that of, spiral and vertical ascents. Systemic chemicals move upward in tree stems along the path of their respective ascents. Crown distribution of water is the most complete by spiral ascent (e.g., red oak), the least effective, by vertical ascent (e.g., white oak) (Rudinski and Vité, 1959). Spiral ascent occurs in a number of species, including conifer xylem (Kozlowski and Winget 1963, Kozlowski et al., 1967).

The size, pattern and distribution of vessels vary in trees. Hardwoods may be grouped as ring- or diffuse- porous; conifers are considered non-porous species (Chaney, 1988). Angiosperm trees have large, wide vessels associated with comparatively high flow rates, while gymnosperms rely solely on very small diameter tracheids to move water. The rate of water flow differs with tree species. Hagen-Poiseulle law describes the rate of flow as a function of the xylem radius to the 4th power (Kramer et al., 1996). Therefore hardwoods (e.g., oaks, elms) move injected liquid at a faster rate than conifers (e.g., pines, hemlocks). In feet per hour, ring porous hardwoods (red oak, ash, elm) move water at 92, 85 and 20; diffuse porous hardwoods (black walnut, maple, beech) move water at 13, 8 and 4; while conifers (pine, hemlock) move water at 6 and 3 (Coder, 1999). Conifers and diffuse porous hardwoods tend to use a larger proportion of sapwood than the ring porous hardwoods for water movement. Drilling more deeply (i.e., 30 rather than 15 mm) in these species serves to access a larger area of sapwood for the injection of systemic chemicals. Sinclair and Larsen investigated wood characteristics that correlated with ease of injection for deciduous trees and suggested the formula, relative frequency of vessels divided by specific gravity (1981).

3. Sapwood composition

Xylem (sapwood) is the conductive tissue of plants, made up of cellulose, lignin and other substances. Cellulose ($C_6H_{10}O_5)_n$ is an organic polymer made up of glucose molecules linked together in long chains (Raven, Evert & Curtis, 1981). Lignin is a complex organic polymer that functions to strengthen wood. Cellulose makes up the cell wall of plants, and is 44.4% carbon (Heukelekian, H. and S.A. Waksman. 1925). When mature, the xylem protoplast dies, leaving only cell wall. It is through the remaining lumen that water conduction occurs. The lumen simultaneously functions as a continuous and extensive conductive and adsorptive structure.

4. Soil and trunk spray applications compared to tree injection

Water soluble insecticides are differentially absorbed by tree roots comparative to insoluble chemistries such as the avermectins (Wislocki, 1989). Imidacloprid and acephate are labeled in the US for soil application, but restricted in areas of ground water concern (for example, Long Island, N.Y., U.S.). In coarse textured, sandy soils and in areas with high precipitation, there is the potential for insecticide leaching. The insecticidal treatment of eastern hemlock (*Tsuga canadensis*) for hemlock woolly adelgid (*Adelges tsugae*) is an example. Eastern hemlock is a riparian species, which grows in moist soils, and near streams and rivers. In these environments, the use of trunk sprays increases the potential for exposure to off target organisms (e.g., aquatic invertebrates, fish). Tree injection of insecticides is an alternative method of application where these conditions exist. Tree injected imidacloprid applied directly to the vascular tissues is conducted upward within those tissues; the procedure reduces the potential for unintended exposures.

5. Pros of tree injection

Canopy sprays are used to control defoliating insects, but drift and limited reach are issues in very tall (>15 meters) trees, where coverage from hydraulic sprayers is inadequate. Employing tree injections resolves these issues; the chemistries move within the vascular system into the canopy for systemic activity. Systemic injections are used to effectively control borers that feed under the bark, where active ingredients sprayed onto the surface of trees may not penetrate in biologically active concentrations. Soil applications are also used, but have a number of limitations. For example, they may be slower acting, require higher amounts of product or repeated applications, may migrate off-target, and be subject to microbial degradation. Finally, tree injections may be more economical to use. Although hydrolysis occurs within the plant, systemically injected chemistries may provide greater residual activity compared to other methods, (i.e., spray, drench) which are subject to drift, leaching, photolysis or microbial degradation. Repeated spray applications each season are necessary for adequate insect control. Aqueous photolysis and mean aerobic soil half-life of selected chemistries appear in Table 1. Soil applications of systemic insecticides are often made at significantly higher volumes (e.g., 5 to 10x) compared to tree injection in order to compensate for leaching, binding to soil particles, microbial degradation or the vagaries of pH and soil moisture. If there are good reasons to utilize tree injection, why are they not employed more often? The objection most often cited is that the application requires drilling into trees. This concern includes the physical wound, and the tree defenses triggered by the introduced formulation. Wounding in trees needs to be placed within context of other types of wounding against which trees evolved effective survival strategies. Trees are wounded in nature when insects bore into the bark and sapwood and when woodpeckers peck and bore into trees after them. People also create wounds in trees for specific purposes.

Half-lives (days)				
Insecticide	Water Sol (g/L)	$K_{o/c}$*	Aqueous Photolysis	Soil[+]
Acephate	700 (Worthing, 1987)	0.48 (Montgomery, 1993)	stable (Chevron, 1972d)	0.5 (Chevron, 1972g)
Imidacloprid	0.514 (Yen & Wendt, 1993)	300-400 (Cox et al., 1997)	3.98×10^{-2} (Anderson, 1991)	38.9 (Yoshida, 1990)
Emamectin	0.024 (Tomlin, 2004)	>25000 (Mushtaq et al. 1996)	3.6-10.9 (Mushtaq et al., 1998)	193.4 (Chukwudebe et al., 1997a)

*organic carbon adsorption coefficient
[+]mean aerobic

Table 1. Water solubility's, organic carbon adsorptions and half-lives of three chemistries systemically injected into trees.

6. Wood boring insects

Insect borers include species of Lepidoptera, Hymenoptera and Coleoptera. Borers may be further categorized as wood or cambium borers. Most native insects are opportunistic, attacking stressed and declining trees. When conditions favor epidemiology, trees are attacked and killed. Exotic insects are comparatively more aggressive and attack and kill healthy trees.

Lepidoptera: Clear-winged borers (Sessidae) include some serious pests including the ash borer (*Podesia syringae*). *Dioryctria* borers (Pyralidae) attack pines causing large masses of sap to exude. The Zimmerman pine moth (*Dioryctria zimmermani*) is a pest of Austrian and Scotch pines (*Pinus nigra, P. sylvestris*) in ornamental landscapes (Cranshaw & Leatherman, 2006).

Hymenoptera: Horntails (Siricidae) are sawflies that develop in damaged or stressed trees. A recent introduction in the US, the Sirex woodwasp (*Sirex noctilio*), a native of Europe, Asia and northern Africa has the potential to cause significant mortality in native pine stands (Haugen & Hoebeke, 2005).

Coleoptera: Several families of beetles bore into trees, which include the Scolytidae (bark beetles), Cerambycidae (Longhorned beetles or roundheaded borers), and Buprestidae (flat-headed borers). Some species vector spores of destructive pathogens.

Scolytidae: In Lodgepole pine (*Pinus contorta*) a native scolytid mountain pine beetle (*Dendroctonus ponderosae*) vectors *Ophiostoma clavigerum*, a blue staining fungus (Solheim and Krokene, 1998). MPB also infests ponderosa (*P. ponderosa*), sugar (*P. lambertiana*) and white (*P. monticola*) pines (Amman et al., 2002). An epidemic can cause widespread tree mortality. The Smaller European Elm bark beetle (*Scolytus multistriatus*) vectors spores of the bluestain fungus (*Ophiostoma novo-ulmi*) that cause Dutch elm disease, a vascular wilt disease that has devastated the American elm (*Ulmus americana*) in the United States.

Cerambycidae: Locust borer (*Megacyllene robiniae*) is a native that attacks, and can severely damage or kill stressed and healthy black locust (*Robinia pseudoacacia*) (Galford, 1984). The Asian longhorned beetle (*Anoplophora glabripennis*) was introduced from Asia (China) and identified in Brooklyn, New York in 1996. ALB has a broad host range in the US but preferentially infests maple (*Acer*), and birch (*Betula*) trees (Sawyer, 2010).

Buprestidae: Emerald Ash Borer (*Agrilus planipennis*), an exotic introduced from Asia (China) was identified in Detroit, MI in 2002 (McCullough and Siegert, 2007; Anulewicz et al., 2008.). EAB attacks native ash (*Fraxinus*) species, preferentially Green (*F. pennsylvanica*) and Black (*F. nigra*), but also White (*F. americana*) and Blue (*F. quadrangulata*) ashes. EAB mines the phloem, cambium and scores the xylem as an actively developing larva. The vascular disruption reduces water movement upward into the canopy, and photosynthate transport through the phloem; unchecked infestations result in tree death. Unlike maple and birch attacked by ALB, ash trees do not bleed and EAB larvae do not remove frass from their galleries, so there are no visible signs of early infestation. Infestations often go undetected for several years, and symptoms in ash (epicormic sprouts, bark cracks, woodpecker flecks) and signs (d-shaped exit holes) do not occur until the damage has occurred. Goldspotted Oak Borer (*Agrilus coxalis*) is native to Southeastern Arizona, detected in San Diego County, California in 2004. It attacks coast live oak (*Quercus agrifolia*), canyon live oak (*Q. chrysolepis*) and California black oak (*Q. kelloggii*). Regarded as an invasive species in California, larval feeding kills phloem and cambium, which results in crown dieback and tree mortality (Coleman & Seybold, 2008). Other Buprestid borers include the two-lined chestnut borer (*A. bilineatus*) and the bronze birch borer (*A. anxius*). Adult two-lined chestnut borers attack stressed or declining oak trees. The bronze birch borer preferentially attacks European cutleaf birches such as *Betula jacquemontii*, *B. pendula and B. pendula* 'Youngii' (Dirr, 2009).

7. Birds that drill into trees

The yellow-bellied sapsucker (*Sphyrapicus varius*) bores into the bark of trees to obtain sap. More than 250 species of woody plants are known to be attacked, but birch (*Betula* spp.), maple (*Acer* spp.) and hemlock (*Tsuga* spp.) are preferentially attacked (Ostry & Nicholls, 1978). Sapsucker damage is characterized by many closely spaced holes on the tree. The tree responds by proliferating new tissues at the wound sites. Woodpeckers feed primarily on wood boring insects. The Northern flicker (*Colaptes auratus*), Red-bellied woodpecker (*Melanerpes carolinus*), Downy woodpecker (*Picoides pubescens*), Hairy woodpecker (*Picoides villosus*) and Red-headed woodpecker (*Melanerpes erthrocephalus*) drill holes into trees to extract insects or sap (Barnes, 1989). These woodpecker behaviors are generally not regarded as detrimental to trees.

8. People drill into trees

People drill into trees for sap extraction and to apply treatments, including injection. In the northeastern US and Canada, Sugar maples (*Acer saccharum*) are tapped annually for maple syrup production. Healthy trees that are tapped according to established guidelines do not suffer adverse health effects and remain productive (Davenport & Staats, 1998), some for over 100 years. Arborists drill into trees to install cabling and lightning protection (ANSI A300 Part 3, 2006; ANSI A300 Part 4, 2008). Tree care specialists treat by injection to protect trees against destructive pests. In the US, destructive, exotic insects such as hemlock woolly adelgid (USDA/FS 2003), Asian long-horned beetle (USDA/FS 2008) and emerald ash borer (USDA/FS 2008a) have recently renewed interest in tree injection technology as an alternative method of insecticide application (McClure, 1992, Doccola et al., 2007; Smitley et al., 2010). To apply tree injections effectively, one needs a basic understanding of the (1) method of application, (2) the chemistry applied, and (3) tree condition.

9. Tree injection methodology

Systemic tree injections effectively treat destructive insect pests of trees. Examples of the new technologies are the TREE I.V. micro-infusion system and the Air/Hydraulic micro-injector (Arborjet, Inc. Woburn, MA, USA) and the Eco-ject® Micro-injection System (Bioforest Technologies, Inc., Canada). The TREE I.V. micro-infusion system and Air/Hydraulic micro-injector deliver 0.50 and 2.0 liters at injection pressures of 172 to 1379 kPa, respectively. These methods require the insertion of an interface into the sapwood (Arborplug™) to inject a systemic insecticide. The Arborplug has an internal rubber septum which is pierced by an injector needle for liquid delivery. The Arborplug is 15 mm in length and has a diameter of either 7 or 9 mm. Drilling 15 mm deep provides a volumetric capacity of 0.6 to 1.1 cm^3, respectively. The Eco-ject Micro-injection System loads re-usable micro-injection capsules, but does not use a plug. Using such devices, one may deliver a number of systemic chemistries by tree injection. Here we discuss three insecticides which are, (1) acephate, (2) imidacloprid and (3) emamectin benzoate.

9.1 Acephate
Acephate (O,S-dimethyl acetylphosphoramidothioate) is water soluble (700 g/L) and readily absorbed by tree roots for systemic activity (Worthing, 1987; Kidd & James, 1991). It has a low $K_{o/c}$ (organic carbon adsorption coefficient) of 0.48 (Montgomery, 1993); it is only weakly adsorbed in the soil. Acephate is an organo-phosphate insecticide designed for insecticidal activity and quick degradation. Acephate's stability is affected by pH. It has a comparatively shorter half-life (of 16-d, pH 9) in alkaline environments (Chevron, unpublished 1972b). Acephate is particularly mobile in coarse textured soils and has the potential to leach (Yen et al., 2000), but it is quickly degraded by microbial activity. In plants, acephate's half-life is approximately 5 to 10-d. Approximately 5 to 10% of acephate is degraded to methamidophos (which has insecticidal activity), the remainder to salts (of N, P and S) (Chevron, unpublished 1973). Acephate has both translaminar and systemic activity in plants. Acephate is a broad spectrum systemic, used for control of aphids, leaf miners, Lepidopterous larvae, sawflies, and thrips. 97.4% acephate is a soluble granular offered as an implant (Ace-Cap, Creative Sales, Fremont Nebraska) or tree injection formulation (ACE-jet, Arborjet, Inc.).

9.2 Imidacloprid
Imidacloprid (1-[(6-chloropyridin-3-yl) methyl]-N-nitro-4, 5-dihydroimidazol-2-amine) is a chloronicotinyl (neonicotinoid) chemistry with a water solubility of 0.51 g/L (Yen and Wendt, 1993). Imidacloprid has moderate binding activity ($K_{o/c}$ of 300 to 400) to clay and organic matter (Cox et al., 1997), however there is potential for the compound to move through porous, coarse textured soils (Jenkins, 1994). Imidacloprid has translaminar and systemic activity in plants (Buchholz and Nauen, 2002). Imidacloprid controls sucking insects such as adelgids, aphids, thrips, whiteflies, and some beetles, including Cerambycids. Examples of tree injection formulations of imidacloprid are Imicide (JJ Mauget, Arcadia, CA), Xytect (Rainbow Treecare Scientific Advancements, Minnetonka, MN) and IMA-jet (Arborjet, Inc.).

9.3 Emamectin benzoate

Emamectin benzoate is a semi-synthetic compound derived from the fermentation by-product of a soil actinomycete, *Streptomyces avermitilis* (Jansson et al., 1996). Emamectin benzoate is a mixture of the benzoic acid salt of two structurally complex heterocyclic (glycoside) compounds. It occurs as a mixture of ≥90% benzoic acid salts of 4'-epi-methylamino-4'-19 deoxyavermectin B1a and ≤10% 4'-epi-methylamino-4'-deoxyavermectin B1b (Wood, 2010). Emamectin benzoate is poorly (0.024 g/L) soluble in water (Tomlin, 2004). It has a $K_{o/c}$ of >25,000 and is immobile in soils (Mushtaq et al. 1996). Emamectin benzoate has translaminar activity, but limited plant systemic activity when applied to the foliage (Copping, 2004). A novel micro-emulsion formulation (TREE-äge, Syngenta Crop Protection, LLC, Greensboro, NC) used for systemic tree injection is registered for use in the US against specific Coleoptera and Lepidoptera pests.

10. Behaviors of injected chemistries

Injected chemistries differ in their rate of movement in the vascular system, and in their residual activity. In Avocado (*Persea americana*), Acephate peaked in foliage 2 weeks following tree injection, whereas peak imidacloprid residues were not observed for 7-9 weeks following application (Morse et al., 2008). The slow upward movement of imidacloprid may be explained by its comparatively higher carbon adsorption, and may play a role in the extended activity observed in field studies (Doccola et al., 2007; Morse et al., 2008). Studies in green ash (*Fraxinus pennsylvanica* Marsh) and white ash (*F. americana* L.) have demonstrated that imidacloprid accumulates in the canopy, but tree injection could also provide a reservoir for continued systemic activity (Cregg et al., 2005; Tanis et al., 2006, 2007, 2009). Takai et al. (2003), reported 3 years of protection in pine trees against pine wilt nematode after injecting a liquid formulation of emamectin benzoate. In the US, emamectin benzoate was reported to provide 2 or more years of protection against Lepidopterous and Coleoptera pests, including Pine cone worm (*Dioryctria*), Southern pine beetle (*Dendroctonus frontalis*) and Emerald ash borer (*Agrilus planipennis*) (Grosman et al., 2002, Grosman et al., 2009; Smitley et al., 2010).

Injection into plant tissues protects the chemistry from phytolysis and microbial degradation, mechanisms that breakdown the chemistry in the environment relatively quickly. Although hydrolysis occurs within the plant, some of the metabolites have insecticidal activity (for example, olefinic-, dihydroxy- and hydroxy-imidacloprid breakdown products of imidacloprid) (Sangha & Machemer, 1992; Suchail et al., 2001). Residual activity is based on the half-life of the chemistry, but carbon adsorption may also play a role in the activity observed in perennial tissues (such as in twig, branch and stem) over time. Injected formulations that provide multiple years of activity must move (spatially) from the original injection site in the xylem tissue into new vascular tissue in order to be effective against insects that perennially attack and feed in the lateral cambium. Residual activity of an injected insecticide provides protection against insect pests that have extended emergence periods, multiple generations per year, or are epidemic (i.e., increase exponentially over time).

11. When to treat trees

Apply treatments before damage (defoliation, vascular mining) occurs for optimum results. Oak trees defoliated by gypsy moth must use stored carbohydrates for recovery

(Shigo, 1989; Shigo, 1991). Furthermore, native insects are opportunistic: oaks that have been defoliated by insects such as gypsy moth (*Lymantria dispar*) are predisposed to attack by the two-lined chestnut borer (Haack & Acciavatti, 1992). Minimizing defoliation in trees is a sound practice to protect tree health. Rather than resorting to "rescue" treatments to save trees at risk of wood and bark infesting insects, treat them when they still appear visibly healthy. Late insecticide treatments (e.g., >33% canopy dieback, epicormic sprouting, bark cracks, woodpecker flecks, exit holes) are contra-indicated. This approach minimizes negative outcomes, such as canopy dieback, delayed recovery or tree mortality.

As discussed earlier, the upward movement of an injected chemistry is dependent upon plant evapo-transpiration. Therefore, tree injections are most efficiently applied when trees are transpiring. Transpiration is dependent on a number of factors, such as soil moisture, soil and ambient temperature, the relative humidity and time of day. For optimal uptake, apply when the soil is moist, soil temperatures are above 7.2°C (45°F), and during the 24 hour period when transpiration is greatest.

When using insecticides with short-residual activity (an example is acephate), make the application when the pest is active. Application of chemistries with greater residual activity are somewhat less dependent upon insect feeding activity (e.g., imidacloprid, emamectin), but are typically applied 30-d or more of expected pest activity. Fall applications may be applied in some instances. For example, imidacloprid applications in evergreen trees may be applied late in the season. Imidacloprid applications for HWA applications are made in the autumn to coordinate with resumption of sistens nymphal activity following summer aestivation. Imidacloprid activity is retained in hemlock (leaves of 3-6 age classes persist in trees) for extended residual activity (Doccola et al., in press). In addition, systemic insecticides with high adsorption coefficients (>5000) may be applied in the fall (at leaf senescence) for activity in the next growing season. TREE-äge (emamectin benzoate) is an example of a fall application used to protect ash trees against EAB (Smitley et al., 2010).

12. Tree defense responses

When trees are wounded, whether by an insect boring into the tree or by a mechanical drill bit, tree defense mechanisms come into play. These defense reactions and responses were systematically described by Shigo and Marx (1977). Dujesiefken and Liese have elaborated on the (CODIT) model taking into account the role of air exposure and embolism formation in the process of walling the damage in trees (2008). Individual trees may vary considerably in the strength of their response to similar types of wounds depending on genetics or tree health (Shigo, 1999). A discussion of tree wound responses must consider basic tree anatomy, in particular the secondary vascular tissues. Of most interest is the lateral meristem (cambium). This secondary cambium is only a few cells thick and occurs between the sapwood (xylem) and inner bark (phloem). This tissue is embryonic in nature. Periclinal divisions form xylem cells inward and phloem cells outward. The cambium is not transport tissue. Sapwood consists of living (symplast) and non-living (apoplast) cells. The living cells within the sapwood are non-differentiated parenchyma. The parenchyma cells store starch, oils and ergastic substances (Esau, 1977). Parenchyma occurs both as radial and axial tissues. Radial parenchyma extends into the phloem. The conductive xylem is functional when it matures and dies. The side walls of the xylem are pitted. Parenchyma cells sometimes balloon into the lumen of the xylem through the sidewall pits to form a tylose, or a physical barrier. Tyloses may be formed in older wood naturally (e.g., white oak, *Quercus alba*, forms tyloses in second year wood), or are a consequence of trauma (e.g., red oak, *Q. rubra*, forms tyloses in response to wounding) (Shigo, 1999). When a tree is physically injured, both biochemical and structural changes occur. The biochemical reactions (changes of stored carbohydrates to phenolic and terpene defense chemicals) are observed in tree sections in three dimensions. These were named reaction zones (or boundary walls) 1 – 3. Reaction zone 1 occurs in the axial direction (i.e., with the stem axis) and is the least limiting boundary. Reaction zone 2 occurs in the radial direction (i.e., with the tree radius, inward toward the pith), and reaction zone 3 occurs in the tangential direction (i.e., with the tree's circumference), and is the strongest limiting boundary of the three reaction zones. The fourth wall, referred to as the barrier zone occurs after injury, and is the strongest limiting boundary. Meristematic cells (cambium) divide to form callus tissue, which later differentiates into new woundwood (new xylem, cambium and phloem). Native insect attacks to healthy trees are fended off by the biochemistry and by the subsequent physical responses. Emerald ash borer attacks to Asian species of ash (*Fraxinus chinensis, F. manchurica*) do not result in tree mortality: plant defense responses effectively isolate the larva in early stages of attack and limit its progression. In *F. pennsylvanica* (a native), the larvae are compartmentalized via physical boundaries (wall 4), but the biochemistry (phenols, terpene chemistries) does not effectively stop the insect's development. Injection of an insecticidal chemistry to compensate for insufficient tree response is the basis of successful tree protection. EAB research has demonstrated that this strategy is very effective (Smitley et al., 2010).

Tree wound responses are dependent upon a number of intrinsic and extrinsic variables such as tree species, tree health, method of treatment and chemistry applied. Tree wound response is under genetic control (Santamour, 1979). For example, birch (*Betula* spp.) poplar (*Populus* spp.) and willow (*Salix* spp.) are considered weak compartmentalizers, whereas oak (*Quercus* spp.), sycamore (*Platanus* spp.) and linden (*Tilia* spp.) are considered strong compartmentalizers (Dujesiefken and Liese, 2008). Santamour (1986) described fourteen cultivars of maple (*Acer*), ash (*Fraxinus*), oak (*Quercus*) and linden (*Tilia*) that were strong wall 2 compartmentalizers. As a group, trees have evolved to resist assaults and are successful, long-lived perennial plants. Tree health is another variable with numerous contributing factors. These include the age of the tree, soil conditions (texture, structure, moisture, pH, minerals and drainage), and exposure (sun, shade). Trees require light, water and minerals for essential life functions (including defense). Photosynthesis is the basis of carbohydrate synthesis. Woundwood responses utilize energy (carbohydrate, lipid) stores. When injections are made to trees in relatively good health (preventative-early therapeutic treatments) tree woundwood development readily proceeds to close wounds. However, the prognosis for recovery is comparatively lower, when making late therapeutic (rescue) applications, because energy stores are reduced. Optimal wound responses are observed when applications are made early, relative to infestation (Doccola et al., 2011). To further manage wounds in trees, make the fewest number of injection sites to apply the dose, and whenever possible, avoid drilling in the valleys between roots (Shigo and Campana, 1977).

The Wedgle Direct-Inject (ArborSystems, LLC, Omaha, NE) is a method of tree injection that does not require drilling into the sapwood. The system relies on forcing the de-lamination (slippage) of the bark from the sapwood to apply a small amount of a formulation. This method directly exposes the lateral cambium to concentrated solvents. A consequence is phytotoxicity (e.g., hypersensitive reactions, necroses) to the tissues of the lateral meristem (the initials for woundwood development). The small doses and exposures to the lateral cambium by this method offers no clear advantage over drilling into trees for injection. Protection of the lateral cambium is of greater consequence to tree wound response compared to drilling into the sapwood. Further, wound closure rates of trees are positively correlated with trunk growth, and greater callus is produced around larger wounds than around smaller diameter wounds (Neely, 1988). Arborjet, Inc. employs a (7 or 9 mm) diameter drill hole to efficiently deliver higher volumes of insecticides into trees. The larger diameter hole is strongly limited by boundary wall 3 (this strong boundary reduces the likelihood of girdling and is an advantage to tree survival). With this system, a plastic Arborplug is inserted into the drilled hole, which creates the injection interface. The Arborplug from a tree wound defense perspective, reduces exposure of the lateral cambium to the solvent carriers in the injection formulation and minimizes wood exposure to air. Placing backflow preventers into the bark do not function in the same manner. Further, when the Arborplug is set correctly (at the sapwood-bark plane), it provides a flat surface for callus and woundwood development and wound closure. This encapsulation is the survival strategy of trees following injury (Dujesiefken and Liese, 2008).

13. Multiple-year activity

It is possible to make applications that are effective against a persistent and destructive tree pest and not require an annual treatment. The residual activity of tree injected imidacloprid may be due to protection against photolysis and microbial degradation. Foliar half-life of imidacloprid is ~9.8-d (Linn, 1992d, unpublished). Plants metabolize imidacloprid via hydrolysis, but some of the metabolites have insecticidal activity. The predominant metabolites associated with toxicity in insects are olefinic-, dihydroxy- and hydroxy-imidacloprid (Sangha & Machemer, 1992; Suchail et al., 2001). In studies of large (50 cm) diameter hemlock infested with HWA, both soil and tree injections with imidacloprid were made (Doccola et al., in press). Two methods of tree injections were employed, one using low volume micro-injection (QUIK-jet, Arborjet, Inc.) and the second using high volume micro-infusion (TREE I.V., Arborjet, Inc.). The soil applications were made using the Kioritz injector (Kioritz Corporation, 7-2, Suehirocho 1 –Chome, Ohme, Tokyo, 198 Japan). Tree injection administered 0.15 g imidacloprid per 2.5 cm dbh, micro-infusion applied 0.3 g per 2.5 cm dbh whereas soil injection applied 1.45 g per 2.5 cm dbh. In that study, data was collected on HWA infestation, tree growth and imidacloprid residues in the foliage over a three year period. Tree foliage responses were greater in the tree injection treatments. Imidacloprid residues taken annually from 70 to 1165-d were above the LC_{50} value of 0.30 µg/g for HWA (Cowles et al., 2006) for all the imidacloprid treatments. At 1165-d, foliage residues (of 1.35 µg/g) in the lowest dose injections continued to protect trees. This residual activity of imidacloprid was attributed to both the perennial nature (of 3-6 years) of the foliage, and to the slow, upward movement of imidacloprid. Green ash trees treated with emamectin benzoate tree injections were protected from EAB for up to four years (Smitley et al., 2010). A recently completed 3 year study using low dose injections of emamectin benzoate protected trees for three years (Deb McCullough, personal communication). These studies point to efficacy and duration of tree injection methods. The TREE-äge label is approved (by US EPA) for up to two years of control against listed arthropods, including EAB. Injection is a very efficient use of insecticidal chemistry to protect trees.

14. Tree injection as an alternative

Today, tree injection is an alternative method of chemical application with definite advantages: (1) efficient use of chemicals, (2) reduced potential environmental exposure, and (3) useful when soil and foliar applications are either ineffective or difficult to apply (Stipes, 1988; Sanchez-Zamora and Fernandez-Escobar, 2004). Tree injection is used when trees are at risk from attack from destructive or persistent pests. It may be put to good use in tall trees. They are administered in trees growing in environmentally sensitive locations (e.g., near water, in sandy soils). Tree injection does create wounds, however the benefit of the introduced chemistry to protect trees often outweigh the drilling wound. The new paradigm weighs the potential of off target consequences of application to the consequences of the drilled wound made by tree injection. Unintended off target exposures include toxicity to earthworms, fish, aquatic arthropods, pollinators and applicator. Insecticides are by design, toxic, albeit useful, substances. Tree injection is a method to deliver specific toxicants to the injurious pest and to minimize non-intended exposures. In this chapter, three specific insecticides used in tree injection were considered, each with unique attributes for specific applications in trees. Tree injection is an alternative methodology to apply systemic insecticides for tree protection.

15. Acknowledgements

The authors thank David Cox, Ph.D., Syngenta Crop Protection, LLC for his review, edits, and comments of this chapter. The authors also thank Ms. Monica Davis for her review and edits.

16. References

Amman, G.D., M.D. McGregor, and R. E. Dolph, Jr. Updated 2002. Mountain Pine Beetle. Forest Insect & Disease Leaflet 2. US Department of Agriculture Forest Service. (website accessed 4/14/2011).

Anderson, C. 1991. Photodegradation of NTN 33893 in water. Unpublished report study prepared by Nitokuno, ESR, Yuki Institute. 128 pp. In SERA (Syracuse Environmental Research Associates, Inc.). 2005. Imidacloprid – human health and ecological risk assessment – final report. SERA TR 05-43-24-03a. 283 pp.

ANSI A300 Part 3. 2006. Supplemental Support Systems. American National Standards Institute (ANSI) A300 Standards for Tree Care Operations.

ANSI A300 Part 4. 2008. Lightning Protection Systems. American National Standards Institute (ANSI) A300 Standards for Tree Care Operations.

Anulewicz, A.C., D.G.McCullough, D.L. Cappaert and T.M. Poland. 2008. Host range of the Emerald ash borer (*Agrilus planipennis* Fairmaire) (Coleoptera: Buprestidae) in North America: results of multiple-choice field experiments. Environ. Entomol. 37(1): 230-241

Barnes, T.G. 1989. Controlling woodpecker damage. FOR-38. http:www.ca.uky.edu/agc/pubs/for/for38/for38.htm (website accessed 4/15/11).

Buchholz, A. and R. Nauen. 2002. Translocation and translaminar bioavailability of two neonicotinoid insecticides after foliar application to cabbage and cotton. Pest Management Science, 58(1): 10-16.

California EPA. 2004. Summary of toxicological data on imidacloprid. Document processing number (DPN) # 51950. Revised date: 3/30/04.

Chaney, W.R. 1988. Anatomy and physiology related to chemical movement in trees. Journal of Arboriculture, 12(4): 85-91.

Chevron Chemical Co. - Ortho Division, 1972a. Orthene Residue Tolerance Petition: Physical and Chemical Properties. Unpublished report submitted to California Department of Pesticide Regulation. *CDPR Volume Number: 108-163.* #54024.

Chevron Chemical Co. - Ortho Division, 1972b. Hydrolysis of Orthene. Unpublished report submitted to California Department of Pesticide Regulation. *CDPR Volume Number: 108-163.* #54145.

Chevron Chemical Co. - Ortho Division, 1972c. Identification of Orthene Hydrolysis Products. Unpublished report submitted to California Department of Pesticide Regulation. *CDPR Volume Number: 108-163.* #54146.

Chevron Chemical Co. - Ortho Division, 1972d. Stability of Orthene to Sunlight. Unpublished report submitted to California Department of Pesticide Regulation. *CDPR Volume Number: 108-163.* #54149.

Chevron Chemical Co. - Ortho Division, 1972e. Orthene Soil Metabolism - Laboratory Studies (Aerobic). Unpublished report submitted to California Department of Pesticide Regulation. *CDPR Volume Number: 108-163.* #54150.

Chevron Chemical Co. - Ortho Division, 1972f. Orthene Soil Metabolism - Laboratory Studies (Anaerobic). Unpublished report submitted to California Department of Pesticide Regulation. *CDPR Volume Number: 108-163.* #54151.

Chevron Chemical Co. - Ortho Division, 1972g. Comparison of Orthene Soil Metabolism Under Aerobic and Anaerobic Conditions (Aerobic). Unpublished report submitted to California Department of Pesticide Regulation. *CDPR Volume Number: 108-163.* #54153.

Chevron Chemical Co. - Ortho Division, 1972h. Comparison of Orthene Soil Metabolism Under Aerobic and Anaerobic Conditions (Anaerobic). Unpublished report submitted to California Department of Pesticide Regulation. *CDPR Volume Number: 108-163.* #54154.

Chevron Chemical Co. - Ortho Division, 1972i. Orthene Leaching Study. Unpublished report submitted to California Department of Pesticide Regulation. *CDPR Volume Number: 108-163.* #54155.

Chevron Chemical Co. - Ortho Division, 1972j. Comparison of Acephate Soil Leaching and Stability in Wet and Dry Soil. Unpublished report submitted to California Department of Pesticide Regulation. *CDPR Volume Number: 108-163.* #54162.

Chevron Chemical Co. - Ortho Division, 1973. Impact of Orthene on the Environment: plant metabolism, fate and metabolism in soil, fate in water, fate in animals, toxicity and hazard to man, wildlife and other non-target species food chain. Unpublished report submitted to California Department of Pesticide Regulation. *CDPR Volume Number: 108-163.* #54161.

Chevron Chemical Co. - Ortho Division, 1988. Freundlich Soil Adsorption/Desorption Coefficients of Acephate and Soil Metabolites. Unpublished report submitted to California Department of Pesticide Regulation. *CDPR Volume Number: 108-189.* #66325.

Chukwudebe AC; Feely WF; Burnett TJ; Crouch LS; Wislocki PG. 1996b. Uptake of Emamectin Benzoate Residues from Soil by Rotational Crops. Journal of Agricultural and Food Chemistry. 44 (12): 4015-4021.

Chukwudebe AC; Atkins RH; Wislocki PG. 1997a. Metabolic Fate of Emamectin Benzoate in Soil. Journal of Agricultural and Food Chemistry. 45 (10): 4137-4146.

Chukwudebe, A.C., D.L. Cox, S.J. Palmer, L.A. Morneweck, L.D. Payne, D.M. Dunbar, and P.G. Wislocki. 1997. Toxicity of emamectin benzoate foliar dislodgeable residues to two beneficial insects. J. Agric. Food Chem. 45(9): 3689-3693.

Coder, K. 1999. Water movement in trees. Daniel B. Warnell School of Forest Resources, University of Georgia. Extension publication FOR99-007. 4pp.

Coleman, T.W. and S.J. Seybold. 2008. Previously unrecorded damage to oak, Quercus spp., in southern California by the goldspotted oak borer, Agrilus coxalis Waterhouse (Coleoptera: Buprestidae). 84 (4): 288-300.

Copping, L.G. (ed.). 2004. The Manual of Biocontrol Agents. Alton, UK. BCPC.

Cox, L., W.C. Koskinen and P.Y. Yen. 1997. Sorption-desorption of imidacloprid and its metabolites in soils. Journal of Agricultural Food Chemistry. 45(4): 1468-1472.

Cranshaw, W.S. and D.S. Leatherman. 2006. Revised. Shade Tree Borers. No. 5.530. Colorado State University Extension. 4pp.

Cregg, B. D. Mota-Sanchez, D. McCullough, R. Hollingworth and T. Poland. 2005. Distribution and persistence of trunk-injected 14C imidacloprid in ash trees. In Emerald Ash Borer Research and Technology Development Meeting. September 26-27, 2005. FHTET-2005-16. Pp 24-25.

Darvas, J.M., J.C. Toerien, and D.L. Milne. 1984. Control of avocado root rot by trunk injection with phosethyl-Al. Plant Dis. 68:691–693.

Davenport, A.L. and L.J. Staats. 1998. Maple syrup production for the beginner. PDF. http:www.dnr.cornell.edu (website accessed 4/15/2011).

Dirr, M.A. 2009. Manual of Woody Landscape Plants: Their Identification, Ornamental Characteristics, Culture, Propagation and Uses. Stipes Pub. Llc. 1325pp.

Doccola, J.J., D.R. Smitley, T.W. Davis, J.J. Aiken and P.M. Wild. 2011. Tree wound responses following systemic injection treatments in Green ash (Fraxinus pennsylvanica Marsh) as determined by destructive autopsy. Arboriculture & Urban Forestry, 37(1): 6-12.

Downing, E. Environmental fate of Acephate. PDF. Environmental Monitoring and Pest Management. Department of Pesticide Regulation. Sacramento, CA. 11pp.

Dujesiefken, D., and W. Liese, 2008. Das CODIT-Prinzip—von Bäumen lernen für eine fachgereche Baumpflege. Haymarket Media, Braunschweig/Germany, 160 pp.

Esau, Katherine. 1977. The Anatomy of Seed Plants. Second Edition. John Wiley and Sons. New York. 550 pp.

Extoxnet/PIP. Imidacloprid. http://extoxnet.orst.edu/pips/imidaclo.htm (website accessed 4/13/2011).

Farm Chemicals Handbook. 1994. Meister Publishing Co. Willoughby, OH.

Fernández-Escobar, R., D. Barranco, and M. Benlloch.1993. Overcoming iron chlorosis in olive and peach trees using a low-pressure trunk-injection method. HortScience 28(3):192–194.

Fernández-Escobar, R., D. Barranco, M. Benlloch, and J.J. Alegria. 1994. Control of Phytophthora root rot of avocado using prepared injection capsules of potassium phosphite. Adv. Hortic. Sci. 8:157–158.

Fernández-Escobar, R., F.J. Gallego, M. Benlloch, J.Membrillo, J. Infante, and A. Perez de Algaba. 1999. Treatment of oak decline using pressurized injection capsules of antifungal materials. Eur. J. For. Pathol. 29:29–38.

Filer, T.H. Jr. 1973. Pressure apparatus for injecting chemicals into trees. Plant Dis. Report. 57:338–340.

Galford, J.R. 1984. Revised. The Locust Borer. Forest Insect & Disease Leaflet 71. US Department of Agriculture Forest Service. http:www.na.fs.fed.us/spfo/pubs/fidl/locust/locust.htm (website accessed 4/15/2011).

Gregory, G. F., T.W. Jones and P. McWain. 1973. Pressure injection of methyl-2-benzimidazole carbamate hydrochloride solution as a control for Dutch elm disease. USDA Forest Service research note NE_176. Northeastern Forest Experiment Station, Upper Darby, PA. 9 p.

Gregory, G.F. and T.W. Jones. 1975. An improved apparatus for pressure-injecting fluids into trees. USDA Forest Service research note NE_214. Northeastern Forest Experiment Station, Upper Darby, PA. 6 p.

Greulach, V.A. 1973. Plant Function and Structure. Macmillan Publishing Co., New York. 575 pp.

Grosman, D.M., S.R. Clarke, and W.W. Upton. 2009. Efficacy of two systemic insecticides injected into loblolly pine for protection against south pine bark beetles (Coleoptera: Curculionidae). J. Econ. Entomol. 102(3): 1062-1069.

Grosman, D.M., W.W. Upton, F.A. McCook and R.F. Billings. 2002. Systemic insecticide injections for control of cone and seed insects in loblolly pine seed orchards – 2 year results. South. J. Appl. For. 26(3): 146-152.

Guest, D.I., K.G. Pegg, and A.W. Whiley. 1994. Control of Phytophthora diseases of tree crops using trunk injected phosphonates. Hortic. Rev. 17:299–330.

Haack, R.A. and R.E. Acciavatti. 1992. Twolined Chestnut Borer. Forest Inset & Disease Leaflet 168. US Department of Agriculture Forest Service. http:www.na.fs.fed.us/spfo/pubs/fidls/chestnutborer/chestnutborer.htm (website accessed 4/15/2011).

Haugen, D. A. and E. R. Hoebeke. 2005. Sirex woodwasp – Sirex noctilio F. (Hymenoptera: Siricidae). Pest Alert. NA-PA-07-05. USDA Forest Service Northeastern Area. State and Private Forestry. http:www.na.fs.fed.us/spfo/pubs/pest_al/sirex_woodwasp/sirex_woodwasp.ht m (website accessed 4/15/2011).

Haugen, L. and M. Stennes. 1999. Fungicide injection to control Dutch elm diseae: understanding the options. Plant Diagnostic Quarterly 20 (2): 29-38.

Helburg, L.B., M.E. Schomaker, and R.A. Morrow. 1973. A new trunk injection technique for systemic chemicals. Plant Dis. Report. 57:513–514.

Herms, D.A. 2010. Multiyear evaluations of systemic insecticides for control of emerald ash borer. In Emerald Ash Borer Research and Technology Development Meeting. Forest Health technology Enterprise Team. Presented October 20-21, 2009. Pittsburgh, PA. FHTET-2010-01.Pp71-75.

Heukelekian, H. and S.A. Waksman. 1925. Carbon and nitrogen transformations in the decomposition of cellulose by filamentous fungi. Journal of Biological Chemistry, 66(1): 323-342.

Jansson, R.K., R. Brown, B. Cartwright, D. Cox, D.M. Dunbar, R.A. Dybas, C. Eckel, J.A. Lasota, P.K. Mookerjee, J.A. Norton, R.F. Peterson, V.R. Starner and S. White. 1996. Emamectin benzoate: a novel avermectin derivative for control of Lepidopterous pests. In Proceedings of the Third International Workshop, 29, October – 1 November, 1996. Kuala Lumpur, Malaysia. The management of diamondback moth and other crucifer pests. Chemical control. Pp. 171-177.

Jenkins, J.J. 1994. Use of Imidacloprid for Aphid Control on Apples in Oregon. Potential for Ground and Surface Water Contamination. Department of Agricultural Chemistry. Oregon State University, Corvallis, OR.

Johnson, W.T. and H.H. Lyon. 1991. Second edition revised. Insects That Feed on Trees and Shrubs. Comstock Publishing, a Division of Cornell University Press. Ithaca, NY. 556 pp.

Jones, T.W. and G.F. Gregory. 1971. An apparatus for pressure injection of solutions into trees. USDA Forest Service research paper NE_233. Northeastern Forest Experiment Station, Upper Darby, PA. 9 p.

Jones, T.W., G.F Gregory and P. McWain. 1973. Pressure injection of solubilized benomyl for prevention and cure of oak wilt. USDA Forest Service research note NE_171. Northeastern Forest Experiment Station, Upper Darby, PA. 4 p.

Kidd, H. and James, D. R., Eds. The Agrochemicals Handbook, Third Edition. Royal Society of Chemistry Information Services, Cambridge, UK, 1991 (As Updated).10-2

Kidd, H. and James, D. R., Eds. The Agrochemicals Handbook, Third Edition. Royal Society of Chemistry Information Services, Cambridge, UK, 1991 (as updated).5-14

Kielbaso, J.J., H. Davidson, J. Hart, A. Jones, and M.K. Kennedy. 1979. In Kielbaso, J.J., et al. (Eds.). Proceedings of Symposium on Systemic Chemical Treatment in Tree Culture, October 9-11, 1978, East Lansing, MI.

Koehler, C.S., and S.S. Rosenthal. 1967. Bark vs. foliage applications of insecticides for control of Psylla uncatoides on acacia. J. Econ. Entomol. 60:1554–1558.

Kondo, E.S. 1978. Root flare and root injection technique. In: Proc. of the symposium on systemic chemical treatment in tree culture, pp. 133-139.

Kozlowski, T.T., and C.H. Winget. 1963. Patterns of water movement in forest trees. Botanical Gazette 124:301–311.

Kozlowski, T.T., J.F.Hughes and L. Leyton. 1967. Movement of injected dyes in gymnosperm stems in relation to tracheid alignment. Forestry, 40(2): 207-219.

Kramer, P.J., S.G. Pallardy and T.T. Kozlowski. 1996. Physiology of Woody Plants. Second edition. Academic Press. 411pp.

Lanier, G.N. 1987. Fungicides fro Dutch elm disease: comparative evaluation of commercial products. Journal of Arboriculture 13 (8): 189-195.

Lewis, R. Jr. 1979. Control of live oak decline in Texas with Lignasan and Arbotect, pp 239-246. In Kielbaso, J.J., et al. (Eds.). Proceedings of Symposium on Systemic Chemical Treatment in Tree Culture, October 9-11, 1978, East Lansing, MI.

Linn, J. 1992d. Evaluation of the foliar half-life and distribution of NTN 33893 in potatoes. Unpublished study prepared by Miles In. 166p. In SERA (Syracuse Environmental

Research Associates, Inc.). 2005. Imidacloprid – human health and ecological risk assessment – final report. SERA TR 05-43-24-03a. 283 pp.

McClure, M.S. 1992. Effects of implanted and injected pesticides and fertilizers on the survival of *Adelges tsugae* (Homoptera: Adelgidae) and on the growth of *Tsuga canadensis*. J. Econ. Entomol. 85(2):468–472

McCullough, D.G. and N. Siegert. 2007. Estimating potential Emerald ash borer (Coleoptera: Buprestidae) populations using ash inventory data. J. Econ. Entomol. 100(5): 1577 – 1586.

McCullough, D.G., T.M. Poland, A.C. Anulewicz, P. Lewis and J. Molongoski. 2010. Evaluation of emamectin benzoate and neonicotinoid insecticides: two-year control of EAB. In Emerald Ash Borer Research and Technology Development Meeting. Forest Health technology Enterprise Team. Presented October 20-21, 2009. Pittsburgh, PA. FHTET-2010-01. Pp68-70.

McKenzie, N., B. Helson, D. Thompson, G. Otis, J. McFarlane, T. Buscarini and J. Meating. 2010. Azadirachtin: an effective systemic insecticide for control of *Agrilus planipennis* (Coleoptera: Buprestidae). J.Econ. entomol. 103(3): 708-717.

McWain, P. and G. F. Gregory. 1971. Solubilizaation of benomyl for xylem injection in vascular wilt disease control. USDA Forest Service research paper NE_234. Northeastern Forest Experiment Station, Upper Darby, PA. 8 p.

Mohapatra, S., Ahuja, A. K., Sharma, D., Deepa, M., Prakash, G. S. and Kumar, S. (2011), Residue study of imidacloprid in grapes (*Vitis vinifera* L.) and soil. Quality Assurance and Safety of Crops & Foods, 3: 24–27.

Montgomery, J.H. (ed.). 1993. Agrochemicals Desk Reference. Environmental Data. Published by Lewis Publishers. Chelsea, MI.

Morse, J., F. Byrne, N. Toscano, and R. Krieger. 2008. Evaluation of systemic chemicals for avocado thrips and avocado lace bug management. Production Research Report. California Avocado Commission. 10pp.

Mushtaq M; Chukwudebe AC; Wrzesinski C; Allen L RS; Luffer-Atlas D; Arison BH. 1998. Photodegradation of Emamectin Benzoate in Aqueous Solutions. Journal of Agricultural and Food Chemistry. 46 (3): 1181-1191.

Mushtaq M; Feely WF; Syintsakos LR; Wislocki PG. 1996. Immobility of Emamectin Benzoate in Soils. Journal of Agricultural and Food Chemistry. 44 (3): 940-944.

Nauen, R., Ebbinghaus-Kintscher, U., and R. Schmuck. 2001. Toxicity and nicotinic acetylcholine receptor interaction of imidacloprid and its metabolites in *Apis mellifera* (Hymenoptera: Apidae). Pest Manag Sci. 57(7): 577-86.

Navarro, C., R. Fernández-Escobar, and M. Benlloch. 1992. A low-pressure, trunk-injection method for introducing chemical formulations into olive trees. J. Am. Soc. Hortic. Sci. 117(2):357–360.

Neely, D. 1988. Wound closure rates on trees. Journal of Arboriculture 14(10): 250-254.

Newsom, L.D. 1967.Consequences of insecticide use on nontarget organisms. Annual Review of Entomology. 12: 257-286.

Nyland, G., and W.J. Moller. 1973. Control of pear decline with a tetracyclina. Plant Dis. Report. 57:634–637.

Ostry, M.E. and T.H. Nicholls. 1978. How to identify and control sapsucker injury on trees. North Central Forest Experiment Station. Forest Service. US Department of Agriculture. St. Paul Minnesota.

http:www.na.fs.fed.us/spfo/pubs/howtos/ht_sap/sap.htm   (website   accessed 4/15/11).

Raven, P., R.F. Evert and H. Curtis. 1981. Biology of Plants. Third Edition. Worth Publishers, Inc., New York. 686 pp.

Reil, W.O. 1979. Pressure-injecting chemicals into trees. Calif. Agric. 33:16–19.

Reil, W.O., and J.A. Beutel. 1976. A pressure machine for injecting trees. Calif. Agric. 30:4–5.

Roach, W.A. 1939. Plant injection as a physiological method. Ann. Bot. NS 3(9):155–227.

Roe, A.H. 2001. Poplar Borer. Fact Sheet No. 94. Utah State University Cooperative Extension. 3 pp.

Rudinsky, J.A. & Vité, J.P. 1959. Certain ecological and phylogenetic aspects of the pattern of water conduction in conifers. Forest Science 5, 259–266.

Sachs, R.M., G. Nyland, W.P. Hackett, J. Coffelt, J. Debie, and G. Giannini. 1977. Pressurized injection of aqueous solutions into tree trunks. Scientia Hortic. 6:297–310.

Sangha, G., and L. Machemer. 1992. An overview of the toxicology of NTN 33893 and its metabolites. Unpublished study prepared by Miles Inc. 134 p. In SERA (Syracuse Environmental Research Associates, Inc.). 2005. Imidacloprid – human health and ecological risk assessment – final report. SERA TR 05-43-24-03a. 283 pp.

Santamour, F.S. 1979. Inheritance of wound compartmentalization in soft maples. Journal of Arboriculture 5(10): 220-225.

Santamour, F.S. 1986. Wound compartmentalization in tree cultivars: addendum. Journal of Arboriculture 12(9): 227-232.

Sawyer, Alan. February 26, 2010, Revised. Asian Longhorned Beetle: Annotated Host List. USDA-APHIS-PPQ, Center for Plant Health Science and Technology, Otis Laboratory

Shigo, A.L. and R. Campana. 1977. Discolored and decayed wood associated with injection wounds in American elm. Journal of Arboriculture 3(12): 230-235.

Shigo, A.L., R. Campana, F. Hayland and J. Anderson, 1980. Anatomy of elms injected to control Dutch elm disease. Journal of Arboriculture 6(4): 96-100.

Shigo, A.L., W.E. Money and D.I. Dodds. 1977. Some internal effects of Mauget tree injections. Journal of Arboriculture 3(11): 213-220.

Shigo, A.L. 1989. A new tree biology: facts, photos, and philosophies on trees and their problems and proper care. Second Edition. Shigo and Trees, Associates. Durham, NH. 618pp.

Shigo, A.L. 1991. Modern Arboriculture: A systems approach to the care of trees and their associates. Shigo and Trees, Associates, Durham, NH.

Sinclair, W. A. and A.O. Larsen. 1981. Wood characteristics related to "injectability" of trees. Journal of Arboriculture, 7(1): 6-10.

Solheim, H. and P. Krokene. 1998. Growth and virulence of mountain pine beetle associated blue-stain fungi, Ophiostoma clavigerum and Ophiostoma montium. Can. J. Bot. 76: 561-566.

Smitley, D.R., J.J. Doccola and D.L. Cox. 2010. Multiple-year protection of ash trees from emerald ash borer with a single trunk injection of emamectin benzoate, and single-year protection with an imidacloprid basal drench. Arboriculture & Urban Forestry, 36(5): 206-211.

Stipes, R.J. 1988. Glitches and gaps in the science and technology of tree injection. Journal of arboriculture 14 (7): 165-171.

Suchail, S., D. Guez and L.P. Belzunces. 2001. Discrepancy between acute and chronic toxicity induced by imidacloprid and its metabolites in *Apis mellifera*. Environ Toxicol Chem. 20(11): 2482-6.

Takai, K., T. Suzuki and K. Kawazu. 2003. Distribution and persistence of emamectin benzoate at efficacious concentrations in pine tissues after injection of a liquid formulation. Pest Manag Sci 60: 42-48.

Tanis, S.R., B.M. Cregg, D. Mota-Sanchez, D.G. McCullough, T.M. Poland and R.M. Hollingworth. 2006. Distribution of trunk-injected 14C imidacloprid in *Fraxinus* trees: a test of the sectored-flow hypothesis. In Emerald Ash Borer Research and Technology Development Meeting. October 29 – November 2, 2006. FHTET-2007-04. Pp 34-38.

Tanis, S.R., B.M. Cregg, D. Mota-Sanchez, D.G. McCullough, T.M. Poland and R.M. Hollingworth. 2007. Sectored flow and reservoirs: a synthesis of 14C-imidacloprid trunk injection research. In Emerald Ash Borer Research and Technology Development Meeting. October 23-24.2007. FHTET-2008-07. Pp 45-47.

Tanis, S.R., B.M. Cregg, D. Mota-Sanchez, D.G. McCullough and T.M. Poland. 2009. Sectored flow and reservoirs: a synthesis of 14C-imidacloprid trunk injection research. In Emerald Ash Borer Research and Technology Development Meeting. October 20-21, 2009. FHTET-2010-01. Pp 79-80.

Tattar, T.A., Dotson, J.A., Ruizzo, M.S. and V.B Steward. 1998. Translocation of imidacloprid in three tree species when trunk- and soil-injected. Journal of Arboriculture. 24(1): 54-56.

USDA/FS (US Department of Agriculture/Forest Service). 2003. Pest Alert NA-PR-03-94: Hemlock Woolly adelgid.

USDA/FS (US Department of Agriculture/Forest Service). 2008. Revised. Pest Alert NA-PR-01-99GEN: Asian Longhorned Beetle (*Anoplophora glabripennis*): A New Introduction. 2pp.

USDA/FS (US Department of Agriculture/Forest Service). 2008a. Revised. Pest Alert NA-PR-02-04. Emerald Ash Borer. 2pp.

Wislocki, P.G., et al. 1989. Environmental Aspects of Abamectin Use in Crop Protection in W.C. Campbell (ed.). Ivermectin and Abamectin. Springer-Verlag, NY.

Wood, A, 2010. Emamectin benzoate. Compendium of Pesticide Common Names. http://www.alanwood.net/pesticides/derivatives/emamectin%20benzoate.html (website accessed 5/06/2011).

Worthing, C. R. (ed.) 1987. The Pesticide Manual: A World Compendium. Eighth edition. Published by The British Crop Protection Council.

Yen, J; K. Lin; Y. Wang, 2000. Potential of the Insecticides Acephate and Methamidophos to Contaminate Groundwater. *Ecotoxicology and Environmental Safety*. Vol. 45, pp. 79-86.

Yoshida, H. 1990. Photodegradation of NTN 33893 on soil. Unpublished study prepared by Nihon Tokushu Noyaku Siezo K.K. 42 pp. In SERA (Syracuse Environmental Research Associates, Inc.). 2005. Imidacloprid – human health and ecological risk assessment – final report. SERA TR 05-43-24-03a. 283 pp.

Zanne, A.E., K. Sweeney, M. Sharma and C.M. Orians. 2006. Patterns and consequences of differential vascular sectorality in 18 temperate tree and shrub species. Function Ecology, 20: 200-206.

3

The Pyrethroid Knockdown Resistance

Ademir Jesus Martins and Denise Valle
Fundação Oswaldo Cruz/ Instituto Oswaldo Cruz/
Laboratório de Fisiologia e Controle de Artrópodes Vetores
Brazil

1. Introduction

New promising insect control efforts are now being evaluated such as biological alternatives or even transgenic insects and *Wolbachia* based strategies. Although it is increasingly clear that successful approaches must involve integrated actions, chemical insecticides unfortunately still play a central role in pest and vector control (Raghavendra et al., 2011). Development of new safe and effective compounds in conjunction with preservation of those currently being utilized are important measures to insure insecticide availability and efficiency for arthropod control. In this sense, understanding the interaction of insecticides with the insect organism (at physiological and molecular levels), the selected resistance mechanisms and their dynamics in and among natural populations is obligatory.

Pyrethroids are synthetic compounds derived from pyrethrum, present in *Chrysanthemum* flowers. Currently, pyrethroids are the most used insecticides against arthropod plagues in agriculture and livestock as well as in the control of vectors of veterinary and human health importance. They are chemically distinguished as type I (such as permethrin, compounds that lack an alpha-ciano group) and type II (with an alpha-ciano group, like deltamethrin) (T. G. Davies et al., 2007b). Pyrethroid insecticides have been largely adopted against vector mosquitoes through indoor, perifocal or ultra-low volume (ULV) applications. As of yet pyrethroids are the only class of insecticides approved for insecticide treated nets (ITNs), an important tool under expansion against malaria, mainly in the African continent (Ranson et al., 2011). The consequence of intense and uncontrolled pyrethroid use is the extremely rapid selection of resistant populations throughout the world.

Just like DDT, pyrethroids act very fast in the central nervous system of the insects, leading to convulsions, paralysis and eventually death, an effect known as *knockdown*. However, unlike DDT, pyrethroids are not claimed to cause severe risks to the environment or to animal or human health, hence its widespread use. The main pyrethroid resistance mechanism (the knockdown resistance phenotype, *kdr*) occurs due to a point mutation in the voltage gated sodium channel in the central nervous system, the target of pyrethroids and DDT.

Herein we aim to discuss the main mechanism of pyrethroid resistance, the knockdown resistance (*kdr*) mutation, its effect and its particularities among arthropods. The most common methods presently employed to detect the *kdr* mutation are also discussed. Some aspects regarding the other main pyrethroid resistance mechanisms, like alterations in behaviour, cuticle and detoxifying enzymes will be only briefly addressed. The proposal of this chapter is to review knockdown resistance to pyrethroids, nowadays the preferred insecticide class worldwide. This topic discusses aspects of general biology, physiology, biochemistry, genetics and evolution, with focus on disease vector mosquitoes. It is expected that the amount and diversity of material available on this subject may well illustrate insecticide resistance in a broader context.

2. Insecticide resistance mechanisms

Besides the resistance to chemical insecticides caused by modifications in the target site (also called phenotypic resistance), other mechanisms commonly associated are: metabolic resistance, behavioral modification and alterations in the integument. In the first case, endogenous detoxifying enzymes become more efficient in metabolizing the insecticide, preventing it from reaching its target in the nervous system. This occurs due to 1) increase in the number of available molecules (by gene amplification or expression activation) or 2) mutation in the enzyme coding portion of the gene, so that its product metabolizes the insecticide more efficiently. These processes can be very complex and involve three major enzyme superfamilies: Esterases, Multi function Oxidases P450 and Glutathion-S-Transferases (Hemingway & Ranson, 2000; Montella et al., 2007). In contrast, there are few examples in literature regarding insect behavioral changes and tegument alterations.

Resistance to insecticides may be functionally defined as the ability of an insect population to survive exposure to dosages of a given compound that are lethal to the majority of individuals of a susceptible lineage of the same species (Beaty & Marquardt, 1996). Resistance is based on the genetic variability of natural populations. Under insecticide selection pressure, specific phenotypes are selected and consequently increase in frequency. Resistance can result from the selection of one or more mechanisms. In order to elucidate the molecular nature of resistance, many studies report laboratory controlled selection of different species (Chang et al., 2009; Kumar et al., 2002; Paeoporn et al., 2003; Rodriguez et al., 2003; Saavedra-Rodriguez et al., 2007). With selected lineages, it becomes easier to separate the role of each distinct mechanism. In a more direct approach, the current availability of a series of molecular tools enables detection of expression of altered molecules in model organisms so that the effect of the insecticide can be evaluated under specific and controlled circumstances (Smith et al., 1997).

Regardless of the mono or multi-factorial character of resistance, this phenomenon may be didactically divided into four categories: behavioral, cuticular, metabolic and phenotypic resistance. In the first case the insect simply avoids contact with the insecticide through behavioral adaptations, which are presumably related to genetic inheritance (Sparks et al., 1989). Among arthropods, mosquitoes are by far the group most intensely investigated in relation to behavioral resistance (Lockwood et al., 1984). For instance, *Anopheles* malaria vector mosquitoes from the Amazon Region had the habit of resting in the walls after a blood meal. There are registers that some populations changed their behavior after a period of indoor residual application of DDT to the dwelling walls (Roberts & Alecrim, 1991). Behavioral changes that minimize contact between insect and insecticide may cause a severe impact in the insecticide application efficacy, especially if resistance is selected by physiological features (Ranson et al., 2011).

Certain alterations in the insect cuticle may reduce insecticide penetration. However, these effects are unspecific, leading to resistance to a series of xenobiotic compounds. This mechanism is known as reduced penetration or cuticle resistance. It is probably not related to high levels of resistance by itself, but it can interact synergistically with other mechanisms. The physiological processes or molecular pathways which describe this type of resistance remain to be elucidated. With respect to pyrethroid resistance, recent evidences point to an increase in the levels of expression of two cuticle genes in populations of two *Anopheles* species (Awolola et al., 2009; Vontas et al., 2007).

The increased ability to detoxify insecticides is one of the main types of resistance, commonly referred to as metabolic resistance. It takes place when the activity of naturally detoxifying enzymes is enhanced, impeding the insecticide to reach its target. Among these enzymes, Multi function Oxidases (or Monoxigenases P450), Esterases and Glutathion-S-Transferases (GST) (ffrench-Constant et al., 2004; Hemingway & Ranson, 2000) are the major representative families. Although the molecular basis of metabolic resistance has been extensively studied, only few reports have investigated the specific metabolic pathways involved or their location in the insect organism. Many different mutations may be attributed to metabolic resistance, such as those leading to production of more enzymes, via gene duplication events or either increases in gene transcription rates, alterations in the normal tissue/time specificity of expression, point mutations leading to a gain of function or changes in the substrate specificity (ffrench-Constant et al., 2004; Hemingway et al., 2004; Perry et al., 2011). Detoxifying enzymes belong to superfamilies composed of numerous genes (Ranson et al., 2002), and it is not unusual for different enzymes to produce the same metabolites. Additionally, an alteration in one type of enzyme may lead to cross-resistance among different classes of insecticides (Ranson et al., 2011). However, population genetic markers that make feasible a complete diagnostic of the resistance mechanisms or their distribution are not yet available. Current studies are generally based on biochemical assays (Valle et al., 2006) and, to a lesser extent, on *microarray detox chips* (David et al., 2005; Vontas et al., 2007). Due to technical limitations, the most common reports are hence oriented to single gene responses, such as punctual mutations that increase the ability of a specific enzyme in detoxifying an insecticide (Lumjuan et al., 2011; Morin et al., 2008).

Multi function P450 Oxidases are the enzymes most commonly associated to metabolic resistance to pyrethroids. However, despite much indirect evidence of P450 total activity increase or even detection of higher expression of some related genes (*cyp*), little is known about their metabolic activity. For instance, 111 genes code for P450 in *Anopheles gambiae*, but only two (*cyp6p3* and *cyp6m2*) were described to be involved in pyrethroid metabolism (Muller et al., 2008). Surprisingly, metabolic resistance can still vary during the course of the day. This is the case of an *Ae. aegypti* population whose resistance to the pyrethroid permethrin is mediated by the *cyp9M9* gene. Expression of this gene is regulated by transcriptional factors enrolled in the circadian rhythm of the insect, varying along the day (Y. Y. Yang et al., 2010).

Finally, phenotypic or target site resistance is designated by modification of the insect molecule where the insecticide binds, inhibiting its effects. Neurotoxic insecticides have as their ultimate target different molecules from the insect central nervous system: the enzyme Acetylcholinesterase (for organophosphates and carbamates), the gama-aminobutiric acid receptor (for ciclodienes), the nicotinic acetylcholine receptors (for spinosyns and neonicotinoids) and the voltage gated sodium channel (for DDT and pyrethroids). Although the mutated target molecule decreases or even abolishes its affinity for the insecticide, it is essential that this alteration does not result in loss of function regarding the insect physiological processes. Since the classical target molecules are much conserved among animals, few mutations are permissive to guarantee the viability of their carriers (ffrench-Constant et al., 1998; Raymond et al., 2001).

The voltage gated sodium channel (Na$_V$) is the effective target for a number of neurotoxins produced by plants and animals, as components of their predation or defense strategies. Knowledge that mutations in the Na$_V$ gene can endow resistance to both the most popular insecticides of the past (DDT) and nowadays (pyrethroids) is leading to significant progress in the understanding of the physiology, pharmacology and evolution of this channel (ffrench-Constant et al., 1998; O'Reilly et al., 2006).

3. The role of the voltage gated sodium channel (Na$_V$) in the nerve impulse propagation in insects

The membrane of all excitable cells (neurons, myocites, endocrinous and egg cells) have voltage gated ion channels responsible for the generation of action potential. These cells react to changes in the electric potential of the membrane, modifying their permeability status (Alberts et al., 2002; Randall et al., 2001). Voltage gated sodium channels (Na$_V$) are transmembrane proteins responsible for the initial action potential in excitable cells (Catterall, 2000). They are members of the protein superfamily which also includes voltage gated calcium (Ca$_V$) and potassium (K$_V$) channels (Jan & Jan, 1992). Both Na$_V$ and Ca$_V$ channels are constituted of four homologous domains whilst K$_V$ is a tetramer with only one domain. A proposed evolution pathway assumes that Ca$_V$ have evolved from K$_v$ by gene duplication during the evolution of multicelular eukaryotes. Na$_V$ channels are supposed to have evolved from an ancestral Ca$_V$ family (family Ca$_V$3) (Spafford et al., 1999). Accordingly, the four Na$_V$ domains are more similar to their Ca$_V$ counterparts than among themselves (Strong et al., 1993). The sodium channel is completely functional by itself, unless the kinetics of opening and closure of the voltage gated channel can be modified by other proteins, sometimes referred to as complementary subunits (beta subunit in mammals and TipE in *Drosophila*) (Catterall et al., 2003).

Cell action potential starts with the depolarization of the membrane, with the internal side attaining a more positive state (compare Figure 1, pannels A and B). A stimulus that causes the depolarization in a given region of the cell membrane promotes activation (opening) of the Na$_V$ in the vicinity. This process results in the influx of Na$^+$ to the cell, enhancing depolarization of the membrane. The action potential works in a positive feedback, that is, once started there is no need of additional stimuli to progress. However, one millisecond after the channel has been activated, the surrounding membrane reaches the Na$^+$ equilibrium potential, and the channel is deactivated. In this state, the pore is still open, but it assumes a conformation that halts the ion influx into the cell (Figure 1, C). After some further milliseconds, the membrane is repolarized and the channel closes, finally returning to its resting configuration (Figure 1, D). This whole process occurs in consonance with other channels and pumps, such as K$_V$ and sodium/ potassium pumps that restore the original electric potential of the cell (Catterall et al., 2003; Randall et al., 2001). The correct operation of sodium channels is essential for nerve impulse propagation. Hence, if the regular propagation of an impulse is altered, as due to the interaction with an insecticide, the organism suffers paralysis and can eventually die.

The structure of Na$_V$ is organized in four homologous domains (I-IV), each containing six hydrophobic segments (S1-S6) and a *P-loop* between S5 and S6 (Figure 2). The segments S1-S4 work as a voltage sensitive module. Since S4 segments are positively charged and sensitive to voltage changes, they move across the membrane in order to initiate the channel activation in response to membrane depolarization (schematically represented in Figure 1, compare relative position of the Na$_V$ blue domains in the different pannels). The pore forming module is composed of the S5-S6 segments and the loop between them, the latter acting as an ion selective filter in the extracellular entrance of the pore (Catterall et al., 2003; Goldin, 2003; Narahashi, 1992). Additionally, the *P-loop* residues D, E, K and A, respectively from domains I, II, III and IV, are critical for the Na$^+$ sensitivity (Zhou et al., 2004).

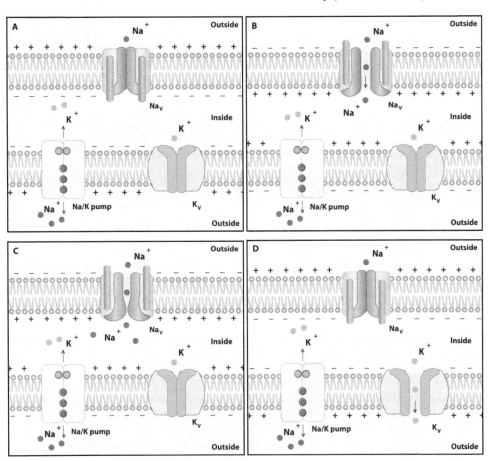

Fig. 1. Propagation of the action potential through a neuronal axon - In the resting potential stage (A) the axon cytoplasm has Na$^+$ and K$^+$ respectively in low and high concentrations compared to the surrounding extracellular fluid. The Na/K pump is constantly expelling three Na+ from the cell for every two K+ it transfers in, which confers a positive charge to the outer part of the membrane. When there is a nervous stimulus, the Na$_V$ opens and the membrane becomes permeable affording the influx of Na$^+$, depolarizing the membrane charge (B). This is the rising phase of the action potential. Soon (~1 millisecond), the Na$_V$ is deactivated, precluding further Na+ entrance to the cell (C), whilst K$^+$ exits the cell through K$_V$ which is now opened, characterizing the falling phase of the action potential (D). The Na/K pump helps to reestablish the initial membrane potential. The action potential generates a wave of sequential depolarization along the axon. Figure based on T. G. Davies et al. (2007b).

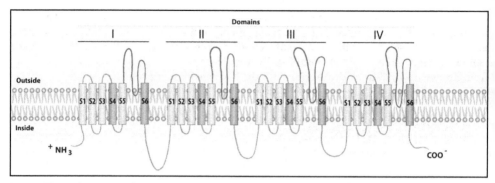

Fig. 2. The voltage gated sodium channel - Scheme representative of the Na$_V$ inserted in a cell membrane, showing its four homologous domains (I-IV), each with six hydrophobic segments (S1-S6). In blue, the voltage sensor segments (S4); in green, the S6 segments, which form the channel pore together with the S5 segments and the link (*P-loop*, in red) between them. Figure adapted from Nelson & Cox (2000).

In the closed state, the putative insecticide contact sites are blocked, corroborating the assumption that pyrethroids and DDT have more affinity to the Na$_V$ channel in its open state when the insecticide stabilizes the open conformation (O'Reilly et al., 2006). These insecticides, therefore, inhibit the channel transition to the non-conducting and deactivated states (T. E. Davies et al., 2008). By interacting with the channel, they form a sort of wedge between segments IIS5 and IIIS6 that restricts displacement of the pore forming helices S5 and S6, preventing closure of the channel. Consequently, the influx of Na$^+$ is prolonged, and the cell is led to work at an abnormal state of hyper-excitability. The amplitude of the Na$^+$ current will not decrease unless the cell's level of hyper-excitability is overcome by its ability to keep the sodium-potassium pump under operation. This process is responsible for the pyrethroid sublethal effect in insects, known as *knockdown* effect, which may lead to paralysis and death if prolonged (T. E. Davies et al., 2008; T. G. Davies et al., 2007b).

Predictive models suggest that DDT and pyrethroids interact with a long and narrow cavity delimited by the IIS4-S5 linker and the IIS5 and IIIS6 helices, accessible to lipophilic insecticides. Moreover, some of the aminoacids belonging to the helices engaged in contact with these insecticides are not conserved among arthropods and other animals, and this could be responsible for the selectivity of pyrethroid effects against insects (O'Reilly et al., 2006). The crystal structure of a Na$_V$ has been recently published (Payandeh et al., 2011), pointing to a better understanding of the channel function and to its interaction with targeted compounds in a near future.

Besides pyrethroids and DDT, other insecticides act on the voltage gated sodium channel, like the sodium channel blocker insecticides (SCBIs) and N-alkylamide inseticides (like BTG 502). There are few reports about these compounds. However, it is known that SCBIs, such as indoxicarb, act by blocking the impulse conduction, an effect opposite to that of DDT and pyrethroids (Du et al., 2011).

4. The knockdown effect and the *kdr* phenotype

In the early 1950s, no sooner had DDT been introduced as an insecticide than resistant strains of houseflies were described. When exposed to DDT, these insects in general did not suffer paralysis followed by death (*knockdown*) but, at most, presented a momentary paralysis followed by complete locomotion recovery, this phenotype being named *kdr* (*knockdown* resistance) (Busvine, 1951; Harrison, 1951; Milani, 1954). Since the introduction of pyrethroids, plenty of insect species exhibiting the *kdr* phenotype have been observed, attributed to previous DDT selection pressure, characterizing cross-resistance between both insecticides (Hemingway & Ranson, 2000). *Kdr* resistance results in 10-20 fold decrease in the sensitivity to the insecticide. However, *kdr* lineages of some species can exhibit up to 100 X increased pyrethroid resistance, an effect denominated *super-kdr*. *Kdr* and *super-kdr* alleles act as recessive traits and hence may persist at low levels in the population in heterozygous individuals (T. G. Davies et al., 2007a).

Over three decades after the description of the *kdr* effect, electrophysiological studies based on neuronal cells and tissues suggested that Na_V had to be the target site for pyrethroids. These reports also indicated that cross-resistance between pyrethroids and DDT must be related to that channel (Pauron et al., 1989). Concomitantly, the gene *paralytic* (*para*) from *Drosophila melanogaster* was cloned and sequenced. This gene is placed in the *locus* related to behavioral changes and paralysis after exposure to high temperatures, similar to the *knockdown* effect produced by DDT and pyrethroids (Loughney et al., 1989). Comparisons within vertebrate nucleotide sequences revealed that *para* is homologous to the voltage gated sodium channel gene (*Na_V*) (Loughney & Ganetzky, 1989). It was then shown, with a DDT resistant housefly lineage, that the *locus* homologous to *para* was in strong linkage with the *kdr* phenotype (Williamson et al., 1993). This evidence was soon extended to other insect species plagues or vectors, such as the tobacco budworm *Heliothis virescens* (Taylor et al., 1993), the cockroach *Blatella germanica* (Dong & Scott, 1994) and the mosquito *Aedes aegypti* (Severson et al., 1997).

Hitherto, Na_V is the only molecule incriminated as the target site for DDT and pyrethroids. Although it has been implied that type II pyrethroids can interact with the GABA receptor (which is the target, for instance, of the insecticide dieldrin), this interaction has not been considered toxically important (Soderlund & Bloomquist, 1989). Research on the molecular interaction between pyrethroids and their target site presently guides a series of approaches towards the development of a great variety of natural and synthetic neurotoxicants acting on the Na_V (Soderlund, 2010).

5. Molecular biology of the insect Na_V and the *kdr* mutation

A great variety of sodium channels have been identified by electrophysiological assays, purification and cloning (Goldin, 2001). In mammals, nine Na_V genes are known, with distinct electrophysiological properties as well as different expression profiles in the tissues and throughout development (Goldin, 2002; Yu & Catterall, 2003), phylogenetic analyses revealing that all are members of only one unique family, deriving from relatively recent gene duplications and chromosome rearrangements. On the other hand, Ca_V and Ka_V have little protein sequence identity and present diverse functions, indicative of more ancient segregation of their coding genes (Catterall et al., 2003).

The Na_V orthologous genes and cDNAs from *D. melanogaster* and *An. gambiae* share, respectively, 56-62% and 82% of nucleotide identity, evidencing a high degree of conservation between these species. The Na_V C-terminal is the most variable region, but as in all dipterans, it is mainly composed of aminoacids of short (Gly, Ala, Ser, Pro) or negative (Asp, Glu) side chains, suggesting a conserved function in this domain (T. G. Davies et al.,

2007a). Concerning size, the voltage gated sodium channel of *Ae. aegypti* (*AaNa$_V$*), for instance, presents 293 Kb of genomic DNA, with 33 exons. Its longer observed transcript has an ORF of 6.4 Kb, coding for 2,147 aminoacids for a protein estimated in 241 KDa (Chang et al., 2009).

The existence of two Na$_V$ evolutionary lines in invertebrates, represented by the genes *para* and *DSC1* in *D. melanogaster*, has been suggested (Spafford et al., 1999). These lines do not correspond to the different genes observed among vertebrates, and they are supposed to have arisen after vertebrate and invertebrate splitting (Goldin, 2002). DSC1 plays a role in the olfactory system (Kulkarni et al., 2002) as it has been found in the peripheral nervous system and also at high density in the synaptic regions. DSC1 is sensitive to tetradotoxin, a specific Na$_V$ blocker (Zhang et al., 2011), while BSC1, its homologous in *B. germanica*, has also been identified as a putative sodium channel, being expressed in the cockroach nerve cord, muscle, gut, fat body and ovary (Liu et al., 2001). Neither *DSC1* nor *BSC1*, however, mapped with any *locus* related to insecticide resistance (Loughney et al., 1989; Salkoff et al., 1987). Actually, these channels probably represent prototypes of a new Ca$_V$ family, highly related to the known Na$_V$ and Ca$_V$ (Zhang et al., 2011; Zhou et al., 2004). On the other hand, in invertebrates, the *D. melanogaster para* gene (or *DmNa$_V$*) and its equivalent in other species actually code for sodium channels and are related to pyrethroid/DDT resistance and to behavioral changes, as aforementioned.

In his review, Goldin (2002) suggested that two to four genes coding for sodium channels should exist in insects and that differences among them would not result from distinct genes but from pos-transcriptional regulation. Accordingly, even after publication of many insect genome sequences, there has been no mention whatsoever of Na$_V$ gene duplication. Furthermore, recent reports attribute the diversity in Na$_V$ sequences to alternative splicing and RNA editing. These modifications seem to be tissue and stage specific and might also have some influence on pyrethroid resistance (Liu et al., 2004; Song et al., 2004; Sonoda et al., 2008).

5.1 Alternative mRNA splicing in the Na$_V$

Briefly, alternative splicing is a post-transcriptional regulated event characterized when certain exons are removed together with introns. This is a common mechanism of gene expression regulation and increment of protein diversity in eukaryotes. The process may occur in different ways: complete exons can be included or excluded (optional exons), splicing sites can be altered and introns can be retained in the mature mRNA. There are also mutually exclusive pairs of exons, when two exons never unite in the same transcript. Alternative mRNA splicing introduces variability in both sequence and size of the Na$_V$ intracellular region, which by itself should have an impact on its operation (T. G. Davies et al., 2007a).

The regulation for excision of an exon, in detriment of others, may be tissue and development specific. In the context of pyrethroid resistance, it is important to know to what extent alternative splicing events compromise the interaction between the insecticide and the channel. It is also necessary to investigate the amount of alternative transcripts in the course of development and their distribution in the different tissues of the insect. The sodium channel genes have alternative exons that potentially synthesize a great number of different mRNAs (Figure 3). There are also mutually exclusive exons that occur in the transmembrane regions of domains II and III (T. G. Davies et al., 2007a). In *D. melanogaster*, many alternative splicing sites have been identified, with seven optional regions and two pairs of mutually exclusive exons (Figure 3) (Olson et al., 2008). These sites are conserved in *M. domestica* (Lee et al., 2002) generating, in both species, 512 potential Na_V transcripts by alternative splicing. However, they are not all necessarily expressed as less than 10 were actually observed in mRNA pools (Soderlund, 2010).

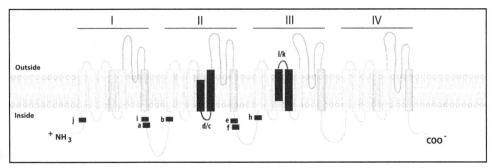

Fig. 3. Alternative splice in the insect voltage gated sodium channel gene. Scheme of Na_V with the sites of alternative exons of $DmNa_V$ indicated in dark color. Exons *a, b, i, j, e* and *f* are optional, while *d/c* and *l/k* are mutually exclusive. Figure adapted from Oslon et al. (2008).

The aminoacid sequences translated from optional exons are conserved and generally consist of intracellular domains of the channel, suggesting functional relevance to these events. Na_V transcript diversity derived from alternative splicing has been investigated in insects of many orders, revealing a high level of conservation, as shown in the cockroach *B. germanica* (Liu et al., 2001; Song et al., 2004), the silk worm *Bombyx mori* (Shao et al., 2009), the moth *Plutella xylostella* (Sonoda et al., 2008) and the mosquitoes *An. gambiae* (T. G. Davies et al., 2007a) and *Ae. aegypti* (Chang et al., 2009). However, in some species not all exons were observed nor their expression detected (see Davies et al., 2007a).

There are two mutually exclusive exons (called c/d) that code for a region between IIS4 and IIS5 segments (Figure 3). The absence of one of these exons might be important for pyrethroid resistance, since the *super-kdr* mutation (Met918Thr) is located in this region, as will be discussed further. In the cockroach *B. germanica*, the mutually exclusive exon pair *k/l* codes for the voltage sensitive region at domain III. The two varieties $BgNa_V1.1a$ and $BgNa_V1.1b$[1], which contain the exons *l* and *k* respectively, exhibit distinct electrophysiological properties. Furthermore, $BgNa_V1.1b$ is 100X more resistant to the pyrethroid deltamethrin than $BgNa_V1.1a$ (Du et al., 2006).

5.2 Sodium channel RNA editing

RNA editing has an important role in the regulation of gene expression and protein diversity. Recent studies implicate RNA editing in the removal of exons in alternative splicing sites, in the antagonism of interference RNA process (iRNA), in the modulation of mRNA processing and in the generation of new exons (for a review see Y. Yang et al., 2008). The basic mechanism of diversity generated by RNA editing includes nucleoside modifications such as C to U or A to I deaminations. Besides, it is possible that non-

---

[1] The genes annotation is in accordance with the nomenclature suggested by Goldin (2000).

templated nucleotides can be inserted in the edited mRNA. This process alters the protein aminoacid constitution so that it differs from the predicted genomic DNA sequence (Brennicke et al., 1999).

Liu et al. (2004) claimed that RNA editing should be the main regulatory mechanism to modulate the insect Na_V function. For instance, no correlation was found between a variety of $DmNa_V$ originated by alternative splicing and the observed changes in gating properties. Therefore it was implied that RNA editing might play a primary role in determining the voltage dependence of activation and deactivation of $DmNa_V$ variants (Olson et al., 2008). At least 10 A/I RNA editing substitutions were observed in the $DmNa_V$ in different points of the Drosophila life cycle indicating developmental regulation (Palladino et al., 2000). These sites are highly conserved in various organisms. Type U/C editing, which is more usual in mitochondria and plastids from higher plants, was also observed in $DmNa_V$ and $BgNa_V$, with electrophysiological alterations in both cases (Liu et al., 2004). Hence, RNA editing should play an important role in the generation of channels with distinct affinities to insecticides. Thus, it seems reasonable to infer that insecticide pressure selects for an adaptive mechanism which might spatially and temporally modulate Na_V mRNA editing. Still, in $Cx.$ $quinquefasciatus$ mosquitoes, diversity based on U/A editing in the sodium channel mRNA was shown to be related to pyrethroid resistance (Xu et al., 2006). In $Ae.$ $aegypti$, however, recent analysis of $AaNa_V$ transcripts from a pyrethroid resistant lineage did not identify any sign of RNA editing (Chang et al., 2009).

5.3 The *kdr* mutation

The very first mutation identified as responsible for the *kdr* trait was a leucine to phenylalanine substitution (Leu1014Phe)[2] in the Na_V IIS6 segment of *M. domestica* (Ingles et al., 1996). Since then, the genomic sequence spanning the region coding for the IIS6 segment has been explored in a vast number of insects, in most of which, the same substitution being found at homologous sites (1014). Besides Phe, Ser is also encountered replacing Leu at the 1014 site in *An. gambiae*. They were initially observed respectively in western and eastern African regions, being commonly referred to as *w-kdr* and *e-kdr* mutations (Pinto et al., 2006). However, nowadays it is known that none of these alleles is restricted to either part of the continent (Ranson et al., 2011). A different substitution in the same 1014 site, Leu1014His, was also associated to pyrethroid resistance in the tobacco budworm *Heliothis virescens* (Park et al., 1999). Many studies identified at least 20 additional substitutions in the Na_V sequence, the majority being placed between segments S4 and S5, or internally to segments S5 or S6 of domain II. However, for most of them, the relationship with pyrethroid resistance is only speculative. Good compilations have recently been presented (T. G. Davies et al., 2007a; Dong, 2007; Du et al., 2009).

It is noteworthy that many of these mutations are not in the precise domain of interaction between insecticide and Na_V (O'Reilly et al., 2006). On the other hand, it is likely that substitutions in these points of interaction could result in the *super-kdr* trait, which has a more pronounced resistance effect (T. G. Davies et al., 2007b). This phenotype was also first described in *M. domestica* (Williamson et al., 1996) and *Haematobia irritans* (Guerrero et al., 1997). In both species, beyond the Leu1014Phe substitution, a Met918Thr mutation (in the IIS4-S5 linker) was disclosed in flies with very high resistant ratios to pyrethroids, referred

---

[2] Number refers conventionally to the position in the voltage gated sodium channel primary sequence of *M. domestica* Vssc1, according to Soderlund & Knipple 2003.

to as the *super-kdr* mutation (Jamroz et al., 1998). However, since it occurs only in association with the Leu1014Phe mutation, its isolated effects are as yet unknown. Although no *super-kdr* mutation has so far been identified in mosquitoes, it was suggested that Leu932Phe, in association with Ile936Val (both also in the IIS4-S5 linker), in *Culex* might play this role, being the first example of *super-kdr* in this group (T. G. Davies et al., 2007a). Accordingly, these sites have proved to be important for the interaction between Na$_V$, in the *D. melanogaster* sodium channel and pyrethroids or DDT (Usherwood et al., 2007).

Substitutions in site 929 are also involved in enhanced pyrethroid resistance, as is the case with the Lepidoptera *Plutella xylostella* mutation Thr929Ile, detected in association with Leu1014Phe (Schuler et al., 1998). However, in the maize weevil *Sitophilus zeamais*, the Thr929Ile was found alone (Araujo et al., 2011). In the louse *Pediculus capitis*, in turn, the Thr929Ile mutation was together with Leu932Phe (Lee et al., 2000). There were other substitutions in the same site: Thr/Cys and/or Thr/Val in the diamondblack moth *Frankliniella occidentalis* (Forcioli et al., 2002) and in the cat flea *Ctenocephalides felis* (Bass et al., 2004).

Ae. aegypti mosquitoes do not present any substitution in the classic 1014 *kdr* site, unlike many other insects or even mosquitoes from other genera, such as *Anopheles* and *Culex*, very likely because the 1014 site of *Ae. aegypti* Na$_V$ is coded by a CTA, in place of the TTA codon present in the majority of other insects. For this reason, two simultaneous nucleotide substitutions would be necessary in order to change from Leu (CTA) to Phe (TTT) or Ser (TCA) (Martins et al., 2009a; Saavedra-Rodriguez et al., 2007). Instead, mutations in different positions have been observed in *Ae. aegypti* populations from Latin America and Southeast Asia, but at least two sites seem to be indeed related to pyrethroid resistance: 1016 (Val to Ile or Gly) and 1534 (Phe to Cys), respectively in the IIS6 and IIIS6 segments (Brengues et al., 2003; Harris et al., 2010; Martins et al., 2009a, b; Saavedra-Rodriguez et al., 2007). Mutations in the vicinity of this site in the IIIS6 segment were also encountered in the southern cattle tick *Rhipicephalus microplus* (He et al., 1999) and in the two-spotted spider mite *Tetranychus urticae* (Tsagkarakou et al., 2009).

Although different Na$_V$ site mutations are known to confer resistance to pyrethroids, their number is quite restricted; additionally, far related taxa present alterations in the same homologous sites. For instance, the Leu1014Phe *kdr* mutation must have arisen at least on four independent occasions in *An. gambiae* (Pinto et al., 2007). Alterations that do not interfere with the endogenous physiological functions of the Na$_V$ must be rare as it is much conserved among animals (ffrench-Constant et al., 1998). As a matter of fact, most of the species studied so far have the *kdr* mutation in the 1014 site, few species proving otherwise due to codon constraints, like *Ae. aegypti* and some anopheline species.

6. Molecular assays for monitoring frequency of *kdr* mutation in insect natural populations

Currently, there are many PCR based diagnostic methods for *kdr* mutation available with elevated sensitivity and specificity. For technique choice, one must consider mainly the laboratory resources, facilities and training of technical personnel, which is as important as establishing an defining localities and frequency of sampling. There is neither consensus nor strict rules suitable for all insect species or even for different populations of the same species. Resistance is a very dynamic process depending upon a series of external factors. Therefore, resistance level as well as the selected mechanisms may fluctuate in a short period of time and space (Kelly-Hope et al., 2008). Moreover, one must be aware about the patterns of distribution and structure of the evaluated populations in order to determine an adequate frequency and sampling size (Ranson et al., 2011).

Allele-specific PCR assays (AS-PCR), as the name suggests, consists of amplification and detection of a specific allele from the DNA of an individual, who is further classified as hetero or homozygous for that allele. Many methodologies based on this strategy have been well succeeded in high-throughput individual diagnostic of *kdr* mutations. Herein, we highlight some PCR based amplifications by allele-specific primers and TaqMan genotyping.

There is ample variation for PCR methods based on allele specific primers. As a first example, one can use two primers (forward and reverse) common for both alleles that amplify a region spanning the mutation site. In this case, additional specific primers, bearing the SNP (single nucleotide polymorphism) at the 3′-end, have opposite orientations in relation to each other (Figure 2-A). The common primers will pair themselves giving rise to a bigger product (that can also be assumed as the positive control reaction) and shorter ones, the consequence of pairing with each allele-specific primer of contrary orientation. The common primers must anneal at sites that result in differently sized products when paring with the specific ones. If both alleles are present (cases when the individual is heterozygous) three products with distinct sizes will be produced (Chen et al., 2010; Harris et al., 2010).

Instead of amplifying a common region for both alleles, it is possible to directly obtain only the specific products (Figure 2-B). This can be accomplished by using only one common primer in one orientation and the two allele specific primers in the opposite sense. However, since the specific primers are at the same orientation and their specificity continues lying upon the 3′-end, something should be incremented in order to obtain distinguishable products. Germer & Higuchi, (1999), later improved by Wang et al. (2005), proposed attaching a GC-tail of different sizes to the 5′-end of the specific primers in a way that the products could be distinguishable by their Tm in a melting curve analysis. In this case the mix reaction contains a fluorescent dye, which lights up when bounded to double strand DNA, carried out in a fluorescence-detecting thermocycler ("Real time PCR"). Additionally, a different mismatch (pirimidine for purine or vice–versa) is added to the third site before the 3′-end of each allele specific primer, in order to strengthen their specificity (Okimoto & Dodgson, 1996). Alternatively, the products can also be distinguishable in a gel electrophoresis.

The second group of techniques is based on the amplification of a region spanning the *kdr* mutation site followed by the detection of the different alleles by specific hybridization with minor groove binding (MGB) DNA fluorescent probes, also known as TaqMan assay (Figure 2-C). Different alleles can be detected in the same reaction, since each probe is attached to a distinct fluorophore. The probe is constituted of an oligonucleotide specific for the SNP with a reporter fluorescent dye in the 5′-end and a non fluorescent quencher in the 3′-end (Araujo et al., 2011; Morgan et al., 2009; Yanola et al., 2011). Bass et al., (2007) concluded that TaqMan probes were the most accurate for *kdr* genotyping among six different evaluated methods.

Other techniques have also been applied. The *Hola* (Heated Oligonucleotide Ligation Assay, see details in Black et al., 2006) revealed high specificity in detecting different Na$_V$ alleles in the 1011 (Ile, Met and Val) and 1016 (Val, Ile and Gly) sites from Thai *Ae. aegypti* populations (Rajatileka et al., 2008) and in the 1014 site of *Cx. quinquefasciatus* from Sri Lanka (Wondji et

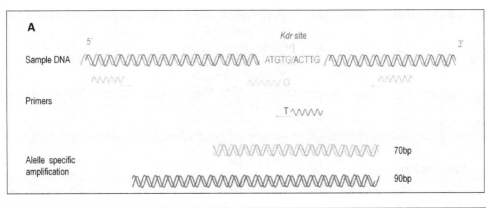

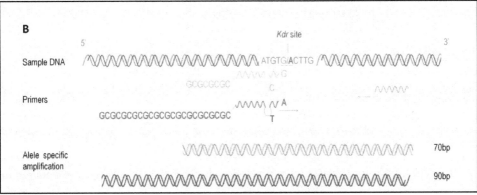

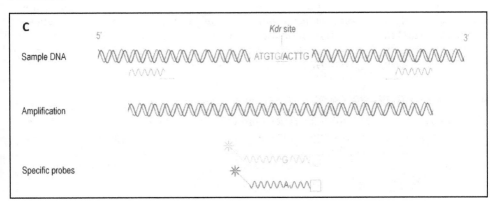

Fig. 4. Examples for kdr genotyping based on PCR methods. A – Allelic specific PCR with specific primers in different orientations; B – Allelic specific PCR with specific primers in the same orientation but with additional and differently sized [GC]ₙ tails, in addition to a mismatch in the 3rd base before the 3'-end; C – TaqMan assay based on specific probes with a different luminescence for each allele. Figure adapted from Yanola et al. (2011).

al., 2008). However, comparison between *HOLA* and pyrosequencing revealed more specificity for this latter method in the diagnostic of the *kdr* mutation Leu1014Phe in *Cx. quinquefasciatus* (Wondji et al., 2008). Sequencing of regions that encompass the SNP allows a direct visualization of the nucleotide allele sequences, eliminating the problem of unspecific amplification or hybridization of PCR based protocols. Moreover, it enables visualizing potential novel variations that would never be identified by PCR diagnostic SNP techniques. However, sequencing in large scale is much more expensive than the aforementioned genotyping tools. It is also mandatory that the eletropherograms generated have a clean profile, so that the heterozygous individuals can be undoubtedly discriminated.

7. Conclusions

New strategies for arthropod control based on the release of laboratory manipulated insects that would suppress or substitute natural populations are being tested in the field with great prospect. The release of transgenic insects carrying a dominant lethal gene (RIDL) (Black et al., 2011) or of mosquitoes with the intracellular *Wolbachia*, that lead to refractoriness to other parasites (Werren et al., 2008) are currently the most discussed strategies. However, the laboratory handling process has to consider specific and sometimes complex aspects for each insect species, and it may take many years until field control based on this kind of approach can be effectively accomplished. Moreover, field studies that guarantee the environmental safety of releasing manipulated insects may take even longer. Hence, even if these strategies prove to be efficient to reduce, extinguish, or substitute a target insect population, the use of insecticides may still indeed play an essential role for many years to come, especially during periods of high insect or disease incidence.

Pyrethroids are largely the most adopted insecticide class in agriculture and for public health purposes. Their use tends to increase, since pyrethroids are the only safe compound to impregnate insecticide treated nets (ITNs), a strategy under expansion against mosquitoes. Advances regarding knowledge of its target, the voltage gated sodium channel, can contribute to the design of new compounds as well as the rapid identification of resistance related mutations. The continuous monitoring of insecticide resistance status, and its mechanisms, in natural populations has proven to be an important tool in the preservation of these compounds.

8. Acknowledgements

We thank Andre Torres for his illustrations presented in this work, the Instituto de Biologia do Exército (IBEx) and Instituto Nacional de Ciência e Tecnologia em Entomologia Molecular (INCT-EM). English review and revision by Mitchell Raymond Lishon, native of Chicago, Illinois, U.S.A – U.C.L.A, 1969. Financial support: Fiocruz, Pronex-dengue/CNPq, Faperj, SVS/MS and CAPES.

9. References

Alberts, B., Johnson, A., Lewis, J., Raff, M., Roberts, K., & Walter, P. (2002). *Molecular Biology of the Cell* (4th), Garland Science, New York.

Araujo, R.A., Williamson, M.S., Bass, C., Field, L.M., & Duce, I.R. (2011). Pyrethroid resistance in *Sitophilus zeamais* is associated with a mutation (T929I) in the voltage-gated sodium channel. *Insect Mol Biol*, Vol.20, No. 4, (Aug), pp. 437-445, ISSN 1365-2583.

Awolola, T.S., Oduola, O.A., Strode, C., Koekemoer, L.L., Brooke, B., & Ranson, H. (2009). Evidence of multiple pyrethroid resistance mechanisms in the malaria vector *Anopheles gambiae* sensu stricto from Nigeria. *Trans R Soc Trop Med Hyg*, Vol.103, No. 11, (Nov), pp. 1139-1145, ISSN 1878-3503.

Bass, C., Nikou, D., Donnelly, M.J., Williamson, M.S., Ranson, H., Ball, A., Vontas, J., & Field, L.M. (2007). Detection of knockdown resistance (kdr) mutations in *Anopheles gambiae*: a comparison of two new high-throughput assays with existing methods. *Malar J*, Vol.6, No. pp. 111, ISSN 1475-2875.

Bass, C., Schroeder, I., Turberg, A., Field, L.M., & Williamson, M.S. (2004). Identification of mutations associated with pyrethroid resistance in the para-type sodium channel of the cat flea, *Ctenocephalides felis*. *Insect Biochem Molec*, Vol.34, No. 12, (Dec), pp. 1305-1313, ISSN 0965-1748.

Beaty, B.J., & Marquardt, W.C. (1996). *The biology of disease vectors*, University Press of Colorado, Colorado.

Black, W.C.t., Alphey, L., & James, A.A. (2011). Why RIDL is not SIT. *Trends Parasitol*, Vol.No. (Jun 7), ISSN 1471-5007.

Black, W.C.t., Gorrochotegui-Escalante, N., & Duteau, N.M. (2006). Heated oligonucleotide ligation assay (HOLA): an affordable single nucleotide polymorphism assay. *J Med Entomol*, Vol.43, No. 2, (Mar), pp. 238-247, ISSN 0022-2585.

Brengues, C., Hawkes, N.J., Chandre, F., McCarroll, L., Duchon, S., Guillet, P., Manguin, S., Morgan, J.C., & Hemingway, J. (2003). Pyrethroid and DDT cross-resistance in *Aedes aegypti* is correlated with novel mutations in the voltage-gated sodium channel gene. *Med Vet Entomol*, Vol.17, No. 1, (Mar), pp. 87-94, ISSN 0269-283X.

Brennicke, A., Marchfelder, A., & Binder, S. (1999). RNA editing. *FEMS Microbiol Rev*, Vol.23, No. 3, (Jun), pp. 297-316, ISSN 0168-6445.

Busvine, J.R. (1951). Mechanism of resistance to insecticides in housefly. *Nature*, Vol.168, No. pp. 193-195,

Catterall, W.A. (2000). From ionic currents to molecular mechanisms: the structure and function of voltage-gated sodium channels. *Neuron*, Vol.26, No. 1, (Apr), pp. 13-25, ISSN 0896-6273.

Catterall, W.A., Chandy, K.G., Clapham, D.E., Gutman, G.A., Hofmann, F., Harmar, A.J., Abernethy, D.R., & Spedding, M. (2003). International Union of Pharmacology: Approaches to the nomenclature of voltage-gated ion channels. *Pharmacol Rev*, Vol.55, No. 4, (Dec), pp. 573-574, ISSN 0031-6997.

Chang, C., Shen, W.K., Wang, T.T., Lin, Y.H., Hsu, E.L., & Dai, S.M. (2009). A novel amino acid substitution in a voltage-gated sodium channel is associated with knockdown resistance to permethrin in *Aedes aegypti*. *Insect Biochem Mol Biol*, Vol.39, No. 4, (Apr), pp. 272-278, ISSN 1879-0240.

Chen, L., Zhong, D., Zhang, D., Shi, L., Zhou, G., Gong, M., Zhou, H., Sun, Y., Ma, L., He, J., et al. (2010). Molecular ecology of pyrethroid knockdown resistance in *Culex pipiens pallens* mosquitoes. *PLoS One*, Vol.5, No. 7, pp. e11681, ISSN 1932-6203.

David, J.P., Strode, C., Vontas, J., Nikou, D., Vaughan, A., Pignatelli, P.M., Louis, C., Hemingway, J., & Ranson, H. (2005). The *Anopheles gambiae* detoxification chip: a highly specific microarray to study metabolic-based insecticide resistance in malaria vectors. *Proc Natl Acad Sci U S A*, Vol.102, No. 11, (Mar 15), pp. 4080-4084, ISSN 0027-8424.

Davies, T.E., O'Reilly, A.O., Field, L.M., Wallace, B., & Williamson, M.S. (2008). Knockdown resistance to DDT and pyrethroids: from target-site mutations to molecular modelling. *Pest Manag Sci*, Vol.64, No. 11, (Nov), pp. 1126-1130, ISSN 1526-498X.

Davies, T.G., Field, L.M., Usherwood, P.N., & Williamson, M.S. (2007a). A comparative study of voltage-gated sodium channels in the Insecta: implications for pyrethroid resistance in Anopheline and other Neopteran species. *Insect Mol Biol*, Vol.16, No. 3, (Jun), pp. 361-375, ISSN 0962-1075.

Davies, T.G., Field, L.M., Usherwood, P.N., & Williamson, M.S. (2007b). DDT, pyrethrins, pyrethroids and insect sodium channels. *IUBMB Life*, Vol.59, No. 3, (Mar), pp. 151-162, ISSN 1521-6543.

Dong, K. (2007). Insect sodium channels and insecticide resistance. *Invert Neurosci*, Vol.7, No. 1, (Mar), pp. 17-30, ISSN 1354-2516.

Dong, K., & Scott, J.G. (1994). Linkage of kdr-type resistance and the para-homologous sodium channel gene in German cockroaches (*Blattella germanica*). *Insect Biochem Mol Biol*, Vol.24, No. 7, (Jul), pp. 647-654, ISSN 0965-1748.

Du, Y., Khambay, B., & Dong, K. (2011). An important role of a pyrethroid-sensing residue F1519 in the action of the N-alkylamide insecticide BTG 502 on the cockroach sodium channel. *Insect Biochem Mol Biol*, Vol.41, No. 7, (Jul), pp. 446-450, ISSN 1879-0240.

Du, Y., Liu, Z., Nomura, Y., Khambay, B., & Dong, K. (2006). An alanine in segment 3 of domain III (IIIS3) of the cockroach sodium channel contributes to the low pyrethroid sensitivity of an alternative splice variant. *Insect Biochem Mol Biol*, Vol.36, No. 2, (Feb), pp. 161-168, ISSN 0965-1748.

Du, Y., Nomura, Y., Luo, N., Liu, Z., Lee, J.E., Khambay, B., & Dong, K. (2009). Molecular determinants on the insect sodium channel for the specific action of type II pyrethroid insecticides. *Toxicol Appl Pharmacol*, Vol.234, No. 2, (Jan 15), pp. 266-272, ISSN 1096-0333.

ffrench-Constant, R.H., Daborn, P.J., & Le Goff, G. (2004). The genetics and genomics of insecticide resistance. *Trends Genet*, Vol.20, No. 3, (Mar), pp. 163-170, ISSN 0168-9525.

ffrench-Constant, R.H., Pittendrigh, B., Vaughan, A., & Anthony, N. (1998). Why are there so few resistance-associated mutations in insecticide target genes? *Philos Trans R Soc Lond B Biol Sci*, Vol.353, No. 1376, (Oct 29), pp. 1685-1693, ISSN 0962-8436.

Forcioli, D., Frey, B., & Frey, J.F. (2002). High nucleotide diversity in the para-like voltage-sensitive sodium channel gene sequence in the western flower thrips (Thysanoptera : Thripidae). *Journal of Economic Entomology*, Vol.95, No. 4, (Aug), pp. 838-848, ISSN 0022-0493.

Germer, S., & Higuchi, R. (1999). Single-tube genotyping without oligonucleotide probes. *Genome Res*, Vol.9, No. 1, (Jan), pp. 72-78, ISSN 1088-9051.

Goldin, A.L. (2001). Resurgence of sodium channel research. *Annu Rev Physiol*, Vol.63, No. pp. 871-894, ISSN 0066-4278.

Goldin, A.L. (2002). Evolution of voltage-gated Na(+) channels. *J Exp Biol*, Vol.205, No. Pt 5, (Mar), pp. 575-584, ISSN 0022-0949.

Goldin, A.L. (2003). Mechanisms of sodium channel inactivation. *Curr Opin Neurobiol*, Vol.13, No. 3, (Jun), pp. 284-290, ISSN.

Goldin, A.L., Barchi, R.L., Caldwell, J.H., Hofmann, F., Howe, J.R., Hunter, J.C., Kallen, R.G., Mandel, G., Meisler, M.H., Netter, Y.B., *et al.* (2000). Nomenclature of voltage-gated sodium channels. *Neuron*, Vol.28, No. 2, (Nov), pp. 365-368, ISSN 0896-6273.

Guerrero, F.D., Jamroz, R.C., Kammlah, D., & Kunz, S.E. (1997). Toxicological and molecular characterization of pyrethroid-resistant horn flies, *Haematobia irritans*: identification of kdr and super-kdr point mutations. *Insect Biochem Mol Biol*, Vol.27, No. 8-9, (Aug-Sep), pp. 745-755, ISSN 0965-1748.

Harris, A.F., Rajatileka, S., & Ranson, H. (2010). Pyrethroid resistance in *Aedes aegypti* from Grand Cayman. *Am J Trop Med Hyg*, Vol.83, No. 2, (Aug), pp. 277-284, ISSN 1476-1645.

Harrison, C.M. (1951). Inheritance of resistance of DDT in the housefly, *Musca domestica* L. *Nature*, Vol.167, No. 4256, (May 26), pp. 855-856, ISSN 0028-0836.

He, H., Chen, A.C., Davey, R.B., Ivie, G.W., Wagner, G.G., & George, J.E. (1999). Sequence analysis of the knockdown resistance-homologous region of the para-type sodium channel gene from pyrethroid-resistant *Boophilus microplus* (Acari: Ixodidae). *J Med Entomol*, Vol.36, No. 5, (Sep), pp. 539-543, ISSN 0022-2585.

Hemingway, J., Hawkes, N.J., McCarroll, L., & Ranson, H. (2004). The molecular basis of insecticide resistance in mosquitoes. *Insect Biochem Mol Biol*, Vol.34, No. 7, (Jul), pp. 653-665, ISSN 0965-1748.

Hemingway, J., & Ranson, H. (2000). Insecticide resistance in insect vectors of human disease. *Annu Rev Entomol*, Vol.45, No. pp. 371-391, ISSN 0066-4170.

Ingles, P.J., Adams, P.M., Knipple, D.C., & Soderlund, D.M. (1996). Characterization of voltage-sensitive sodium channel gene coding sequences from insecticide-susceptible and knockdown-resistant house fly strains. *Insect Biochem Mol Biol*, Vol.26, No. 4, (Apr), pp. 319-326, ISSN 0965-1748.

Jamroz, R.C., Guerrero, F.D., Kammlah, D.M., & Kunz, S.E. (1998). Role of the kdr and super-kdr sodium channel mutations in pyrethroid resistance: correlation of allelic frequency to resistance level in wild and laboratory populations of horn flies (*Haematobia irritans*). *Insect Biochem Mol Biol*, Vol.28, No. 12, (Dec), pp. 1031-1037, ISSN 0965-1748.

Jan, L.Y., & Jan, Y.N. (1992). Tracing the roots of ion channels. *Cell*, Vol.69, No. 5, (May 29), pp. 715-718, ISSN 0092-8674.

Kelly-Hope, L., Ranson, H., & Hemingway, J. (2008). Lessons from the past: managing insecticide resistance in malaria control and eradication programmes. *Lancet Infect Dis*, Vol.8, No. 6, (Jun), pp. 387-389, ISSN 1473-3099.

Kulkarni, N.H., Yamamoto, A.H., Robinson, K.O., Mackay, T.F., & Anholt, R.R. (2002). The DSC1 channel, encoded by the smi60E locus, contributes to odor-guided behavior in *Drosophila melanogaster*. *Genetics*, Vol.161, No. 4, (Aug), pp. 1507-1516, ISSN 0016-6731.

Kumar, S., Thomas, A., Sahgal, A., Verma, A., Samuel, T., & Pillai, M.K. (2002). Effect of the synergist, piperonyl butoxide, on the development of deltamethrin resistance in yellow fever mosquito, *Aedes aegypti* L. (Diptera: Culicidae). *Arch Insect Biochem Physiol*, Vol.50, No. 1, (May), pp. 1-8, ISSN 0739-4462.

Lee, S.H., Ingles, P.J., Knipple, D.C., & Soderlund, D.M. (2002). Developmental regulation of alternative exon usage in the house fly Vssc1 sodium channel gene. *Invert Neurosci*, Vol.4, No. 3, (Apr), pp. 125-133, ISSN 1354-2516.

Lee, S.H., Yoon, K.S., Williamson, M.S., Goodson, S.J., Takano-Lee, M., Edman, J.D., Devonshire, A.L., & Clark, J.M. (2000). Molecular analysis of kdr-like resistance in permethrin-resistant strains of head lice *Pediculus capitis*. *Pestic Biochem Physiol*, Vol.66, No. pp. 103-143.

Liu, Z., Chung, I., & Dong, K. (2001). Alternative splicing of the BSC1 gene generates tissue-specific isoforms in the German cockroach. *Insect Biochem Mol Biol*, Vol.31, No. 6-7, (Apr 27), pp. 703-713, ISSN 0965-1748.

Liu, Z., Song, W., & Dong, K. (2004). Persistent tetrodotoxin-sensitive sodium current resulting from U-to-C RNA editing of an insect sodium channel. *Proc Natl Acad Sci U S A*, Vol.101, No. 32, (Aug 10), pp. 11862-11867, ISSN 0027-8424.

Lockwood, J.A., Sparks, T.C., & Story, R.N. (1984). Evolution of insect resistance to insecticides: a reevaluation of the roles of physiology and and behavior. *Bull Entomol Soc Am*, Vol.30, No. pp. 41-51.

Loughney, K., & Ganetzky, B. (1989). The *Para* Locus Encodes a Protein Homologous to the Vertebrate Sodium-Channel. *J Neurogenet*, Vol.5, No. 4, pp. 262-262, ISSN 0167-7063.

Loughney, K., Kreber, R., & Ganetzky, B. (1989). Molecular Analysis of the Para Locus, a Sodium-Channel Gene in *Drosophila*. *Cell*, Vol.58, No. 6, (Sep 22), pp. 1143-1154, ISSN 0092-8674.

Lumjuan, N., Rajatileka, S., Changsom, D., Wicheer, J., Leelapat, P., Prapanthadara, L.A., Somboon, P., Lycett, G., & Ranson, H. (2011). The role of the *Aedes aegypti* Epsilon glutathione transferases in conferring resistance to DDT and pyrethroid insecticides. *Insect Biochem Mol Biol*, Vol.41, No. 3, (Mar), pp. 203-209, ISSN 1879-0240.

Martins, A.J., Lima, J.B., Peixoto, A.A., & Valle, D. (2009a). Frequency of Val1016Ile mutation in the voltage-gated sodium channel gene of *Aedes aegypti* Brazilian populations. *Trop Med Int Health*, Vol.14, No. 11, (Nov), pp. 1351-1355, ISSN 1365-3156.

Martins, A.J., Lins, R.M., Linss, J.G., Peixoto, A.A., & Valle, D. (2009b). Voltage-gated sodium channel polymorphism and metabolic resistance in pyrethroid-resistant *Aedes aegypti* from Brazil. *Am J Trop Med Hyg*, Vol.81, No. 1, (Jul), pp. 108-115, ISSN 1476-1645.

Milani, R. (1954). Comportamento medeliano della resistenza alla azione del DDT e correlazione tra abbattimento e e mortalitá in *Musca domestica* L. *Revista di Parassitologia*, Vol.15, No. pp. 513-542.

Montella, I.R., Martins, A.J., Viana-Medeiros, P.F., Lima, J.B., Braga, I.A., & Valle, D. (2007). Insecticide resistance mechanisms of Brazilian *Aedes aegypti* populations from 2001 to 2004. *Am J Trop Med Hyg*, Vol.77, No. 3, (Sep), pp. 467-477, ISSN 0002-9637.

Morgan, J.A., Corley, S.W., Jackson, L.A., Lew-Tabor, A.E., Moolhuijzen, P.M., & Jonsson, N.N. (2009). Identification of a mutation in the para sodium channel gene of the cattle tick *Rhipicephalus (Boophilus) microplus* associated with resistance to synthetic pyrethroid acaricides. *Int J Parasitol*, Vol.No. (Jan 17), ISSN 1879-0135.

Morin, S., Alon, M., Alon, F., & Nauen, R. (2008). Organophosphates' resistance in the B-biotype of *Bemisia tabaci* (Hemiptera: Aleyrodidae) is associated with a point mutation in an ace1-type acetylcholinesterase and overexpression of carboxylesterase. *Insect Biochem Molec*, Vol.38, No. 10, (Oct), pp. 940-949, ISSN 0965-1748.

Muller, P., Warr, E., Stevenson, B.J., Pignatelli, P.M., Morgan, J.C., Steven, A., Yawson, A.E., Mitchell, S.N., Ranson, H., Hemingway, J., *et al.* (2008). Field-caught permethrin-resistant *Anopheles gambiae* overexpress *CYP6P3*, a P450 that metabolises pyrethroids. *PLoS Genet*, Vol.4, No. 11, (Nov), pp. e1000286, ISSN 1553-7404.

Narahashi, T. (1992). Nerve membrane Na+ channels as targets of insecticides. *Trends Pharmacol Sci*, Vol.13, No. 6, (Jun), pp. 236-241, ISSN 0165-6147.

Nelson, D.L., & Cox, M.M. (2000). *Lehninger - Principles of Biochemistry* (3rd), Worth Publishers, New York.

O'Reilly, A.O., Khambay, B.P., Williamson, M.S., Field, L.M., Wallace, B.A., & Davies, T.G. (2006). Modelling insecticide-binding sites in the voltage-gated sodium channel. *Biochem J*, Vol.396, No. 2, (Jun 1), pp. 255-263, ISSN 1470-8728.

Okimoto, R., & Dodgson, J.B. (1996). Improved PCR amplification of multiple specific alleles (PAMSA) using internally mismatched primers. *Biotechniques*, Vol.21, No. 1, (Jul), pp. 20- 26, ISSN 0736-6205.

Olson, R.O., Liu, Z., Nomura, Y., Song, W., & Dong, K. (2008). Molecular and functional characterization of voltage-gated sodium channel variants from *Drosophila melanogaster*. *Insect Biochem Mol Biol*, Vol.38, No. 5, (May), pp. 604-610, ISSN 0965-1748.

Paeoporn, P., Komalamisra, N., Deesin, V., Rongsriyam, Y., Eshita, Y., & Thongrungkiat, S. (2003). Temephos resistance in two forms of *Aedes aegypti* and its significance for the resistance mechanism. *Southeast Asian J Trop Med Public Health*, Vol.34, No. pp. 786-792.

Palladino, M.J., Keegan, L.P., O'Connell, M.A., & Reenan, R.A. (2000). A-to-I pre-mRNA editing in Drosophila is primarily involved in adult nervous system function and integrity. *Cell*, Vol.102, No. 4, (Aug 18), pp. 437-449, ISSN 0092-8674.

Park, Y., Taylor, M.F.J., & Feyereisen, R. (1999). Voltage-gated sodium channel genes hscp and hDSC1 of *Heliothis virescens* F-genomic organization. *Insect Molecular Biology*, Vol.8, No. 2, (May), pp. 161-170, ISSN 0962-1075.

Pauron, D., Barhanin, J., Amichot, M., Pralavorio, M., Berge, J.B., & Lazdunski, M. (1989). Pyrethroid Receptor in the Insect Na+ Channel - Alteration of Its Properties in Pyrethroid-Resistant Flies. *Biochemistry*, Vol.28, No. 4, (Feb 21), pp. 1673-1677, ISSN 0006-2960.

Payandeh, J., Scheuer, T., Zheng, N., & Catterall, W.A. (2011). The crystal structure of a voltage-gated sodium channel. *Nature*, Vol.475, No. 7356, (Jul 21), pp. 353-358, ISSN 1476-4687.

Perry, T., Batterham, P., & Daborn, P.J. (2011). The biology of insecticidal activity and resistance. *Insect Biochem Mol Biol*, Vol.41, No. 7, (Jul), pp. 411-422, ISSN 1879-0240.

Pinto, J., Lynd, A., Elissa, N., Donnelly, M.J., Costa, C., Gentile, G., Caccone, A., & do Rosario, V.E. (2006). Co-occurrence of East and West African kdr mutations suggests high levels of resistance to pyrethroid insecticides in *Anopheles gambiae* from Libreville, Gabon. *Med Vet Entomol*, Vol.20, No. 1, (Mar), pp. 27-32, ISSN 0269-283X.

Pinto, J., Lynd, A., Vicente, J.L., Santolamazza, F., Randle, N.P., Gentile, G., Moreno, M., Simard, F., Charlwood, J.D., do Rosario, V.E., *et al.* (2007). Multiple origins of knockdown resistance mutations in the Afrotropical mosquito vector *Anopheles gambiae*. *PLoS One*, Vol.2, No. 11, pp. e1243, ISSN 1932-6203.

Raghavendra, K., Barik, T.K., Reddy, B.P., Sharma, P., & Dash, A.P. (2011). Malaria vector control: from past to future. *Parasitol Res*, Vol.108, No. 4, (Apr), pp. 757-779, ISSN 1432-1955.

Rajatileka, S., Black, W.C.t., Saavedra-Rodriguez, K., Trongtokit, Y., Apiwathnasorn, C., McCall, P.J., & Ranson, H. (2008). Development and application of a simple colorimetric assay reveals widespread distribution of sodium channel mutations in Thai populations of *Aedes aegypti*. *Acta Trop*, Vol.108, No. 1, (Oct), pp. 54-57, ISSN 1873-6254.

Randall, D., Burggren, W., & French, K. (2001). *Eckert Animal Physiology: Mechanisms and Adaptations* (5th), W.H. Freeman and Company.

Ranson, H., Claudianos, C., Ortelli, F., Abgrall, C., Hemingway, J., Sharakhova, M.V., Unger, M.F., Collins, F.H., & Feyereisen, R. (2002). Evolution of supergene families associated with insecticide resistance. *Science*, Vol.298, No. 5591, (Oct 4), pp. 179-181, ISSN 1095-9203.

Ranson, H., N'Guessan, R., Lines, J., Moiroux, N., Nkuni, Z., & Corbel, V. (2011). Pyrethroid resistance in African anopheline mosquitoes: what are the implications for malaria control? *Trends Parasitol*, Vol.27, No. 2, (Feb), pp. 91-98, ISSN 1471-5007.

Raymond, M., Berticat, C., Weill, M., Pasteur, N., & Chevillon, C. (2001). Insecticide resistance in the mosquito *Culex pipiens*: what have we learned about adaptation ? *Genetica*, Vol.112, No. (Nov), pp. 287-296, ISSN 0016-6707.

Roberts, D.R., & Alecrim, W.D. (1991). [Response of *Anopheles darlingi* to spraying with DDTin amazonas, Brazil]. *Bol Oficina Sanit Panam*, Vol.110, No. pp. 480-488.

Rodriguez, M.M., Bisset, J.A., Diaz, C., & Soca, L.A. (2003). [Cross resistance to pyrethroids in *Aedes aegypti* from Cuba induced by the selection with organophosphate malathion]. *Rev Cubana Med Trop*, Vol.55, No. 2, (May-Aug), pp. 105-111, ISSN 0375-0760.

Saavedra-Rodriguez, K., Urdaneta-Marquez, L., Rajatileka, S., Moulton, M., Flores, A.E., Fernandez-Salas, I., Bisset, J., Rodriguez, M., McCall, P.J., Donnelly, M.J., *et al.* (2007). A mutation in the voltage-gated sodium channel gene associated with pyrethroid resistance in Latin American *Aedes aegypti*. *Insect Mol Biol*, Vol.16, No. 6, (Dec), pp. 785-798, ISSN 0962-1075.

Salkoff, L., Butler, A., Wei, A., Scavarda, N., Giffen, K., Ifune, C., Goodman, R., & Mandel, G. (1987). Genomic organization and deduced amino acid sequence of a putative sodium channel gene in *Drosophila*. *Science*, Vol.237, No. 4816, (Aug 14), pp. 744-749, ISSN 0036-8075.

Schuler, T.H., Martinez-Torres, D., Thompson, A.J., Denholm, I., Devonshire, A.L., Duce, I.R., & Williamson, M.S. (1998). Toxicological, electrophysiological, and molecular characterisation of knockdown resistance to pyrethroid insecticides in the diamondback moth, *Plutella xylostella* (L.). *Pestic Biochem Phys*, Vol.59, No. 3, (May), pp. 169-182, ISSN 0048-3575.

Severson, D.W., Anthony, N.M., Andreev, O., & ffrenchConstant, R.H. (1997). Molecular mapping of insecticide resistance genes in the yellow fever mosquito (*Aedes aegypti*). *J Hered*, Vol.88, No. 6, (Nov-Dec), pp. 520-524, ISSN 0022-1503.

Shao, Y.M., Dong, K., Tang, Z.H., & Zhang, C.X. (2009). Molecular characterization of a sodium channel gene from the *Silkworm Bombyx* mori. *Insect Biochem Mol Biol*, Vol.39, No. 2, (Feb), pp. 145-151, ISSN 1879-0240.

Smith, T.J., Lee, S.H., Ingles, P.J., Knipple, D.C., & Soderlund, D.M. (1997). The L1014F point mutation in the house fly Vssc1 sodium channel confers knockdown resistance to pyrethroids. *Insect Biochem Mol Biol*, Vol.27, No. 10, (Oct), pp. 807-812, 0965-1748 (Print) 0965-1748 (Linking).

Soderlund, D.M. (2010). State-Dependent Modification of Voltage-Gated Sodium Channels by Pyrethroids. *Pestic Biochem Physiol*, Vol.97, No. 2, (Jun 1), pp. 78-86, ISSN 0048-3575.

Soderlund, D.M., & Bloomquist, J.R. (1989). Neurotoxic actions of pyrethroid insecticides. *Annu Rev Entomol*, Vol.34, No. pp. 77-96, ISSN 0066-4170.

Song, W., Liu, Z., Tan, J., Nomura, Y., & Dong, K. (2004). RNA editing generates tissue-specific sodium channels with distinct gating properties. *J Biol Chem*, Vol.279, No. 31, (Jul 30), pp. 32554-32561, ISSN 0021-9258.

Sonoda, S., Igaki, C., & Tsumuki, H. (2008). Alternatively spliced sodium channel transcripts expressed in field strains of the diamondback moth. *Insect Biochem Mol Biol*, Vol.38, No. 9, (Sep), pp. 883-890, ISSN 0965-1748.

Spafford, J.D., Spencer, A.N., & Gallin, W.J. (1999). Genomic organization of a voltage-gated Na+ channel in a hydrozoan jellyfish: insights into the evolution of voltage-gated Na+ channel genes. *Receptors Channels*, Vol.6, No. 6, pp. 493-506, ISSN 1060-6823.

Sparks, T.C., Lockwood, J.A., Byford, R.L., Graves, J.B., & Leonard, B.R. (1989). The Role of Behavior in Insecticide Resistance. *Pestic Sci*, Vol.26, No. 4, pp. 383-399, ISSN 0031-613X.

Strong, M., Chandy, K.G., & Gutman, G.A. (1993). Molecular evolution of voltage-sensitive ion channel genes: on the origins of electrical excitability. *Mol Biol Evol*, Vol.10, No. 1, (Jan), pp. 221-242, ISSN 0737-4038.

Taylor, M.F.J., Heckel, D.G., Brown, T.M., Kreitman, M.E., & Black, B. (1993). Linkage of Pyrethroid Insecticide Resistance to a Sodium-Channel Locus in the Tobacco Budworm. *Insect Biochem Molec*, Vol.23, No. 7, (Oct), pp. 763-775, ISSN 0965-1748.

Tsagkarakou, A., Van Leeuwen, T., Khajehali, J., Ilias, A., Grispou, M., Williamson, M.S., Tirry, L., & Vontas, J. (2009). Identification of pyrethroid resistance associated mutations in the para sodium channel of the two-spotted spider mite *Tetranychus urticae* (Acari: Tetranychidae). *Insect Mol Biol*, Vol.18, No. 5, (Oct), pp. 583-593, ISSN 1365-2583.

Usherwood, P.N., Davies, T.G., Mellor, I.R., O'Reilly, A.O., Peng, F., Vais, H., Khambay, B.P., Field, L.M., & Williamson, M.S. (2007). Mutations in DIIS5 and the DIIS4-S5 linker of *Drosophila melanogaster* sodium channel define binding domains for pyrethroids and DDT. *FEBS Lett*, Vol.581, No. 28, (Nov 27), pp. 5485-5492, ISSN 0014-5793.

Valle, D., Montella, I.R., Medeiros, P.F.V., Ribeiro, R.A., Martins, A.J., & Lima, J.B.P. (2006). *Metodologia para quantificação de atividade de enzimas relacionadas com a resistência a inseticidas em Aedes aegypti/ Quantification metodology for enzyme activity related to insecticide resistance in Aedes aegypti*, Editora MS, ISBN: 853341291-6 1., DF: Brasília.

Vontas, J., David, J.P., Nikou, D., Hemingway, J., Christophides, G.K., Louis, C., & Ranson, H. (2007). Transcriptional analysis of insecticide resistance in *Anopheles stephensi* using cross-species microarray hybridization. *Insect Mol Biol*, Vol.16, No. 3, (Jun), pp. 315-324, ISSN 0962-1075.

Wang, J., Chuang, K., Ahluwalia, M., Patel, S., Umblas, N., Mirel, D., Higuchi, R., & Germer, S. (2005). High-throughput SNP genotyping by single-tube PCR with Tm-shift primers. *Biotechniques*, Vol.39, No. 6, (Dec), pp. 885-893, ISSN 0736-6205.

Werren, J.H., Baldo, L., & Clark, M.E. (2008). *Wolbachia*: master manipulators of invertebrate biology. *Nat Rev Microbiol*, Vol.6, No. 10, (Oct), pp. 741-751, ISSN 1740-1534.

Williamson, M.S., Denholm, I., Bell, C.A., & Devonshire, A.L. (1993). Knockdown resistance (*kdr*) to DDT and pyrethroid insecticides maps to a sodium channel gene locus in the housefly (*Musca domestica*). *Mol Gen Genet*, Vol.240, No. 1, (Jul), pp. 17-22, ISSN 0026-8925.

Williamson, M.S., Martinez-Torres, D., Hick, C.A., & Devonshire, A.L. (1996). Identification of mutations in the housefly para-type sodium channel gene associated with knockdown resistance (kdr) to pyrethroid insecticides. *Mol Gen Genet*, Vol.252, No. 1-2, (Aug 27), pp. 51-60, ISSN 0026-8925.

Wondji, C.S., Priyanka De Silva, W.A., Hemingway, J., Ranson, H., & Parakrama Karunaratne, S.H. (2008). Characterization of knockdown resistance in DDT- and pyrethroid-resistant *Culex quinquefasciatus* populations from Sri Lanka. *Trop Med Int Health*, Vol.13, No. 4, (Apr), pp. 548-555, ISSN 1365-3156.

Xu, Q., Wang, H., Zhang, L., & Liu, N. (2006). *Kdr* allelic variation in pyrethroid resistant mosquitoes, *Culex quinquefasciatus* (S.). *Biochem Biophys Res Commun*, Vol.345, No. 2, (Jun 30), pp. 774-780, ISSN 0006-291X.

Yang, Y., Lv, J., Gui, B., Yin, H., Wu, X., Zhang, Y., & Jin, Y. (2008). A-to-I RNA editing alters less-conserved residues of highly conserved coding regions: implications for dual functions in evolution. *RNA*, Vol.14, No. 8, (Aug), pp. 1516-1525, ISSN 1469-9001.

Yang, Y.Y., Liu, Y., Teng, H.J., Sauman, I., Sehnal, F., & Lee, H.J. (2010). Circadian control of permethrin-resistance in the mosquito *Aedes aegypti*. *J Insect Physiol*, Vol.56, No. 9, (Sep), pp. 1219-1223, ISSN 1879-1611.

Yanola, J., Somboon, P., Walton, C., Nachaiwieng, W., Somwang, P., & Prapanthadara, L.A. (2011). High-throughput assays for detection of the F1534C mutation in the voltage-gated sodium channel gene in permethrin-resistant *Aedes aegypti* and the distribution of this mutation throughout Thailand. *Trop Med Int Health*, Vol.16, No. 4, (Apr), pp. 501-509, ISSN 1365-3156.

Yu, F.H., & Catterall, W.A. (2003). Overview of the voltage-gated sodium channel family. *Genome Biol*, Vol.4, No. 3, ISSN 1465-6914.

Zhang, T., Liu, Z., Song, W., Du, Y., & Dong, K. (2011). Molecular characterization and functional expression of the DSC1 channel. *Insect Biochem Mol Biol*, Vol.41, No. 7, (Jul), pp. 451-458, ISSN 1879-0240.

Zhou, W., Chung, I., Liu, Z., Goldin, A.L., & Dong, K. (2004). A voltage-gated calcium-selective channel encoded by a sodium channel-like gene. *Neuron*, Vol.42, No. 1, (Apr 8), pp. 101-112, ISSN 0896-6273.

4

Photoremediation of Carbamate Residues in Water

Anđelka V. Tomašević and Slavica M. Gašić
Institute of Pesticides and Environmental Protection,
Belgrade-Zemun,
Serbia

1. Introduction

Pesticides are extensively world wide used for agriculture and for non-agricultural purposes. The major environmental concern of used pesticides is their ability to leach down to subsoil and contaminate the ground water, or, if they immobile, they could persist on the top soil and become harmful to microorganisms, plants, animal and people (Jha & Mishra 2005; Radivojević et al., 2008). Harmfull pesticide residues can contaminate the environment and accumulate in ecosystems than entering the human food chain (Đurović et al., 2010; Gašić et al., 2002a; Gevao et al., 2000). Pesticides have various characteristics that determine how act once in soil where it could accumulate to toxic level. Generally, soil and groundwater pollution are the major consequences environmental effects of pesticides application. Pesticides can reach water through surface runoff from treated plants and soil. Pesticide sprays usually directly hit non-target vegetation or can drift or volatilize from the treated areas that contaminate air, soil, and non-target plants. Finally, using of pesticides has resulted in acute and chronic ecological damage either by direct injury such as birds and fish or by indirect.

Carbamates are large group of pesticides which have been extensively used in almost sixty years. In this chapter an attempt is made to give the available data of the carbamates used as pesticides, their physico-chemical and toxicological characteristics, behaviour and fate in the environment, types of formulations which exist on the market as well as photochemical degradation for the certain members. Owing to widespread use in agriculture and relatively good solubility in water carbamate compounds can contaminate surface and ground waters and therefore carries a risk to various consumers, as well as the environment.

In this chapter we will also discuss some very important photocatalytic methods for remediation of water containing carbamate residues: direct photodegradation (photolysis), photosensitized degradation and photocatalytic degradation (including heterogeneous TiO_2 and ZnO processes and photo-Fenton and Fenton-like processes).

2. Carbamates

Carbamates were developed into commercial pesticides in the 1950s. It is a very huge family which members are effective as insecticides, herbicides, and fungicides, but they are most commonly used as insecticides. More than 50 carbamates are known. The most often used members of carbamate group are: aldicarb, asulam, bendiocarb, carbaryl, carbetamid, carbofuran, carbosulfan, chlorpropham, desmedipham, ethiofencarb, formetanate, furatiocarb, fenoxycarb, isoprocarb, methiocarb, methomyl, oxamyl, phenmedipham, pirimicarb, promecarb, propamocarb and propoxur.

Carbamates are N-substituited esters of carbamic acid. Their general formula is:

$$R_1NH - \overset{\overset{\textstyle O}{\|}}{C} - OR_2$$

Fig. 1. General carbamate structure, where R_2 is an aromatic or aliphatic moiety, if R_1 is a methyl group it is carbamate insecticide, if R_1 is an aromatic moiety it is carbamate herbicide and if R_1 is a benzimidazole moiety it is carbamate fungicide (WHO, 1986).

Pesticide activity	Chemical structure	Common or other names
Insecticide	$CH_3 - NH - \overset{\overset{\textstyle O}{\|}}{C} - O - aryl$	aldoxycarb, allyxycarb, aminocarb, BPMC, bendiocarb, bufencarb, butacarb, carbanolate, carbaryl, carbofuran, cloethocarb, dimetilan, dioxacarb, ethiofencarb, formetanate, hoppcide, isoprocarb, trimethacarb, MPMC, methiocarb, metolcarb, mexacarbate, pirimicab, promacyl, promecarb, propoxur, MTMC, XMC, xylylcarb
	$CH_3 - NH - \overset{\overset{\textstyle O}{\|}}{C} - O - N - alkyl$	aldicarb, methomyl, oxamyl, thiofanox, thiodicarb
Herbicide	$aryl - NH - \overset{\overset{\textstyle O}{\|}}{C} - O - alkyl$	asulam, barban, carbetamide, chlorbufam, desmedipham, phenmedipham, swep
	$alkyl - NH - \overset{\overset{\textstyle O}{\|}}{C} - O - aryl$	dichlormate, karbutilate, terbucarb
Herbicide and sprout inhibitors	$aryl - NH - \overset{\overset{\textstyle O}{\|}}{C} - O - alkyl$	propham, chlorpropham
Fungicide	$aryl - NH - \overset{\overset{\textstyle O}{\|}}{C} - O - alkyl$	benomyl, carbendazim, thiophanate-methyl, thiophanate-ethyl

Table 1. Relationship of chemical structure and pesticide activity of carbamates (WHO, 1986).

2.1 Carbamates physical and chemical properties

It is known that esters or N-substituted derivates of carbamic acid are unstable compounds, especially under alkaline conditions. Decomposition under this conditions takes place and the compounds as alcohol, phenol, ammonia, amine and carbon dioxide are formed.

Derivates of carbamic acid as salts or esters are more stable than carbamic acid. This enhanced stability is the basis for synthesis of many derivates that are biologically active pesticides.

Carbamate ester derivates are crystalline solids of low vapor pressure with variable, but usually low water solubility. They are moderately soluble in solvents such as benzene, toluene, xylene, chloroform, dichloromethane and 1,2-dichloromethane. Generally, they are poorly soluble in nonpolar organic solvents such as petroleum hydrocarbons but highly soluble in polar organic solvents such as methanol, ethanol, acetone, dimethylformamide, etc (WHO, 1986).

2.2 Carbamates mode of action and toxicity

Most carbamates are active inhibitors of acetylholinesteraze (AChE), but some carbamates as benzimidazole have no acetylcholinesterase activity. Carbamates toxicity to insects, nematodes, and mammals is based on inhibition of acetylcholinesterase, which is the enzyme responsible for the hydrolysis of acetycholine into choline and acetic acid. Acetylcholine (ACh) is a substance that transmits a nerve impulse from a nerve cell to a specific receptor such as another nerve cell or a muscle cell. Acetylcholine, in essence, acts as a chemical switch. When it is present (produced by nerve cell) it turns the nerve impulse on. When it is absent, the nerve impulse is discontinued. The nerve transmission ends when the enzyme aceylcholinesterase breaks down the acetylcholine into choline and acetic acid. Without the action of this enzyme acetylcholine builds up at the junction of nerve cell and the receptor site, and the nerve impulse continues. Carbamate insecticides block (or inhibit) the ability of this enzyme, acetylcholinesterase, to break down the acetylcholine and the nerve impulse (Kamrin, 1997; Machemer & Pickel, 1994).

In mammals, cholinesterase inhibition caused by carbamates is labile, reversible process. Estimates of the recovery time in humans range from immediate up to four days, depending on the dose, the specific pesticide and the method of exposure. The breakdown of carbamate compounds within an organisms is a complex process and is depended on the specific pesticide structure. The rapid degradation of carbamates *in vivo* by mammals occurs by hydrolysis, oxidation and conjugation. The end products include amines, alcohols or phenol derivates. The urinary route is the main excretory route (Machemer & Pickel, 1994).

Inhibition of acetylholinesteraze (AChE) by carbamates causes toxic effects in animals and human beings that result in variety of poisoning symptoms. Carbamates acute toxicity and poisoning are dose related. Acute poisoning occurs rapidly after exposure. Ingestion of carbamate insecticides at low doses can cause excessive salivation and an increase in the rate of breathing within 30 min. At higher doses this is followed by excessive tearing, urination, no control defecation, nausea and vomiting. At the highest doses, symptoms can include those listed above along with violent intestinal movements, muscle spasm and convulsions. Death has occurred in a few instances, usually due to respiratory failure resulting from paralysis of the respiratory muscles (Kamrin, 1997; WHO, 1986).

While the insecticidal carbamate produce the typical symptoms of cholinesterase inhibition, they don't appear to induce a delayed neurotoxic reaction similar to that seen with some organophosphourus compound. Chronic exposure to carbamate compounds may cause adverse effects on organs or acetylcholinesterase levels. These effects are unlikely to occur in humans at expected exposure levels (Kamrin, 1997).

The acute toxicity of different member of carbamates ranges from highly toxic to only slightly toxic. The LD_{50} for the rats ranges from less than 1 mg/kg to over 5000 mg/kg body weight. The acute dermal toxicity of carbamates is generally low to moderate except aldicarb which is very toxic. The carbamates in short term and long term toxicity studies showed different toxicity. Some carbamates are very toxic and others less. Carbamate pesticides are transformed metabolically by variety of chemical reactions in more water soluble molecules which can be excreted via the urine. Rats eliminate carbamate compounds rapidly in that way. Most metabolites are excreted within 24 h of exposure and therefore carbamate residues don't accumulate in animals (Kamrin, 1997; WHO, 1986). In the study of carbofuran toxicity on rats during subhronic exposure the histopathological changes in liver and kidneys were observed but there was cell regeneration in all test groups as well (Brkić et al., 2008).

Aldicarb is the most toxic among the carbamates and establish acceptable daily intake (ADI) for humans is 0.001 mg/kg/body weight. The other carbamates have ADIs values in range of 0.001-0.1 mg/kg/ body weight (WHO, 1986). According to the European Food Safety Authority (2009), the lowest ADI has carbofuran 0.00015 mg/kg/body weight (EU Pesticide Database, 2011).

Many carbamates have been studied for reproductive, teratogenic, mutagenic and carcinogenic effects and the results of this is that a few members of this family has been banned by the regulatory bodies worldwide.

2.3 Environmental fate

Generally, carbamates remain active for a few hours to a few month in soils and crop, but they may leave residues in agricultural products (Takino et al., 2004). The rate of degradation in soil depends on soil type, soil moisture, adsorption, pH, soil temperature, concentration of pesticide, microbial activity and photodecomposition. The higher the organic content, the greater the binding to soil and thus the greater the persistence. Also, the higher the soil acidity, the longer it takes for carbamates to be degraded. Carbamate insecticides are mainly applied on the plants, but can reach the soil, while carbamate nematocides and herbicides are applied directly to the soil. Generally, in soil carbamates degraded by chemical hydrolysis and microbial processes. Microorganisms that have capability to degrade carbamate pesticides play a significant role in the break dawn and elimination of them from environment. Because the different carbamates have different properties, it is clear that each of them should be evaluated on its own merits, and no extrapolation of results can be made from one carbamate to another. One carbamate may be easily decomposed, while another may be strongly adsorbed on soil. Some leach out easily and may reach groundwater. In these processes, the soil type and water solubility are of great importance. Furthermore, it should be recognized that this not only concerns the parent compound but also the breakdown products or metabolites (Kamrin, 1997; WHO, 1986).

Persistence of carbamate herbicides is increased by application to dry soil surface or by soil incorporation. Environmental factors which increasing microbial activity in soil generally decrease the persistence of carbamate herbicides. In most of degradation reactions the initial cleavage of the molecule occurs at ester linkage. Enzymatic hydrolysis of some carbamates can be correlated to soil acidity, and rate differences explained by consideration of certain steric and electronic properties of the carbamates. The carbamate derivates with herbicidal action are substantially more stable to alkaline hydrolysis than the methyl carbamate derivatives, which have an insecticidal action (Kaufman, 1967).

Carbamate compounds degrade through chemical hydrolysis and this is the first step in the metabolic degradation. The hydrolysis products will be further metabolized in soil and plant. Chemical degradation does not appear to have much influence in the total degradation of pesticides in soil. Carbamate compounds are adsorbed and translocated through plants and treated crops. In most cases, carbamates will break down quickly in plants and the residues in plants will last not very long .

Finally carbamates are metabolized by microorganisms, plants and animals or broken down in water or soil. In water carbamates degraded by chemical hydrolysis, but photodegradation and aquatic microbes may also contribute degradation. Generally, in alkaline water and under sunlight carbamate compounds will decompose more rapidly (WHO, 1986).

2.4 Formulations

Carbamate products come in variety of solid and liquid formulations on the market. They contain beside carbamate compounds inert ingredients which could be toxic, flammable or reactive. Examples of inert ingredients are wetting agents, spreaders, dispersing agents, solvents, solubilizers, carriers, ticker, surfactants and so on. A surfactant is a substance that reduced surface tension of a system, allowing oil-based and water-based substances to mix more readily. A common groups of non-ionic surfactants are the alkylphenol polyethoxylates or alcohol ethoxylates which may be used in pesticide formulations. Nonyl phenols, one of the members of above mention alkylphenol surfactant has been linked to endocrine-disrupting effects in aquatic animals and should be substituted by less hazardous alternatives. Commonly used formulation types include liquid and dry formulations as emulsifiable concentrates (EC), soluble concentrates (SL), suspension concentrates (SC), than wettable powders (WP), water dispersible granules (WG), granules (GR), etc, and they are signed by international coding system (CropLife, 2008).

Pesticides are very often formulated as emulsifiable concentrates (EC) which produce emulsions when dissolved in water. The first problem in defining this formulation is the selection of an adequate surfactants (emulsifiers) for the intended purposes (Gašić et al., 1998a, 1998b, 2002b; Shinoda & Friberg, 1986). Recently there is increasing interest in the effect of emulsifiers on toxicity to mammals and fish. These effects can be due to inherent toxicity of the surfactant itself or to the enhancing effect that the emulsifiers may have on toxicity of active ingredient. So, the formulation type can have implications for product efficacy and exposure to humans and other non-target organisms (Knowles, 2005, 2006; Sher, 1984).

While the toxicity of the active ingredient of a pesticide is property which can not be changed, the acute toxicity effects of the formulation are strongly influenced by the way in which the active ingredient is formulated. While pesticide formulations are influenced by both the physical and chemical properties of the active ingredient and the economic pressures of the marketplace, there are formulation choices which will increase the safety of pesticide formulations (Mollet & Grubenmann, 2001).

The type of pesticide formulations and in, some cases, the choice of product of the same formulation type can significantly affect the results obtained in practical use. Safety, efficacy, residual life, cost, availability and ease of use must all be considered in selecting formulation. The ways in which pesticides are formulated considerably influence their persistence. Formulations in order of increasing persistence on plants are prepared in the way that more readly adsorbed on the soil fractions and not appreciably degradated (Edwards, 1975).

3. Photodegradation processes for carbamates wastewater treatments

3.1 Photolysis

Photolysis (direct photodegradation reaction) is photodegradation process without any catalysts and use light only for degradation of different organic molecules, including pesticides and related compounds. Direct irradiation will lead to the promotion of the pesticides to their excited singlet states and such excited states can then undergo among homolysis, heterolysis or photoionization processes (Burrows et al., 2002). Direct photodegradation by solar light is limited and various lamps have been used for irradiation of contaminated water solutions. The photolysis of contaminants (including pesticides) in aqueos solution depends on the different reaction parameters such as type of light, lamp distance, temperature, initial concentration of pesticides, type of water, pH, the presence of humic and fulvic acids, the presence of O_2, O_3, O_2/O_3 and H_2O_2, the presence of inorganic ions and organic matter dissolved in water (Burrows et al., 2002; Tomaševic at al., 2010a).

3.2 Photosensitized degradation

The photosensitized reaction is based on the absorption of light by a molecule of the sensitizer and includes an energy transfer from molecul excited state to the pesticides. The most famous sensitizers are aceton, rose Bengal, methylene blue and humic and fulvic acids (Burrows et al., 2002).

3.3 Advanced oxidation processes

Advanced Oxidation Processes (AOPs) include catalytic and photochemical methods and have H_2O_2, O_3 or O_2 as oxidant. The principal active species in this system is the hydroxyl radical $^{\bullet}OH$, which is an extremely reactive and non-selective oxidant for organic contaminants (Legrini at al., 1993; Sun & Pignatello, 1993). The main advantage of these processes is a complete mineralization of many organic pollutants (Andreozzi at al., 1999; Neyens & Baeyens, 2003). Several of AOPs are currently employed for the elimination of pesticides from water: heterogeneous photocatalytic reactions with semiconductor oxides TiO_2 (Malato et al., 2002a, 2002b; Tomaševic at al., 2010a) or ZnO (Tomaševic at al., 2010a) as photocatalysts, photo-Fenton (Malato et al., 2002a; Tamimi et al, 2008; Tomaševic at al., 2010b) and photo-assisted Fenton processes (Huston & Pignatello, 1999). Electro-photo-Fenton (Kesraoui Abdessalem et al., 2010) and electrochemical oxidation processes (Tomašević et al., 2009a) have been seldom studied.

Heterogeneous photocatalysis is combination of semiconductor particles (TiO_2, ZnO, Fe_2O_3, CdS, ZnS), UV/solar light and different oxidants (H_2O_2, $K_2S_2O_8$, KIO_4, $KBrO_3$). The main equations of the heterogeneous photocatalysis are (Andreozzi et al., 1999; Daneshvar et al., 2003; Karkmaz et al., 2004; Legrini at al., 1993):

$$C + h\nu \rightarrow C\ (e^- + h^+) \tag{1}$$

$$h^+ + H_2O \rightarrow {}^{\bullet}OH + H^+ \tag{2}$$

$$e^- + O_2 \rightarrow O_2{}^{\bullet-} \tag{3}$$

Among AOP$_s$, heterogeneous photocatalysis using TiO_2 as photocatalyst appears as the most emerging destructive technology. The following mechanism of the TiO_2 photocatalysis has been proposed (Daneshvar et al., 2003; Gomes da Silva & Faria, 2003; Karkmaz et al., 2004; Tomaševic at al., 2010a):

a) absorption of efficient photons by titania ($h\nu \geq Eg=3.2$ eV):

$$TiO_2 + h\nu\,(UV) \rightarrow e_{CB}{}^- + h_{VB}{}^+ \tag{4}$$

b) oxygen ionosorption:

$$(O_2)_{ads} + e_{CB}{}^- \rightarrow O_2{}^{\bullet-} \tag{5}$$

c) neutralization of OH- groups into $^\bullet OH$ by photoholes:

$$(H_2O \leftrightarrow H^+ + OH^-)_{ads} + h_{VB}{}^+ \rightarrow H^+ + {}^\bullet OH \tag{6}$$

d) oxidation of the organic reactant via successive attacks by $^\bullet OH$ radicals:

$$R + {}^\bullet OH \rightarrow R'{}^\bullet + H_2O \tag{7}$$

e) or by direct reaction with holes:

$$R + h^+ \rightarrow R^{\bullet+} \rightarrow \text{degradation products} \tag{8}$$

ZnO is also frequently used as a catalyst in heterogeneous photocatalytic reactions. The biggest advantage of ZnO in comparison to TiO_2 is that it absorbs over a larger fraction of the UV spectrum and the corresponding threshold wavelength of ZnO is 387 nm. Upon irradiation, valence band electrons are promoted to the conduction band leaving a hole behind. These electron-hole pairs can either recombine or interact separately with other molecules. The holes at the ZnO valence band can oxidize adsorbed water or hydroxide ions to produce hydroxyl radicals. Electron in the conduction band at the catalyst surface can reduce molecular oxygen to superoxide anion. This radical may form organic peroxides or hydrogen peroxide in the presence of organic scavengers. The hydroxyl radical attacks organic compounds (R) and intermediates (Int) are formed. These intermediates react with hydroxyl radicals to produce the final products (P). The mechanism of heterogeneous photocatalysis in the presence of ZnO can be given by the following reactions (Behnajady et al., 2006; Daneshvar et al., 2004, 2007; Pera-Titus at al,. 2004; Tomaševic et al., 2010a):

$$ZnO + h\nu\,(UV) \rightarrow e_{CB}{}^- + h_{VB}{}^+ \tag{9}$$

$$e_{CB}{}^- + h_{VB}{}^+ \rightarrow \text{heat} \tag{10}$$

$$h_{VB}{}^+ + H_2O_{ads} \rightarrow H^+ + {}^\bullet OH_{ads} \tag{11}$$

$$h_{VB}{}^+ + {}^-OH_{ads} \rightarrow {}^\bullet OH_{ads} \tag{12}$$

$$e_{CB}{}^- + O_2 \rightarrow O_2{}^{\bullet-} \tag{13}$$

$$O_2{}^{\bullet-} + HO_2{}^\bullet + H^+ \rightarrow H_2O_2 + O_2 \tag{14}$$

$$O_2^{\bullet-} + R \to R\text{-}OO^{\bullet} \tag{15}$$

$$^{\bullet}OH_{ads} + R \to Int. \to P \tag{16}$$

Fenton's processes belong to AOPs and utilize H_2O_2 activation by iron salts. The classic Fenton's reagent is a mixture of ferrous ion and H_2O_2 in acidic solution or suspension (Neyens & Baeyens, 2003; Tamimi at al.,2008):

$$Fe^{2+} + H_2O_2 \to Fe^{3+} + OH^- + {}^{\bullet}OH \tag{17}$$

Equation (17) presents the most important steps of a Fenton reaction and involves electron transfer between H_2O_2 and Fe(II) with oxidation of Fe(II) to Fe(III) and the resulting production of highly reactive hydroxyl radical $^{\bullet}OH$ and potentially reactive ferryl species. The degradation of pesticides by Fenton's reagent can be strongly accelerated upon UV or UV-visible light. This process is the photo-Fenton reaction (Malato et al., 2002a, 2002b; Tamimi et al, 2008; Tomaševic at al., 2010b). Equation (17) is the key of photo-Fenton processes. The obtained Fe^{3+} ion or its $Fe(OH)^{2+}$ complexes act as light absorbing species, that produce another hydroxyl radical, while the initial Fe^{2+} ion is regained:

$$Fe(OH)^{2+} + hv \to Fe^{2+} + {}^{\bullet}OH \tag{18}$$

The main advantage of the photo-Fenton process is light sensitivity up to a wavelength of 600 nm (Malato et al., 2002a).

4. Photodegradation of carbamate pesticides

4.1 Aldicarb

Aldicarb (IUPAC name: 2-methyl-2-(methylthio)propionaldehyde O-methylcarbamoy-oxime) is a systemic oxime carbamate pesticide, effective against a variety of insects, mites, and nematodes. It is sold commercially only in granular form (GR). Aldicarb is applied on a variety of crops, including cotton, sugar beet, sugarcane, citrus fruits, potatoes, sweet potatoes, peanuts, beans (dried beans), soybeans, pecans, and ornamental plants. Home and garden use is not permitted in many countries. The current regulation status of this active ingredient under directive 91/414/EEC is not included in Annex 1 (EU Pesticide Database, 2011; Tomlin, 2009).

The complete conversion of 38 mg/L of aldicarb and 62% reduction in TOC content using the photo-Fenton reaction (Fe(III)/H_2O_2/UV) within 120 min in acidic aqueous solution (pH 2.8) at 25 °C with fluorescent blacklight irradiation (300-400 nm) has been considered (Huston & Pignatello, 1999). They also observed the formation of sulfate and nitrate ions during the photo-Fenton process.

4.2 Asulam

Asulam (IUPAC name: methyl sulanylcarbamate) is selective systemic herbicide, which is used for control of annual and perennial grasses and broad-leaved weeds in spinach, oilseed poppies, alfalfa, some ornamentals, sugar cane, bananas, coffee, tea, cocoa, coconuts, rubber, fruit trees and bushes, and forestry. It could be found only as soluble concentrate (SL) on the market. The current regulation status of this active ingredient under directive 91/414/EEC is not included in Annex 1 (EU Pesticide Database, 2011; Tomlin, 2009).

The degradation of asulam was studied in homogeneous aqueous solution in the presence of molecular oxygen at pH 3.0-3.4, by irradiation at 365 nm and by solar irradiation (Catastini et al., 2002a). When the iron(III) aquacomplexes was photoreduced to iron(II) ions and hydroxyl radicals the degradation of asulam in the presence of oxygen continud to completion. The Fe^{2+} ions are oxidized back to Fe^{3+} ions through various pathways such as photooxidation and oxidation by H_2O_2 generated within the system, where another $^{\bullet}OH$ forms. Their experimental results indicate that the presence of Fe^{3+}, Fe^{2+} and molecular oxygen accelerate the mineralization of asulam. Also, less than 10% conversion of asulam was observed when the irradiation was performeds in the presence of 0.01 M 2-propanol, used as hydroxyl radical scavenger. Complete conversion and nearly complete TOC reduction of 23 mg/L of asulam was achieved with 16.7 mg/L of Fe^{3+} ions, within 17 h (at 365 nm) and 28-30 h (under solar light). In this process intermediates or degradation by-products of asulam were not identified. The photodegradation of the herbicide asulam in aqueous solution (1.0×10^{-4} M or 23 mg/L) has been investigated with and without Fe(III) (Catastini et al., 2002b).The asulam disappearance were monitored by photolysis at 254 nm as a functuion of pH and oxygen concentration and no complete transformation of organic carbon into CO_2 was observed. In the presence of Fe(III) at 365 nm the complete mineralization of asulam has been achieved.

4.3 Bendiocarb

Bendiocarb (IUPAC name: 2,3-isopropyldenedioxyphenyl methylcarbamate, 2,2-dimethyl-1,3-benzodioxol-4-yl methylcarbamate) is systemic insecticide with contact and stomach action. It is active against many public health, industrial and storage pest. This active ingredient is especially useful inside buildings, due to its low odor and lack of corrosive and staining properties. It comes in variety formulations type as DP, FS, GR, SC, WP on the market. The current regulation status of this active ingredient under directive 91/414/EEC is not included in Annex 1 (EU Pesticide Database, 2011; Tomlin, 2009).

Evaluation of different pathway (photolysis, photo-Fenton, H_2O_2/UV and electro-Fenton) of bendiocarb (112-188 mg/L) photodegradation have been proposed (Aaron & Oturan, 2001). The conversion of insecticide was apparently much faster in the H_2O_2/UV and photo-Fenton proces (λ = 254 nm, 68 mg/L of H_2O_2 and 55.8 mg/L of Fe^{3+}) than in the other processes. Also, the degradation mechanism of bendiocarb has been proposed. The photolysis of aqueous bendiocarb (3.3×10^{-3} M, 4 h, room temperature, 125 W medium-pressure mercury lamp) has been examined by GC-MS (Climent & Miranda, 1996). Upon irradiation the only one photo-product (corresponding phenol) was detected and 30% conversion of bendiocarb was achieved.

4.4 Carbaryl

Carbaryl (IUPAC name: 1-naphthyl methylcarbamate) is insecticide with contact and stomach action and has slight systemic properties. It is used for control of chewing and sucking insects on more than 120 different crops, including vegetables, tree fruit (including citrus), mangoes, bananas, strawberries, nuts, vines, olives, okra, cucurbits, peanuts, soya beans, cotton, rice, tobacco, cereals, beet, maize, sorghum, alfalfa, potatoes, ornamentals, forestry, etc, than for control earthworms in turf and as a growth regulator for fruit thinning of apples. Also it is used against an animal ectoparasiticide. Carbaryl can be found formulated as DP, GR, OF, RB, SC, TK and WP. The current regulation status of this active ingredient under directive 91/414/EEC is not included in Annex 1 (EU Pesticide Database, 2011; Tomlin, 2009).

The degradation of carbaryl under UV light using a continuous flow of TiO_2 slurry shown that the degradation proceeds through a multi-step process involving the attack of the substrate by •OH radicals (Peris at al., 1993). The studies on the degradation of carbaryl under simulated solar light in aqueous TiO_2 dispersions showed that the reaction follows pseudo-first-order kinetics and the complete mineralization (to CO_2, nitrate and ammonium ions) is achieved in less 30 min (Pramauro et al., 1997). The effect of ionic and non-ionic aliphatic surfactants (constitute an important ingredient of pesticide formulations and can influence the degradation of pesticide) on the degradation of aqueous carbaryl solutions (20 mg/L) containing 500 mg/L of TiO_2 (anatase) in the presence of simulated solar light (1500 W Xenon lamp with 340 nm cut-off filter) was investigated (Bianco Prevot at al., 1999). Depending on the surfactant and on the initial pH of the solution, an inhibition ot the photodegradation rate was observed. Also, mineralization of the carbaryl to CO_2, nitrate and ammonium ions was evidence in the presence of added surfactants, suggesting the feasibility of photocatalytic treatment of aqueous pesticide wastes.

4.5 Carbetamid

Carbetamid (IUPAC name: (R)-1-(ethylcarbamoyl)ethyl carbamilate) is selective herbicide, absorbed by the roots, and also by the leaves. It is used for control of annualgrasses and some broad-leaved weeds, alfalfa, sainfoin, brassicas, field beans, peas, lentils, sugar beast, oilseed rape, chicory, endive, sunflowers, caraway, strawberries, wines, and fruit orchards. Formulations types for this active ingredient are EC and WP. The current regulation status of this active ingredient under directive 91/414/EEC is included in Annex 1, expiration of inclusion: 31/05/2021 (EU Pesticide Database, 2011; Tomlin, 2009).

Photodegradation of herbicide carbetamide with ultraviolet light ($\lambda > 290$ nm) in the presence of TiO_2, H_2O_2 and ozone was studied in the aqueous solutions (Mansour et al., 1992). Using spectrometric methods several photoproducts were isolated and identified, suggesting that photodegradation pathways of carbetamide in the presence of TiO_2 and H_2O_2 are hydroxylations of the aromatic ring. Also, UV-ozonation rapidly oxydized carbetamide to water, ammonia and CO_2. The kinetics of photodegradation of carbetamide in water in the presence of TiO_2 (Degussa P 25 grade, surface area 50.0 m²/g) or ZnO (surface area 9.5 m²/g) were examined upon $\lambda \geq 310$ nm (Percherancier et al., 1995). The effects of various parameters, such as the kind of semiconductor, mass of TiO_2, initial concentration of pesticide, radiation flux and quantum yield were studied. The degradation with ZnO is faster than that with TiO_2 in spite of the lager surface area of the later catalyst. Also, the mechanism of the carbetamide photocatalytic degradation has been proposed.

4.6 Carbofuran

Carbofuran (IUPAC name: 2,3-dihydro-2,2-dimethylbenzofuran-7-yl methylcarbamate) is systemic insecticide with predominantly contact and stomach action. It is used for control of soil-dwelling and foliar-feeding insects and nematodes in vegetables, ornamentals, beet, maize, sorghum, sunflowers, oilseed rape, potatoes, alfalfa, peanuts, soya beans, sugar cane, rice, cotton, coffee, cucurbits, tobacco, lavender, citrus, wines, strawberries, bananas, mushrooms and other crops. This active ingredient is prepared as FS, GR, SC and WP formulation. The current regulation status of this active ingredient under directive 91/414/EEC is not included in Annex 1 (EU Pesticide Database, 2011; Tomlin, 2009).

Various carbofuran photodegradation processes (by ozon, UV photolysis, Fenton, O_3 + UV, UV + H_2O_2 and photo-Fenton) upon polychromatic UV irradiation were evaluated (Benitez et al., 2002). For all these reactions, the apparent pseudo-first–order rate constants are evaluated in order to compare the efficiency of each process. The most effective process in removing carbofuran from water was the photo-Fenton system (UV + Fe^{2+} + H_2O_2) with rate constants k from 17.2 x 10^{-4}/s to >200.0 x 10^{-4}/s. The degradation of pure carbofuran and commercial product Furadan 4F in acidic aqueous solution upon polychromatic light (300-400 nm) by photo-assisted Fenton process has been studied (Huston & Pignatello, 1999). The complete conversion of 2.0 x 10^{-4} M of pure carbofuran and more than 90% TOC reduction in the water solution within 120 min has been achieved. Nitrate and oxalate ions were detected as organic ionic species after the treatment. Also, the results show that the adjuvants in Furadan 4F have little or no influence on degradation of carbofuran nor of TOC mineralization. Two different Advanced oxidation processes (photo- and electro-Fenton) have been used for photodegradation of carbofuran in water (Kesraoui Abdessalem et al., 2010). For the photo-Fenton process TOC removal ratio was influenced by the initial concentration of the pesticides and the amout of Fe^{3+} and H_2O_2. The TOC measurement indicate an efficient mineralization of 93 and 94% respectively, for photo- and electro-Fenton processes after 480 min of treatment. Carbofuran could not be mineralized on AlFe-PILC and Fe-ZSM-5 zeolite catalysts in the heterogeneous photo-Fenton reactions at 575.6 nm, even in the catalytic reaction promoted at high temperature (Tomašević et al., 2007a, 2007b).

4.7 Ethiofenocarb

Ethiofencarb (IUPAC name: $\propto$-ethythio-$\propto$-tolyl methylcarbamate) is systemic insecticide with contact and stomach action. It is applied for control of aphids on pome fruit , stone fruit and soft fruit, than vegetables, ornamentals and sugar beet. Formulations types which can be found on the market are: emulsifieble concentrate (EC), emulsions oil in water (EW) and granules (GR). The current regulation status of this active ingredient under directive 91/414/EEC is not included in Annex 1 (EU Pesticide Database, 2011; Tomlin, 2009).

Solar photodegradation of ethiofencarb was examined in pure water, natural water and in the pure water containing 10mg/L of humic acids (Vialaton & Richard, 2002). Photosensitized reactions are main degradation pathway of pesticide in natural water and in the presence of humic acids. Photosensitized transformations were shown to be largely due to photoreactants other than singlet oxygen and hydroxyl radicals. A comparative photolysis reactions of ethiofencarb in water and non-water media were performed in the presence of simulated solar light (Sanz-Asensio et al., 1999). The studies showed that the photolysis reaction follows pseudo-first-order kinetics and that the degradation kinetics depend on the solvent polarity. In the water media the reaction of pesticide degradation was completed for 30 h. Also, the photoproducts are dependent on the solvent and the main photoproduct in water was 2-(methyl)phenyl-N-methylcarbamate. The photolysis of aqueous ethiofencarb (3.3 x 10^{-3} M, 4 h, room temperature, 125 W medium-pressure mercury lamp) has been examined by GC-MS (Climent & Miranda, 1996). Upon irradiation three photoproducts were detected and 66% conversion of ethiofencarb was achieved. The main product was 2-methylphenyl methylcarbamate, and two corresponding phenols also were registered.

4.8 Formetanate

Formetanate (IUPAC name: 3-dimethylaminomethyleneaminophenyl methylcarbamate) is acaricide and insecticide with contact and stomach action. It is used for control of spider mites and some insects on ornamentals, pome fruit, stone fruit, citrus fruit, vegetables and alfalfa. It is sold commercially only as soluble powder (SP). The current regulation status of this active ingredient under directive 91/414/EEC is included in Annex 1, expiration of inclusion: 30/09/2017 (EU Pesticide Database, 2011; Tomlin, 2009).

The solar driven photo-Fenton process using pilot-scale compound parabolic collector was applied to the degradation of formetanate in the form of AgrEvo formulated product Dicorzol (Fallman et al., 1999). The results shown that a good conversion of formetanate was achieved (about 25 min was a TOC half-life and about 70 min was the time necessary for degradation of 80% of TOC). The heterogeneous photocatalysis with TiO_2 (200 mg/L) and homogeneous photocatalysis by photo-Fenton (0.05 mM of $FeSO_4$ x $7H_2O$) of 50 mg/L of formetanate have been studied (Malato et al., 2002b). In the presence of 2.8 mg/L of Fe^{2+} complete conversion of formetanate and more than 90% TOC reduction was demonstrated in pilot-scale solar reactor. The kinetics of formetanate degradation by the TiO_2 solar photocatalysis and by the solar photo-Fenton process were also investigated (Malato et al., 2002b, 2003).

4.9 Methomyl

Methomyl (IUPAC name: S-methyl N-(methylcarbamoyloxy)thioacetimidate) is systemic insecticide and acaricide with contact and stomach action. It is used for control of a wide range of insects and spider mites in fruit, vines, olives, hops, vegetables, ornamentals, field crops, cucurbits, flax, cotton, tobacco, soya beans, etc. Also it can be used for control of flies in animal and poultry houses and dairies. Formulations types for this active ingredient are SL, SP, WP. The current regulation status of this active ingredient under directive 91/414/EEC is included in Annex 1 expiration of inclusion: 31/08/2019 (EU Pesticide Database, 2011; Tomlin, 2009).

The solar driven homogeneous photo-Fenton and heterogeneous TiO_2 processes for methomyl detoxification in water have been evaluated (Malato et al., 2002b, 2003). According to TOC removal, the photo-Fenton process was more efficient in degrading 50 mg/L of methomyl than was the TiO_2 process. The both processes were capable of mineralizing more than 90% of the insecticide (Malato et al., 2002b). The photodegradation of methomyl by Fenton and photo-Fenton reactions were investigated (Tamimi et al., 2008). The degradation rate and the effect of reaction parameters (initial concentration of pesticide, pH, ferrous and H_2O_2 dosage, etc) were monitored. The photo-Fenton was more efficient than Fenton, both for methomyl degradation and TOC removal. The catalytic wet peroxide oxidation of methomyl at 575.6 nm (photo-Fenton reaction) with two types of heterogeneous iron catalysts (Fe-ZSM-5 zeolite and AlFe-pillared montmorillonite) were performed (Lazar et al., 2009; Tomašević et al., 2007c, 2009c, 2010a, 2010b; Tomašević, 2011). The effect of catalyst type on the reaction is shown in Fig. 2. The photolysis of 16.22 mg/L of methomyl in different types of water (deionized, disstiled and sea water) at 254 nm was performed (Tomašević et al., 2009c, 2010a; Tomašević, 2011) and the influence of reaction parameters to degradation of pesticide were investigated. The studies showed that the photolysis reactions depend on the lamp distance (Fig. 3), water type (Fig. 4), reaction temperature and pH. The photocatalytic removal of the methomyl from aqueous solutions upon UV/Vis (366 and 300-400 nm) and natural solar light in the presence of TiO_2 and ZnO has been examined

(Tomašević et al., 2009b, 2010a; Tomašević, 2011) and the influence of reaction conditions (initial concentration of methomyl, catalysts type and concentration, pH, presence of Cl- ions) were studied. The results (Table 2) showed that the degradation of methomyl was much faster with ZnO than with TiO₂. The IC results confirmed that mineralization of methomyl led to the formation of sulfate, nitrate, and ammonium ions during the all investigated processes (Tomašević et al., 2010a, 2010b; Tomašević, 2011).

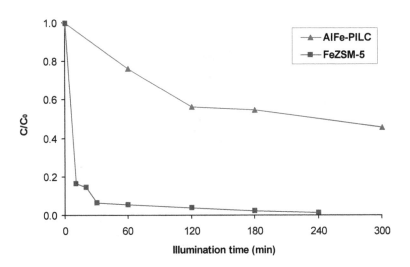

Fig. 2. Photodegradation of methomyl with 5 g/L of catalysts (Tomašević, 2011).

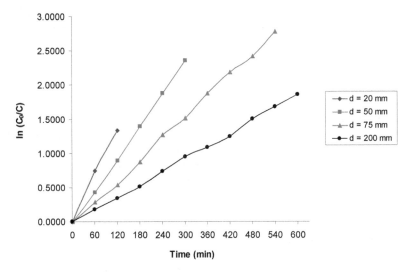

Fig. 3. The effect of lamp distance on the photolysis rate of methomyl (Tomašević, 2011).

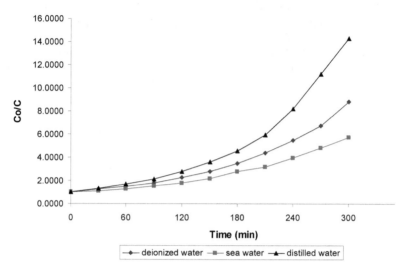

Fig. 4. The effect of the type of water on the photolysis rate of methomyl (Tomašević, 2011).

Technical methomyl	Parameters	Water type Deionized
With 2.0 g/L of TiO$_2$	k (min^{-1})	0.0058
	R	0.9880
	t$_{1/2}$ (min)	119.51
With 2.0 g/ L of ZnO	k (min^{-1})	0.0120
	R	0.9915
	t$_{1/2}$ (min)	57.76

Table 2. Kinetics of methomyl photodegradation at 366 nm (Tomašević, 2011).

4.10 Oxamyl

Oxamyl (IUPAC name: N,N-dimethyl-2-methylcarbamoyoxyimino-2-(methylthio) acetamide) is contact and systemic insecticide, acaricide and nematocide. It is used for control of chewing and sucking insects, spider mites and nematodes in ornamentals, frut trees, vegetables, cucurbits, beet, bananas, pineapples, peanuts, cotton, soya beans, tobacco, potatoes, and other crops. It could be found only as soluble concentrate (SL) on the market. The current regulation status of this active ingredient under directive 91/414/EEC is included in Annex 1, expiration of inclusion: 31/07/2016 (EU Pesticide Database, 2011; Tomlin, 2009).

An pre-industrial solar treatmen is used to prevent pollution of waters with commercial pesticide Vydate L, containing 24% oxamyl (Malato et al., 2000). Oxamyl is completely photodegraded, but mineralization is slow with illuminated TiO$_2$ only. The use of additional oxidants such as peroxydisulphate enhanced the degradation rate by a factor of 7 compared to TiO$_2$ alone. Solar photodegradation in aqueous solution of oxamyl in the presence of two photocatalysts TiO$_2$ and sodium decatungstate Na$_4$W$_{10}$O$_{32}$ is reported (Texier et al., 1999).

For pure compounds TiO_2 was a better catalyst than $Na_4W_{10}O_{32}$, concerning the rate of photodegradation and mineralization. When the pesticide is used as formulation product, the decatungstate anion becomes as efficien or even more efficient than TiO_2. This difference of reactivity is accounted for by the different nature of the active species during both photodegradation processes. The solar driven photo-Fenton process was applied to the degradation of oxamyl in the form of DuPont formulated product Vydate (Fallman et al., 1999). The obtained results shown that oxamyl was relatively recalcitrant (about 100 min was a TOC half-life and about 160 min was the time necessary for degradation of 80% of TOC).

4.11 Pirimicarb

Pirimicarb (IUPAC name: 2-dimethylamino-5,6-dimethylpyrimidin-4-yl dimethylcarbamate) is selective systemic insecticide with contact , stomach, and respiratory action. It is used as a selective aphicide for control a wide range of crops, including cereals, oil seeds, potatoes and other vegetables, ornamentals, and other non-food uses. Formulations types for this active ingredient are AE, DP, EC, FU, WG and WP. The current regulation status of this active ingredient under directive 91/414/EEC is included in Annex 1, expiration of inclusion: 31/07/2017 (EU Pesticide Database, 2011; Tomlin, 2009).

Photolysis of pirimicarb upon simulated solar light in natural water and in different aqueous solutions was investigated (Taboada et al., 1995). Aceton strongly increased degradation of pesticide, while methanol did not have any significant effect. The rate of pesticide degradation in the presence of river water was 4.5 times slower than in distilled water, and the half-life of pirimicarb in presence of dissolved humic and fulvic acids was 2-10 times longer than in distilled water. In all studied solutions the degradation reaction followed a first-order kinetics. The solar light and simulated sunlight were used for the photolysis of pirimicarb in water (Romero et al., 1994). The photodegradation mechanism seemed to be similar under both conditions, but the half-life of pirimicarb was found to be about three times longer under natural than under simulated conditions. Also, four main products were isolated and identified by spectroscopic methods. The photolysis of aqueous pirimicarb (3.3×10^{-3} M, 4 h, room temperature) has been examined by GC-MS (Climent & Miranda, 1996). Upon irradiation with 125 W medium-pressure mercury lamp three main photoproducts were detected.

4.12 Promecarb

Promecarb (IUPAC name: 3-methyl-5-methylphenyl methylcarbamate) is an obsolete carbamate insecticide once used to combat foliage and fruit eating insects. It is systemic insecticide. Promecarb is highly toxic by ingestion and is adsorbed through the skin. Formulations type is EC. The current regulation status of this active ingredient under directive 91/414/EEC is not included in Annex 1 (EU Pesticide Database, 2011; Tomlin, 2009).

The photolysis of promecarb in water solution (3.3×10^{-3} M, 4 h, room temperature, 125 W medium-pressure mercury lamp) has been examined by GC-MS (Climent & Miranda, 1996). Upon irradiation, 24% conversion of promecarb was achieved and photolysis of promecarb led to the phenol derivative (22%) as major product. Also, minor amounts of two compounds (isomers arising from photo-Fries rearrangement) were also obtained.

4.13 Propamocarb

Propamocarb (IUPAC name: propyl 3-(dimethylamino)propylcarbamate) is systemic fungicide with protective action. It is used for specific control of Phycomycetes. Also it is used against of wide variety of pest on tomatoes and potatoes, lettuce, cucumber, cabbages, ornamentals, fruit, vegetables, and vegetable seedbeds. Formulations types on the market are SC and SL. The current regulation status of this active ingredient under directive 91/414/EEC is included in Annex 1 expiration of inclusion: 30/09/2017 (EU Pesticide Database, 2011; Tomlin, 2009).

The application of solar photo-Fenton process for degradation of DuPont commercial product Previcur (Fallman et al., 1999) confirmed that propamocarb was one of the hardest pesticides to degrade by process (106 min was a TOC half-life and more than 200 min was the time necessary for degradation of 80% of TOC).

4.14 Propoxur

Propoxur (IUPAC name: 2-isopropoxyphenyl methylcarbamate) is non-systemic insecticide with contact and stomach action. It is used for control of cockroaches, flies, fleas, mosquitoes, bugs, ants, millipedes and other insect pests in food storage areas, houses, animal houses, etc. Also it is used for control of sucking and chewing insects (including aphids) in fruit, vegetables, ornamentals, vines, maize, alfalfa, soya beans, cotton, sugar cane, rice, cocoa, forestry, etc, and against migratory locusts and grasshoppers. There are a lot of different formulations with this active ingredient as AE, DP, EC, FU, GR, RB, SL, UL, WP and Oil spray. The current regulation status of this active ingredient under directive 91/414/EEC is not included in Annex 1 (EU Pesticide Database, 2011; Tomlin, 2009).

An study of the photodegradation of aerated aqueous propoxur solution is given very interesting data (Sanjuan et al., 2000). Photolysis of 1.0×10^{-3} M solution (pH 6.8) with 125 W medium-pressure mercury lamp leads to an almost complete degradation of pesticide and the formation of photo-Fries rearrangement products, but with a relatively minor degree of mineralization. Photocatalyzed degradations in the presence of TiO_2 (40 mg) or with 150 mg of triphenylpyrylium-Zeolite Y (TPY) were shown the same degree of propoxur mineralization. Laser flash photolysis (266 nm) has shown that the degradation could be initiated by a single electron transfer between excited 2,4,6-triphenylpyrylium cation and propoxur to form the corresponding 2,4,6-triphenylpyrylium radical and propoxur radical cation.

5. Conclusion

The reviewed literature reflects that in case of carbamate pesticides the most of the studies have been reported using photo-Fenton processes, photolysis and heterogeneous catalysis with TiO_2 as a catalyst. This photodegradation processes have been proposed as an effective and attractive techniques for degradation of carbamate residues in water. The kinetics of all photodegradation processes depend on several main parameters such as the nature of pesticides, type of light, initial concentration of pesticides (and catalysts), pH of solution, temperature, and presence of oxidant. The AOPs provide an excellent opportunity to use solar light as an energy source. Photocatalytic processes can lead to the mineralization of toxic and hazardous carbamate pesticides into carbon dioxide, water and inorganic mineral salts.

6. Acknowledgment

The authors are grateful to the Ministry of Education and Science of the Republic of Serbia for financial support (Project No. III 46008). The authors wish to thank also the DuPont de Nemours and FMC, USA companies for kindly support with the analytical standards. We would like to express thanks to Mr. Aleksandar F. Tomaši for technical assistance.

7. References

Andreozzi, R.; Caprio, V.; Insola, A. & Marotta, R. (1999). Advanced oxidation processes (AOP) for water purification and recovery. *Catalysis Today*, Vol.53, No.1, pp. 51-59, ISSN 0920-5861

Aaron, J.J. & Oturan, M.A. (2001). New Photochemical and Electrochemical Methods for the Degradation of Pesticides in Aqueous Media. *Turkish Journal of Chemistry*, Vol.25, No. 4, pp. 509-520, ISSN 1300-0527

Behnajady, M.A.; Modirshahla, N. & Hamzavi, R. (2006). Kinetic study on photocatalytic degradation of C.I. Acid Yellow 23 by ZnO photocatalyst. *Journal of Hazardous Materials* B, Vol.133, No.1-3, pp. 226-232, ISSN 0304-3894

Benitez, F.J.; Acero, J.L. & Real. F.J. (2002). Degradation of Carbofuran by Using ozone, UV Radiation and Advanced Oxidation Processes. *Journal of Hazardous Materials B*, Vol.89, No.1, pp. 51-65, ISSN 0304-3894

Burrows, H.D.; Canle, M.L.; Santaballa, J.A. & Steenken, S. (2002). Reaction pathways and Mechanisms of Photodegradation of Pesticides. *Journal of Photochemistry and Photobiology B: Biology*, Vol.67, pp. 71-108, ISSN 1011-1344

Bianco Prevot, A.; Pramauro, E. & de la Guardia, M. (1999). Photocatalytic Degradation of Carbaryl in Aqueous TiO$_2$. Suspensions Containing Surfactants. *Chemosphere*, Vol.39, No.3, pp. 493-502, ISSN 0045-6535

Brkić, D.; Vitorović, S.; Gašić, S. & Nešković, N. (2008). Carbofuran in Water: Subchronic Toxiticity to Rats. *Environmental Toxicology and Pharmacology*, Vol.25, No.3, pp. 334-341, ISSN 1382-6689

Catastini, C.; Sarakla, M.; Mailhot, G. & Bolte, M. (2002a). Iron (III) Aquacomplexes as Effective Photocatalysts for the Degradation of Pesticides in Homogeneous Aqueous Solutions. *The Science of the Total Environment*, Vol.298, No.1-3, pp. 219-228, ISSN 0048-9697

Catastini, C.; Sarakla, M. & Mailhot, G. (2002b). Asulam in Aqueous solutions: Fate and Removal Under Solar Irradiation. *International Journal of Environmental Analytical Chemistry*, Vol.82, No.8-9, pp. 591-600, ISSN 0306-7319

Climent, M.H. & Miranda, M.A. (1996). Gas Chromatographic-Mass Spectrometric Study of Photodegradation of Carbamate Pesticides. *Journal of Chromatography A*, Vol.738, No.2, pp. 225-231, ISSN 0021-9673

CropLife Tehnical Monograph No.2 (2008). http//www.croplife.org/en-us/technical_ monographs

Daneshvar, N.; Salari, D. & Khataee, A.R. (2003). Photocatalytic Degradation of Azo Dye Acid red 14 in Water: Investigation of the Effect of Operational Parameters. *Journal of Photochemistry and Photobiology A: Chemistry*, Vol.157, No.1, pp. 111-116, ISSN 1010-6030

Daneshvar, N.; Salari, D. & Khataee, A.R. (2004). Photocatalytic Degradation of Azo Dye Acid red 14 in Water on ZnO as an Alternative Catalyst to TiO₂. *Journal of Photochemistry and Photobiology A: Chemistry*, Vol.162, No.2-3, pp. 317-322, ISSN 1010-6030

Deneshvar, N.; Aber, S.; Seyed Dorraji, M. S.; Khataee, A.R. & Rasuolifard, M.H. (2007). Photocatalytic Degradation of the Insecticide Diazinon in the Presence of Prepared Nanocrystalline ZnO Powders Under Irradiation of UV-C Light. *Separation and Purification Technology*, Vol.58, No.1, pp. 91-98, ISSN 1383-5866

Đurović, R.; Đorđević, T.; Šantrić, Lj.; Gašić, S. & Ignjatović, Lj. (2010). Headspace Solid Phase Microextraction Method for Determination of Triazine and Organophosphorus Pesticides in Soil. *Journal of Environmental Science and Health, Part B*, Vol.45, No.7, pp. 626-632, ISSN 0360-1234

Edwards, C.A. (1975). Factors that Affect the Persistence of Pesticides in Plants and Soils. *Pure and Applied Chemistry*, Vol.42, No. 1-2, pp. 39-56, ISSN 0033-4545

EU Pesticide Database: http//ec.europa.eu/sanco_pesticides/public/

Fallmann, H.; Krutzler, T.; Bauer, R., Malato, S. & Blanco, J. (1999). Applicability of the Photo-Fenton Method for Treating Water Containing Pesticides. *Catalysis Today*, Vol.54, No.2-3, pp. 309-319, ISSN 0920-5861

Gašić, S.; Jovanović, B. & Jovanović, S. (1998a). Emulsion Inversion Point (EIP) as a Parameter in the Selection of Emulsifier. *Journal of the Serbian Chemical Society*, Vol.63, No.7, pp. 529-536, ISSN 0352-5139

Gašić, S.; Jovanović, B. & Jovanović, S. (1998b). Phase Inversion Temperature (PIT) as a Parameter for the Selection of an Appropriate Nonionic Emulsifier. *Journal of the Serbian Chemical Society*, Vol.63, No.10, pp.763-771, ISSN 0352-5139

Gašić, S.; Budimir, M.; Brkić, D. & Nešković, N. (2002a). Residues of Atrazine in Agricultural Areas of Serbia. *Journal of the Serbian Chemical Society*, Vol.67, No.12, pp. 887-892, ISSN 0352-5139

Gašić, S.; Jovanović, B. & Jovanović, S. (2002b). The Stability of Emulsions in the Presence of Additives. *Journal of the Serbian Chemical Society*, Vol.67, No.1, pp. 31-39, ISSN 0352-5139

Gevao, B.; Semple, K.T. & Jones, K.C. (2000). Bound Pesticide Residues in Soils: a Review. *Environmental Pollution*, Vol.108, No.1, pp. 3-14, ISSN 0269-7491

Gomes da Silva, C. & Faria, J.L. (2003). Photochemical and Photocatalytic Degradation of an Azo Dye in Aqueous Solution by UV Irradiation. *Journal of Photochemistry and Photobiology A: Chemistry*, Vol.155, No.1-3, pp. 133-136, ISSN 1010-6030

Huston, P.L. & Pignatello, J.J. (1999). Degradation of Selected Pesticide Active Ingredients and Commercial Formulations in Water by the Photo-assisted Fenton Reaction. *Water Research*, Vol.33, No.5, pp. 1238-1246, ISSN 0043-1354

Jha, M.N. & Mishra, S.K. (2005). Decrease in Microbial Biomass due to Pesticide Application/Residues in Soil Under Different Cropping Systems. *Bulletin of Environmental Contamination and Toxicology*, Vol.75, No.2, pp. 316-323, ISSN: 0007-4861

Kamrin, M. A. (Ed.) (1997). *Pesticide profiles*, CRC Lewis Publishers. pp 53-87, ISBN 0-56670-190-2, New York, USA

Kaufman, D.D. (1967). Degradation of carbamate herbicides in soil. *Journal of Agricultural and Food Chemistry*, Vol.15, No.4, pp. 582-591, ISSN 0021-8561

Karkmaz, M.; Puzenat, E.; Guillarad, C. & Herrmann, J.M. (2004). Photocatalytic Degradation of the Alimentary Azo Dye Amaranth. Mineralization of the Azo Group to Nitrogen. *Applied Catalysis. B: Environmental*, Vol.51, No.3, pp. 183-194, ISSN 0926-3373

Kesraoui Abdessalem, A.; Bellakhal, N.; Oturan, N.; Dachraoui, M. & Oturan, M. A. (2010). Treatment of a Mixture of Three Pesticides by Photo- and Electro-Fenton processes. *Desalination,*Vol.250, No.1, pp. 450-455, ISSN 0011-9164

Knowles, A. (2005). *New Developments in Crop Protection Product Formulation*, T&F Informa UK Ltd. DS243, Agrow Reports, pp. 33-52, www.agrowreports.com

Knowles, A. (2006). *Adjuvants and Additives*, T&F Informa UK Ltd., DS256 Agrow Reports, pp. 33-62, www.agrowreports.com

Lázár, K.; Tomašević, A.; Bošković, G. & Kiss, E. (2009). Comparison of FeAl-PILC and Fe-ZSM-5 Catalysts Used for Degradation of Methomyl. *Hyperfine Interactions,* Vol.192, No.1-3, pp. 23-29, ISSN 0304-3843

Legrini, O.; Oliveros, E. & Braun, A.M. (1993). Photochemical Processes for Water Treatment. *Chemical Reviews*, Vol.93, No.2, pp. 671-698 ISSN 0009-2665

Machemer, H.M. & Pickel, M. (1994). Chapter 4, Carbamate Insecticides. *Toxicology*, Vol. 91, pp. 29-36, ISSN 0300-483X

Malato, S.; Blanco, J.; Richter, C., Fernandez, P. & Maldonado, M.I. (2000). Solar Photocatalytic Mineralization of Commercial Pesticides: Oxamyl. *Solar Energy Materials & Solar Cells*, Vol.64, No.1, pp. 1-14, ISSN 0927-0248

Malato, S.; Blanco, J.; Vidal, A. & Richter, C. (2002a) Photocatalysis with Solar Energy at a Pilot-plant Scale: an Overview. *Applied Catalysis. B: Environmental*, Vol.37, pp. 1-15, ISSN 0926-3373

Malato, S.; Blanco, J.; Caceres, J.; Fernandez-Alba, A.R.; Agüera, A. & Rodriguez, A. (2002b). Photocatalytic Treatment of Water-Soluble Pesticides by Photo-Fenton and TiO$_2$ Using Solar Energy. *Catalysis Today*, Vol.76, No.2-4, pp. 209-220, ISSN 0920-5861

Malato, S.; Blanco, J.; Vidal, A.; Alarcon, D.; Maldonado, M.I.; Caceres, J. & Gernjak, W. (2003). Applied Studies in Solar Photocatalytic Detoxification: an Overview. *Solar Energy*, Vol.75, No.4, pp. 329-336, ISSN 0038-092X

Mansour, M.; Schmitt, Ph. & Mamouni, A. (1992). Elimination of Metoxuron and Carbetamide in the Presence of Oxigen Species in Aqueous Solutions. *The Science of the Total Environment*, Vol.123-124, pp. 183-193, ISSN 0048-9697

Mollet, H. & Grubenmann, A. (2001). *Formulation Technology*, Wiley-Vch, pp.389-397, ISBN 3-527-30201-8, Weinheim Federal Repablic of Germany.

Neyens, E. & Baeyens, J. (2003). A Review of Classic Fenton·s Peroxidation as an Advanced Oxidation Technique. *Journal of Hazardous Materials* B, Vol.98, pp. 33-47, ISSN 0304-3894

Pera-Titus, M.; Garcia-Molina, V.; Banos, M.A.; Gimenez, J. & Esplugas, S. (2004) Degradation of Chlorophenols by Means of Advanced Oxidation Processes: a

General Review. *Applied Catalysis. B: Environmental*, Vol.47, No. 4, pp. 219-256, ISSN 0926-3373

Percherancier, J.P.; Chapelon, R. & Pouyet, B. (1995). Semiconductor-Sensitized Photodegradation of Pesticides in Water: The Case of Carbetamide. *Journal of Photochemistry and Photobiology A: Chemistry*, Vol.87, No.3, pp. 261-266, ISSN 1010-6030

Peris, C.E.; Terol, J.; Mauri, A.R.; de la Guardia, M. & Pramauro, E. (1993). Continuous Flow Photocatalytic Degradation of Carbaryl in Aqueous Media. *Journal of Environmental Science and Health, Part B*, Vol.28, No.4, pp. 431-440, ISSN 0360-1234

Pramauro, E.; Bianco Prevot, A.; Vincenti, M. & Brizzolesi M. (1997). Photocatalytic Degradation of Carbaryl in Aqueous Solutions Containing TiO_2 Suspensions. *Environmental Science and Technology*, Vol.31, No.11, pp. 3126-3131, ISSN 0013-936X

Radivojević, Lj.; Gašić, S.; Šantrić, Lj. & Stanković-Kalezić R. (2008). The Impact of Atrazine on Several Biochemical Propertes of Chernozem Soil. *Journal of the Serbian Chemical Society*, Vol.73, No.10, pp. 951-959, ISSN 0352-5139

Romero, E; Schmitt, Ph. & Mansour, M. (1994). Photolysis of Pirimicarb in Water Under Natural and Simulated Sunlight Conditions. *Pesticide Science*, Vol.41, No.1, pp. 21-26, ISSN 0031-613X

Sanjuan, A.; Aguirre, G.; Alvaro, M., Garcia, H. & an Scaiano, J.C. (2000). Degradation of Propoxur in Water Using 2,4,6-triphenylpyrylium-Zeolite Y as Photocatalyst. *Applied Catalysis. B: Environmental*, Vol.25, No.4, pp. 257-265, ISSN 0926-3373

Sanz-Asensio, J.; Plaza-Medina, M.; Martinez-Soria M.T. & Perez-Clavijo, M. (1999). Study of Photodegradation of the Pesticide Ethiofencarb in Aqueous and Non-Aqueous Media, by Gas-Chromatography-Mass-Spectrometry. *Journal of Chromatography A*, Vol.840, No.2, pp. 235-247, ISSN 0021-9673

Sher, H.B. (Ed.). (1984). *Advances in Pesticide Formulation Technology*, In: ACS Symposium Series 254, American Chemical Society, pp. 1-7, 141-151. ISBN: 0-8412-0840-9, Washington DC, USA

Shinoda, K. & Friberg. S. (1986). *Emulsions and Solubilization*, A Wiley-Interscience Publication, John Wiley&Sons, pp. 55-91, ISBN 0-471-03646-3, New York, USA

Sun, Y. & Pignatello, J.J. (1993). Photochemical Reactions Involved in the total Mineralization of 2,4-D by $Fe^{3+}/H_2O_2/UV$. *Environmental Science and Technology*, Vol. 27, No. 2, pp. 304-310, ISSN 0013-936X

Taboada, R.E; Schmitt, Ph. & Mansour, M. (1995). Photodegradation of Pirimicarb in Natural Water and in Different Aqueous Solutions Under Simulated Sunlight Conditions. *Fresenius Environmental Bulletin*, Vol.4, No.11, pp. 649-654, ISSN 1018-4619

Takino, M.; Yamaguchi, K. & Nakahara, T. (2004). Determination of Carbamate Pesticide in Vegetables and Fruits by Liquid Chromatography - Atmospheric Pressure Chemical Ionization-Mass Spectrometry. *Journal of Agricultural and Food Chemistry*, Vol.52, No.4, pp. 727-735, ISSN 0021-8561

Tamimi, M.; Qourzal, S.; Barka, N.; Assabbane, A. & Ait-Ichou, Y. (2008). Methomyl Degradation in Aqueous Solutions by Fenton's Reagent and the Photo-Fenton System. *Separation and Purification Technology*,Vol. 61, No.1, pp. 103-108, ISSN 0376-7388

Texier, I.; Giannotti, C.; Malato, S.; Richter, C. & Delaire, J. (1999). Solar Photodegradation of Pesticides in Water by Sodium Decatungstate. *Catalysis Today*, Vol.54, No.2-3, pp. 297-307, ISSN 0920-5861

Tomašević, A.; Bošković, G.; Mijin, D.; Đilas, S. & Kiss, E. (2007a). The Extremely High Stability of Carbofuran Pesticide in Acidic Media. *Acta Periodica Technologica*,Vol.38, pp. 97-103, YU ISSN 1450-7188

Tomašević, A.; Bošković, G.; Mijin, D. & Kiss, E. (2007b). Degradation of Carbamates over AlFe-PILC Catalyst, *Proceedings of PSU-UNS International Conference of Engineering and Environment, ICEE-2007 and PEC-5*, pp. 124-127, Phuket, Thailand, May 10-11, 2007

Tomašević, A.; Bošković, G.; Mijin, D. & Kiss, E. (2007c). Decomposition of Methomyl Over Supported Iron Catalysts. *Reaction Kinetics and Catalysis Letters*, Vol.91, No.1, pp. 53-59, ISSN 0133-1736

Tomašević, A.V.; Avramov Ivić, M.L.; Petrović, S.D.; Jovanović, M. B. & Mijin, D.Ž. (2009a). A Study of the Electrochemical Behaviour of Methomyl on a Gold Electrode in a Neutral Electrolyte. *Journal of the Serbian Chemical Society*, Vol.74, No.5, pp. 573-579, ISSN 0352-5139

Tomašević, A.; Đaja, J.; Petrović, S:, Kiss, E. E. & Mijin, D. (2009b). A Study of the Photocatalytic Degradation of Methomyl by UV Light. *Chemical Industry & Chemical Engineering Quarterly*, Vol.15, No.1, pp. 17-19, YU ISSN 1451-9372

Tomašević, A.; Mijin, D. & Kiss, E. (2009c). Photochemical Behaviour of the Insecticide Methomyl Under Different Conditions, *Proceedings of 2nd European Conference on Environmental Applications of Advanced Oxidation Processes*, pp. 1-6, ISBN 978-9963-689-09-5, Nicosia, Cyprus, September 9-11, 2009

Tomašević, A.; Mijin, D. & Kiss, E. (2010a). Photochemical Behavior of the Insecticide Methomyl Under Different Conditions. *Separation Science and Technology*, Vol.45, No.11, pp. 1617-1627, ISSN 1520-5754

Tomašević, A.; Kiss, E.; Petrovic, S. & Mijin, D. (2010b). Study on the Photocatalytic Degradation of Insecticide Methomyl in Water, *Desalination*, Vol.262, No.1-3, pp. 228-234, ISSN 0011-9164

Tomašević, A. (2011). Contribution to the Studied of the Photodegradation Mechanism of Carbamate Pesticides. *PhD Dissertation*, pp. 1-202, University of Belgrade, Faculty of Technology and Metalurgy, Belgrade, Serbia

Tomlin, C.D.S. (Ed.) (2009). *The Pesticide Manual*, A World Compendium (15th Edition). British Crop Production Council (BCPC), Omega Park, Alton, Hampshire, GU34 2QD, UK, ISBN 9781901396188

Vialaton, D. & Richard, C. (2002). Phototransformations of Aromatic Pollutants in Solar Light: Photolysis Versus Photosensitized Reactions Under Natural Water Conditions. *Aquatic Sciences*, Vol.64, No.2, pp. 207-215, ISSN 1015-1621

World Health Organization (1986). Carbamate pesticides: a general introduction. Environmental Health Criteria (EHC) No.64, World Health Organization (WHO), ISBN 9241542640, Geneva, Switzerland

5

Adverse Effect of Insecticides on Various Aspects of Fish's Biology and Physiology

Mahdi Banaee

Department of Aquaculture, Natural Resource and Environmental Faculty,
Behbahan University of Technology,
Iran

1. Introduction

Today, water quality management faces greater problems than at any time in its history. In addition to natural pollutants, varied contaminants exist in surface waters including multiple chemical compounds and different products of industrial and agricultural revolution. The insecticides constitute one group of these pollutants, both synthetic and natural, which contribute to the environmental problems. At present, it seems that the problem is more conspicuous in developing countries, where lately there has been an increase in the use of insecticides as a means of increasing agricultural productivity, without much concern to the consequences of indiscriminate application. There are many pathways by which insecticides leave their sites of application and distribute throughout the environment and enter the aquatic ecosystem. The major route of insecticides to water ecosystems in urban areas is through rainfall runoff and atmospheric deposition. Another source of water contamination by insecticides is from municipal and industrial dischargers. Most insecticides ultimately find their way into rivers, lakes and ponds (Tarahi Tabrizi, 2001; Honarpajouh, 2003; Bagheri, 2007; Shayeghi *et al.*, 2007; Vryzas *et al.*, 2009; Werimo *et al.*, 2009; Arjmandi *et al.*, 2010) and have been found to be highly toxic to non-target organisms that inhabit natural environments close to agricultural fields. The contamination of surface waters by insecticides is known to have ill effects on the growth, survival and reproduction of aquatic animals. In the past few years, the increase of mortality among the fish in various streams, lakes and ponds of around the world has drawn scholars' attention to the problems caused by insecticides and pesticides runoff associated with intense agricultural practices. Different concentrations of insecticides are present in many types of wastewater and numerous studies have found them to be toxic to aquatic organisms especially fish species (Talebi, 1998; Üner *et al.*, 2006; Banaee *et al.*, 2008). Fishes are particularly sensitive to the environmental contamination of water. Hence, pollutants such as insecticides may significantly damage certain physiological and biochemical processes when they enter into the organs of fishes (John, 2007; Banaee *et al.*, 2011). Authors found out that different kinds of insecticides can cause serious impairment to physiological and health status of fishes (Begum, 2004; Monteiro *et al.*, 2006; Siang *et al.*, 2007; Banaee *et al.*, 2009). Since fishes are important sources of proteins and lipids for humans and domestic animals, so health of fishes is very important for human beings. Recently, many studies have been conducted to determine the mechanisms of insecticides in fishes, with the ultimate goal of monitoring, controlling and possibly intervening in xenobiotics exposure and its effects on the aquatic ecosystem. This chapter presents further information concerning the toxic effects of different concentrations of insecticides on various aspects of fish's biology and physiology. In other words, this chapter depicts the effects of insecticides on the survival chance, blood biochemical parameters, tissues and organs, reproduction, development and growth, nervous system, behavior, genetic and immune system of fish. The information given in this part facilitates the evaluation of potential toxic hazard resulting from exposure to different levels of these compounds.

2. Biokinetics and bioteransformation

After exposure to difference concentrations of insecticides in water, the fish absorbs them in its gills, skin or gastrointestinal tract. In the other words, Due to their lipophilicity, most insecticides easily permeate the biological membranes and it increases the sensitivity of fish to aqueous insecticides. Then, these compounds are rapidly metabolized and extracted, and may be bio-concentrated in various tissues of fish. In other words, bio-accumulation occurs if the insecticides is slowly metabolized or excreted from the body. As the amount of insecticide increases, it becomes more harmful to the consumer or animal. Accumulated insecticide can cause death or long-term damage. Ballesteros et al. (2011) showed that during the initial 24 h of exposure, insecticides may be transformed in various tissues of fish. However, some differences exist among tissues relating to metabolism rates, leading to different distribution models of the original compounds and their metabolites. For example, the low biodegradation and the high lipid solubility of some insecticides such as organochlorine insecticides have led to problems with the bio-concentrations of these compounds in different tissues of fish. In addition, since some fish are lower on the food chain, bioaccumulation of insecticides may increase in tissues of their predators and consumers, such as humans and thus affecting their health and survival. So, the bioaccumulation of these contaminants in fish and the potential biomagnification in humans are perceived as threats (Favari et al., 2002). Bioaccumulation rate of insecticides in fish depends on the species, life stages, the amount of fat reservation in different tissues and diet of fish, chemical and physical properties of insecticides and the rate of water pollution.

In order to facilitate the elimination and detoxification of toxic compounds, fishes have developed partly complex detoxification mechanisms including the release of several enzymes collectively termed xenobiotic metabolizing enzymes. Enzymatic biotransformation of insecticides can potentially alter their activity and toxicity. Enzymes participating in the biotransformation of insecticides are classified into phase I and phase II enzymes. The phase I enzymes, cytochrome P450 enzymes including CYP1A and CYP3A, are generally involved in the biotransformation of exogenous and endogenous compounds; thereby creating a more polar and water soluble compound. A great diversity of cytochrome P450 enzymes in fish has been recognized (Stegeman and Hahn, 1994), and CYP1A, CYP2B, CYP2E1, CYP2K1 and CYP3A have been recently identified in liver of some freshwater fish (Nabb et al., 2006) which play an important role in the detoxification of organophosphate and carbamate insecticides (Ferrari et al., 2007). The common pathways of biotransformation of different kinds of insecticides include three cytochrome P450 (CYP) mediated reactions: O-dealkylation, hydroxylation, and epoxidation of insecticides (Soldano et al., 1992; Keizer et al., 1995; Kitamura et al., 2000; Straus et al., 2000; Behrens & Segner, 2001; Nebbia, 2001). In phase II reactions, metabolites produced in phase I detoxification often conjugate with glutathione, uridyl-diphosphate glucose (UDPG), uridyl-diphosphate-glucuronic acid (UDPGA), amino acid derivatives and sulfate derivatives and can readily excrete from the fish body (Iannelli *et al.*, 1994; Keizer *et al.*, 1995; Kitamura *et al.*, 2000; Straus *et al.*, 2000; Behrens & Segner, 2001; Nebbia, 2001). In other words, final metabolites may also be excreted from the body of fish through the skin, gills, genital products, urine as sulphated and glucuronidated metabolites and stool as glutathione conjugated metabolites (Kitamura *et al.*, 2000; Straus *et al.*, 2000; Behrens & Segner, 2001; McKim & Lein, 2001; Nebbia, 2001). Since metabolites produced during detoxification process may be more dangerous than parental compounds, these metabolites can cause serious damage in fish. Furthermore, the production of reactive oxygen species (ROS) during detoxification process can induce oxidative damage and may be a mechanism of toxicity for aquatic organisms living in environments receiving insecticides (Monteiro *et al.*, 2009). ROS can indiscriminately attack and react with susceptible vital macromolecules -lipids, proteins and DNA- in living cells, inducing cytotoxicity and can result in serious disturbances in physiological cell processes (Dogan *et al.*, 2011; Jin *et al.*, 2011). Lipid peroxidation, the major contributor to the loss of cell function, DNA damage, enzyme inactivation, and hormone oxidation are bio-indicators of oxidative cell damage and examples of toxic mechanisms of insecticide induced ROS being involved in pathological processes and in the etiology of many fish diseases (Üner *et al.*, 2006; Dogan *et al.*, 2011).

3. Acute toxicity

In acute toxicity, sudden and intense mortality may be observed in a fish population exposed to toxic materials. The most apparent symptoms of insecticides' acute poisoning in fishes include lethargy, forward extension fins, pallor or blur parts of body, severe reaction to external stimuli and muscle spasms and sudden fast swimming in circles. The main clinical internal sings that can lead to death of fishes include neurological disorder and disruption of nerve functions, respiratory dysfunction and suffocation (Banaee *et al.*, 2011).

Acute toxicity testing is widely used in order to identify the dose or exposure concentration and the time associated with death of 50 percent of the fish exposed to toxic materials which is expressed as LC_{50} in parts per million (ppm) or milligrams per liter (mg/L). In addition, we can use the LC_{50} value in the classification of insecticides based on potential toxicity for fishes. Furthermore, the relative acute toxicity of chemicals to fish can be categorized as follows:

Toxicity rating	96 hr LC_{50}
Slightly toxic	10-100 ppm
Moderately toxic	1-10 ppm
Highly toxic	0.1-1.0 ppm
Extremely toxic	Less than 0.1 ppm

Our literature reviews demonstrate that different fish species, even from the same family, show differences in the sensitivity to high concentrations of insecticides in water. Acute toxicity of different insecticides is influenced by the age, sex, genetic properties and body size of fish, water quality and its physicochemical parameters, and purity and formulation of insecticides.

The eight tables, which give relative acute toxicity of some insecticides to fishes, can be used to determine the potential toxicity to fish of using these compounds in farms around surface waters and to select products which less likely to cause problems. The data are derived from laboratory studies and are given only as a guideline and not absolute data of the acute toxicity of the insecticides to different species of fish (Table 1-11).

Insecticide	species	Range of 96h LC$_{50}$	Reference
Akton	Channel catfish, Bluegill, Rainbow trout, Fathead minnow	0.17-1370 ppb	Jonson & Finley, 1980
Aldicarb	Fathead minnow	1.3-2.4 ppm	Pant et al., 1987
Aldrin	Chinook salmon, Rainbow trout, Fathead minnow, Black bullhead, Bluegill, Largemouth bass	2.3-53 ppb	Jonson & Finley, 1980
Allethrin	Rainbow trout, Bluegill	19-56 ppb	Jonson & Finley, 1980
Aminocrab	Cutthroat trout, Rainbow trout, Bluegill, Atlantic salmon, Fathead minnow, Channel catfish, Largemouth bass, Yellow perch, Brook trout	3.1-31 ppm	Jonson & Finley, 1980
Azinphos ethyl	Rainbow trout, Bluegill	1.1-20 ppb	Jonson & Finley, 1980
Azinphos methyl	Gilthead seabream, Coho salmon, Rainbow trout, Bluegill, Atlantic salmon, Brown trout, pike, Goldfish, Carp, Fathead minnow, Black bullhead, Channel catfish, Green sunfish, Largemouth bass, Black crappie, Yellow perch	2.1-4270 ppb	Arufe et al., 2007; Jonson & Finley, 1980
Azodrin	Rainbow trout, Bluegill, Channel catfish. Fathead minnow	4.9-50 ppm	Jonson & Finley, 1980

Table 1. Summary of acute toxicity.

Insecticide	species	Range of 96h LC$_{50}$	Reference
Bensulide	Rainbow trout, Bluegill, Black bullhead, Common eel, Guppy, Sheephead minnow, Crucian carp	0.1-30 ppm	McAllister *et al.*, 1986a, b;
Benzene hexachloride	Rainbow trout, Bluegill, Cutthroat trout, Goldfish, fathead minnow, Channel catfish, Largemouth bass	9-348 ppb	Jonson & Finley, 1980
Carbaryl	Coho salmon, Chinook salmon, Cutthroat trout, Atlantic salmon, Brown trout, Brook trout, Lake trout, Carp, Channel Catfish, Fathead Minnow, Rainbow Trout, Bluegill, Goldfish, Black bullhead, Green sunfish, Largemouth bass, Black crappie, Yellow perch	0.9-39 ppm	Jonson & Finley, 1980
Carbofuran	Walked catfish, *Poecilia reticulate*, Chubs	0.22-23 ppm	Dobšíková, 2003; Begum, 2004
Carbophenthion	Channel catfish, Bluegill, Green sunfish	0.02-6 ppm	Jonson & Finley, 1980
Carbosulfan	Bluegill, Cutthroat trout, Rainbow trout, Lake trout, Channel catfish	2.4-280 ppb	Yi *et al.*, 2006; Jonson & Finley, 1980

Table 2. Summary of acute toxicity.

Insecticide	species	Range of 96h LC$_{50}$	Reference
Chlordane	Coho salmon, Cutthroat trout, Rainbow trout, Brown trout, Fathead minnow, Channel catfish, Bluegill, Largemouth bass	3-115 ppb	Jonson & Finley, 1980
Chlorethoxyphos	Cutthroat trout, Rainbow trout, Fathead minnow, Channel catfish, Bluegill, Largemouth bass	0.72-11.9 ppm	Jonson & Finley, 1980
Chlorfenapyr	Bluegill, Rainbow trout, channel catfish	6.5-14.7 ppb	Rand, 2004
Chlorpyrifos	Mosquito fish, Bluegill, Fathead minnow, Rainbow trout, Nile tilapia, Goldfish	0.57-3270 ppb	Davey *et al.*, 1976; Holcombe *et al.*, 1982; Bowman, 1988a, b; Gül, 2005; Wang *et al.*, 2009
Coumaphos	Cutthroat trout, Rainbow trout, Lake trout, Channel catfish, Bluegill, Largemouth bass	0.34-1.2 ppm	Jonson & Finley, 1980
Cryolite	Rainbow trout, Bluegill	47-400 ppm	Jonson & Finley, 1980

Cypermethrin	Sheepshead minnow, Rainbow trout, Bluegill, Freshwater catfish	0.39-0.95 ppb	Jaber & Hawk, 1981; Sousa, 1998; Mishra et al., 2005
DDD	Rainbow trout, Fathead minnow, Channel catfish, Largemouth bass, Walleye	14-4400 ppb	Jonson & Finley, 1980
DDE	Rainbow trout, Atlantic salmon, Bluegill	32-240 ppb	Jonson & Finley, 1980

Table 3. Summary of acute toxicity.

Insecticide	species	Range of 96h LC$_{50}$	Reference
DDT	Coho samon, Rainbow trout, Fathead minnow, Channel catfish, Bluegill, Largemouth bass, Black bullhead, Yellow perch	1.5-21.5 ppb	Jonson & Finley, 1980
Deltamethrin	Guppies, Channa punctatus,	1.5-5.13 ppb	Viran et al., 2003; Sayeed et al., 2003
Diazinon	Cutthroat trout, Rainbow trout, Lake trout, Fathead minnow, Carp, Bluegill	0.9-2.6 ppm	Calmbacher, 1978a, b; Banaee et al., 2011; Banaee et al., 2008; Jonson & Finley, 1980
Dichlorvos	Lake Trout, Sheephead minnow	0.18-7.5 ppm	Mayer & Ellersieck 1986; Jones & Davis, 1994
Dicrotophos	Bluegill, Rainbow trout, Channel catfish	6.3-24.2 ppm	Jonson & Finley, 1980
Dieldrin	Cutthroat trout, Rainbow trout, Goldfish, Fathead minnow, Channel catfish, Bluegill, Largemouth bass	1.2-19 ppb	Jonson & Finley, 1980
Diflubenzuron	Cutthroat trout, Rainbow trout, Brook trout, Fathead minnow, Channel catfish, Bluegill, Yellow perch	25-240 ppm	Jonson & Finley, 1980
Dimethoate	Rainbow trout, Bluegill	6-9.3 ppm	Jonson & Finley, 1980
Dimethrin	Fathead minnow, Channel catfish, Yellow perch, Bluegill,	28-1275 ppb	Jonson & Finley, 1980
Dinitrocresol	Rainbow trout, Bluegill	66-360 ppb	Jonson & Finley, 1980

Table 4. Summary of acute toxicity.

Insecticide	species	Range of 96h LC_{50}	Reference
Dioxathion	Cutthroat trout, Rainbow trout, Largemouth bass	22-110 ppb	Jonson & Finley, 1980
Disulfoton	Rainbow trout, Fatheah minnow, Channel catfish, Bluegill, Largemouth bass	60-4700 ppb	Jonson & Finley, 1980
d-Trans Allethrin	Coho salmon, Steelhead, Lake trout, pike, Fathead minnow, Channal catfish, Largemouth bass, Yellow perch	2.6-66 ppb	Jonson & Finley, 1980
Endosulfan	Striped bass, Bluegill, Rainbow trout, Fathead minnows, Asian swamp eel, Milk fish, Zebra fish	0.1-20 ppb	Mayer & Ellersieck, 1986; Siang *et al.*, 2007; Capkin *et al.*, 2006; Magesh & Kumaraguru, 2006, Velasco-Santamaría *et al.*, 2011
Endrin	Rainbow trout, Goldfish, fathead minnow, black bullhead, Channel catfish, Mosquito fish, Bluegill, Largemouth bass, Yellow perch, carp	0.15-1.8 ppb	Jonson & Finley, 1980
EPN	Cutthroat trout, Rainbow trout, Channel catfish, Bluegill, Largemouth bass, Walleye	110-420 ppb	Jonson & Finley, 1980
Ethion	Cutthroat trout, Rainbow trout, Channel catfish, Bluegill, Largemouth bass, Fathead minnow	0.17-7.6 ppm	Jonson & Finley, 1980

Table 5. Summary of acute toxicity.

Insecticide	species	Range of 96h LC_{50}	Reference
Ethyl Parathion	Coho salmon, Cutthroat trout, Rainbow trout, Brown trout, Goldfish, Carp, Fathead minnow, Channel catfish, Bluegill, Black bullhead, Largemouth bass, Yellow perch	0.4-3.52 ppm	Jonson & Finley, 1980
Fenitrothion	Coho salmon, Cutthroat trout, Rainbow trout, Brown trout, Brook trout, Atlantic salmon, Goldfish, Bluegill, Channel catfish, Fathead minnow, Carp	1.7-12 ppm	Johnson & Finley, 1980; Woodward & Mauck, 1980; Jonson & Finley, 1980
Fenthion	Coho salmon, , Rainbow trout, Brown trout, Brook trout, Atlantic salmon, Goldfish, Yellow perch, Bluegill, Channel catfish, Green sunfish, Fathead minnow, Largemouth bass, Carp	1.1-3.4 ppm	Jonson & Finley, 1980

Fenvalerate	Zebra fish	3.5-193 ppb	Ma *et al.*, 2009
Heptachlor	Rainbow trout, Northern pike, Fathead minnow, Black bullhead, Channel catfish, Redear sunfish, Bluegill, Largemouth bass,	5.3-63 ppb	Jonson & Finley, 1980
Isoprocarb	Goldfish	4.61 ppm	Wang *et al.*, 2009

Table 6. Summary of acute toxicity.

Insecticide	species	Range of 96h LC_{50}	Reference
Kepone	Rainbow trout, Channel catfish, Bluegill, redear sunfish	30-225 ppb	Jonson & Finley, 1980
Leptophos	Rainbow trout, Lake trout, Fathead minnow, Bulegill	0.03-30 ppm	Jonson & Finley, 1980
Lindane	Eel, Tilapia, African Catfish, Coho salmon, Rainbow trout, Brown trout, Goldfish, Carp, Fathead minnow, Black bullhead, Green sunfish, Largemouth bass, Yellow perch	0.03-1.29 ppm	Ferrando *et al.*, 1988, Feltz, 1971; Lawson *et al.*, 2011; Jonson & Finley, 1980
Linuron	Bluegill, Rainbow trout	3-16.2 ppm	Wetzel, 1986; Mayer & Ellersieck,1986
Malathion	Coho salmon, Cutthroat trout, Rainbow trout, Brown trout, Lake trout, Goldfish, Carp, Fathead minnow, Black bullhead, Bluegill, Green sunfish, Largemouth bass, Yellow perch, Redear sunfish	4-12900 ppb	Mayer & Ellersieck,1986; Jonson & Finley, 1980
Methamidophos	Rainbow trout, Fathead minnow, Channel catfish, Bluegill	1.6-100 ppm	Jonson & Finley, 1980
Methomyl	Cutthroat trout, Rainbow trout, Atlantic salmon, Brook trout, Fathead minnow, Channel catfish, Bluegill, Largemouth bass	0.3-6.8 ppm	Jonson & Finley, 1980; Yi et al., 2006

Table 7. Summary of acute toxicity.

Insecticide	species	Range of 96h LC_{50}	Reference
Methoxychlor	Rainbow trout, Atlantic salmon, Cutthroat trout, Brook trout, Lake trout, pike, Goldfish, Largemouth bass, Bluegill, Yellow perch, Fathead minnow, Channel catfish	15-64 ppb	Jonson & Finley, 1980
Methyl Parathion	Freshwater characid fish, Coho salmon, Cutthroat trout, Rainbow trout, Brown trout, Lake trout, Goldfish, Carp, Fathead minnow, Channel catfish, Bluegill, Black bullhead, Green sunfish, Largemouth bass, Yellow perch	0.25-9 ppm	Mayer & Ellersieck,1986; Monteiro et al., 2006; Jonson & Finley, 1980
Methyl Trithion	Cutthroat trout, Rainbow trout, Channel catfish, Bluegill, Largemouth bass	0.76-2.8 ppm	Jonson & Finley, 1980
Mexacarbate	Coho salmon, Cutthroat trout, Rainbow trout, Atlantic salmon, Lake trout, Carp, Fathead minnow, Channel catfish, Bluegill, Black bullhead, Largemouth bass, Yellow perch	0.32-23 ppm	Jonson & Finley, 1980
Mirex	Rainbow trout, Brown trout, Fathead minnow, Channel catfish, Bluegill, Largemouth bass, Yellow perch, Walleye	100 < ppm	Jonson & Finley, 1980

Table 8. Summary of acute toxicity.

Insecticide	species	Range of 96h LC_{50}	Reference
Monocrotophos	Tilapia, Mosquito fish	11.5-20.5 ppm	Rao, 2006; Kavitha & Rao, 2007
Naled	Cutthroat trout, Rainbow trout, Lake trout, Fathead minnow, Channel catfish, Bluegill, Largemouth bass	0.13-3.3 ppm	Jonson & Finley, 1980
Oxydemeton-methyl	Rainbow trout, Channel catfish, Bulegill, Largemouth bass, Walleye	13-31.5 ppm	Jonson & Finley, 1980
Permethrin	Brook trout	1.4-7.9 ppb	Jonson & Finley, 1980
Phorate	Cutthroat trout, Rainbow trout, Northern pike, Largemouth bass, Channel catfish, Bluegill	2-110 ppb	Jonson & Finley, 1980

Phosmet	Coho salmon, Rainbow trout, Fathead minnow, Channel catfish, bluegill, Smallmouth bass, Largemouth bass	0.15-10.6 ppm	Jonson & Finley, 1980
Phosphamidon	Rainbow trout, Fathead minnow, Channel catfish, Bluegill	3.4-100 ppm	Jonson & Finley, 1980
Phoxim	Coho salmon, Atlantic salmon, Rainbow trout, Brown trout, Brook trout, Northern pike, Fathead minnow, Channel catfish, bluegill	0.11-2.9 ppm	Jonson & Finley, 1980
Propoxur	Goldfish, Rainbow trout, Fathead minnow, Bluegill	4.8-36.2 ppm	Wang et al., 2009; Jonson & Finley, 1980

Table 9. Summary of acute toxicity.

Insecticide	species	Range of 96h LC$_{50}$	Reference
Pyrethrum	Coho salmon, Atlantic salmon, Brown trout, Lake trout, Channel catfish, bluegill	13-65 ppb	Jonson & Finley, 1980
Resmethrin	Cho salmon, Lake trout, Fathead minnow, Channel catfish, bluegill	1.7-9.9 ppb	Jonson & Finley, 1980
Ronnel	Rainbow trout, Channel catfish, bluegill, Cutthroat trout, Lake trout	0.6-1.6 ppm	Jonson & Finley, 1980
Rotenone	Rainbow trout, Channel catfish, bluegill	2.6-36 ppb	Jonson & Finley, 1980
RU-1169	Coho salmon, Atlantic salmon, Lake trout, fathead minnow, White sucker, Bluegill	0.3-28 ppb	Jonson & Finley, 1980
S-Bioallethrin	Fathead minnow, Channel catfish, Bluegill, Yellow perch	7.8-90 ppb	Jonson & Finley, 1980
SD-17250	Coho salmon, Rainbow trout, bluegill	1.5-5.7 ppm	Jonson & Finley, 1980
Strobane	Bluegill, Rainbow trout	8.7-12 ppb	Jonson & Finley, 1980
TEPP	Rainbow trout, Fathead minnow, Bluegill	240-980 ppb	Jonson & Finley, 1980

Table 10. Summary of acute toxicity.

Insecticide	species	Range of 96h LC50	Reference
Temephos	Cutthroat trout, Rainbow trout, Atlantic trout, Brook trout, Lake trout, Fathead minnow, Channel catfish, Bluegill, Largemouth bass	1.44-34 ppm	Jonson & Finley, 1980
Thanite	Rainbow trout, Channel catfish, Bluegill	1.2-3.7 ppm	Jonson & Finley, 1980
Thiodicarb	Bluegill, Rainbow trout	1.4-3.3 ppm	Yi et al., 2006
Toxaphene	Coho salmon, Rainbow trout, Brown trout, Goldfish, Carp, Fathead minnow, Black bullhead, Channel catfish, Bluegill, Largemouth bass, Yellow perch	2-18 ppb	Jonson & Finley, 1980
Trichlorfon	Eel, Rainbow trout, Cutthroat trout, Atlantic salmon, Brown trout, Brook trout, Lake trout, Fathead minnow, Channel catfish, Bluegill, Largemouth bass	0.36-9.2 ppm	Lopes et al., 2006; Jonson & Finley, 1980

Table 11. Summary of acute toxicity.

4. Sub-lethal toxicity

Sub-lethal toxicity testing was planned based on one tenth or more of LC_{50} dose in moderate periods. In sub-lethal toxicity, the organs or biological systems which may be affected at such exposure can be respiratory, hepatic, haematopoietic, nervous, cardiovascular, and reproductive and immune systems. Different biomarkers of fish exposed to insecticides are usually evaluated in these experiments. Insecticides may lead to changes in the blood biochemical parameters and haematological profile of fish which can be investigated as biomarker in pollution monitoring (Mushigeri & David, 2005; Banaee et al., 2008; Kavitha & Rao, 2009). In fact, these compounds may induce alterations in the activities or levels of a number of different enzyme systems, including those necessary for biochemical reactions in cells (Banaee et al., 2011). Decreased rate of growth, reproductive disorder, spinal deformities, histopathological changes (Benli & Özkul, 2010) in gills, liver, haematopoietic tissue such as spleen, head of kidney, and renal tubules, endocrine tissues as well as brain, neurological and behavioral disorder and genetic defects are other biological indicators of exposure to insecticides which are described in details in the following sections.

5. Chronic toxicity

Chronic toxicity tests commonly include the measurement of long term effects of low concentrations of insecticides on the survival, growth, reproduction, nervous system and other biological and physiological aspects of fishes. Type of injury to fish in chronic toxicity is similar to sub-lethal toxicity damage, but the frequency and intensity injury and lesion resulting from chronic toxicity may be more or even less than damage of sub-lethal toxicity. Therefore, this experiment is important in insecticides toxicology.

6. Side effect of insecticides on various aspects of fish's biology and physiology

6.1 Alterations in blood biochemical parameters

Insecticides can cause serious impairment to physiological and health status of fish. Therefore, biochemical tests are routine laboratory tests useful in recognizing acute or chronic toxicity of insecticides (Banaee *et al.*, 2008; Al-Kahtani, 2011) and can be a practical tool to diagnose toxicity effects in target organs and to determine the physiological status in fish. Blood biochemistry test gives indicates what is happening in the body of fish exposed to insecticides. When different tissues are injured, the damaged cells release specific enzymes into plasma and we can recognize their abnormality levels in blood. Then it helps locate the lesions. Also, if certain organs are not eliminating certain waste products or not synthesizing certain important materials, this can tell us they are not functioning properly. In some cases due to the severity of the damage to tissues, particularly liver, synthesis of many biochemical parameters may reduce significantly in cells, which can decrease some biochemical factors in blood of fish exposed to insecticides. These changes were observed in *Channa punctatus* (Agrahari *et al.*, 2007), *Oreochromis niloticus* (Velisek *et al.*, 2004), *O.mossambicus* (Arockia and Mitton, 2006; Matos *et al.*, 2007), *Heteropneustes fossilis* (Saha & Kaviraj, 2009), *Cirrhinus mrigala* (Prashanth & Neelagund, 2008) *Clarias batrachus* (Begum, 2005; Ptnaik, 2010), *Cyprinus carpio* (Banaee *et al.*, 2008), *Oncorhynchus mykiss* (Banaee *et al.*, 2011), *Colisa fasciatus* (Singh *et al.*, 2004) which were exposed to monocrotophos, bifenthrin, carbaryl, dimethoate, cypermethrin, sevin, diazinon, and malathion respectively.

6.2 Tissue and organ damage

Histopathological investigations on different tissues of fish are valuable tools for toxicology studies and monitoring water pollutions. In histopathology, we can provide information about the health and functionality of organs. Tissues injuries and damages in organs can result in the reduced survival, growth and fitness, the low reproductive success or increase of susceptibility to pathological agents. Frequency and intensity of tissue lesions depend on the concentrations of insecticides and the length of the period fish are exposed to toxins. Nevertheless, many insecticides cause specific or non-specific histopathological damage (Fanta *et al.*, 2003). For example, histopathological lesions in the liver tissue of freshwater fish (*Cirrhinus mrigala*) (Velmurugan *et al.*, 2009) and common carp carp (*Cyprinus carpio*) (Banaee *et al.*, In press) were observed after 10 and 30 days exposure to sublethal concentrations of dichlorvos and diazinon insecticides, respectively. Other researchers reported the same histopathological alterations in different tissues of fish treated with diazinon (Duttaa & Meijer, 2003; Banaee *et al.*, 2011), deltamethrin (Cengiz, 2006; Cengiz & Unlu, 2006), fenitrothion (Benli & Özkul, 2010).

6.3 Reproductive dysfunction

Reproduction guarantees the survival of fish population. Any changes in environmental parameters or physiological conditions of fish can affect its reproductive success. Since fish may be exposed to environmental pollutants, including insecticides, herbicides, heavy metals and other xenobiotics, disorders may occur in their natural reproductive process. Recent researches showed the dysfunction in the reproductive systems of fishes exposed to insecticides. Insecticides' effects on reproductive biology of fishes are numerous, and include decreased fecundity, testicular and ovarian histological damage (Duttaa & Meijer,

2003; Banaee *et al.*, 2009), vitellogenesis process impairment (Haider and Upadhyaya, 1985;), and disruption in steroidogenesis process (Zaheer Khan & Law, 2005), delay in gonads maturation (Skandhan *et al.*, 2008), alter in reproductive and parental behavior (Jaensson *et al.*, 2007), impairment in olfactory response and disorder in reproductive migrations (Scholz *et al.*, 2000), as well as disruption in coordinating courtship behavior of male and female fish and time of spawning (Jaensson *et al.*, 2007).

Some insecticides are known as endocrine disrupting chemicals (EDC) which can interfere with the normal functioning of endocrine system in fish. Adverse effects of insecticides on the hypothalamus-pituitary-gonads axis can also play a significant role in causing reproductive failures in fish. In fishes, chronic toxicity of insecticides can change sex steroid hormone levels in plasma. While the mechanism is not exactly known, it is possible that insecticides and their metabolites disrupt reproductive systems through activation or inhibition of key enzymes which participated in the steroid hormone biosynthesis in fishes. For example, DDT, endosulfan, methoxychlor and some other insecticides possess estrogenic properties and are probably capable of disrupting functions of endocrine system and causing disorder in the reproductive system of fish (Arukwe, 2001). These compounds may directly or indirectly interact with natural hormones, changing the hormone functions and thus altering physiological cellular response or mutate the natural patterns of hormone synthesis and metabolism. Impact of organophosphate insecticides such as malathion, diazinon and fenitrothion on the hypothalamus-pituitary-gonads axis and also disturbance in hormones associated with reproductive systems were studied by Kapur & Toor, (1978); Singh and Singh, (1987); Maxwell & Dutta, (2005); Skandhan *et al.* (2008).

Insecticides can also cause adverse effects on gonad histology, morphology and its growth. In addition, there are significant relationships between blood sex steroid hormone concentrations, sperm or oocytes quality, rate of fecundity and histopathological alterations in ovary and testis of fish exposed to different insecticides (Duttaa & Meijer, 2003; Maxwell & Dutta, 2005). Banaee *et al.*, (2008) reported that diazinon inhibits steroidogenesis in testis of male carp by histopatological alterations. Research results showed that direct toxic effects of insecticides on seminiferous tubules or Leydig cells may be the most important parameter for the low quality of sperms in fish (Fadakar Masouleh *et al.*, 2011). Similar results were reported in walking catfish (*Clarias batrachus*), freshwater eel (*Monopterus albus*), and Atlantic salmon (*Salmo salar*) that were exposed to different insecticides (Singh & Singh, 1987; Singh, 1989; Moore & Waring, 1996; 2001).

Exposure of fish eggs and milt to insecticides also reduced the level of fertilization, hatching rate and the larval survivability. Further studies on bluegills (*Lepomis macrochirus*), atlantic salmon (*Salmo salar* L.) demonstrated that the gametes and fertilized eggs were sensitive to the insecticides (Tanner & Knuth, 1996; Moore & Waring, 2001) suggesting a further toxic impact of these toxicants on the fish reproduction. In addition, the waste of energy in the fish exposed to insecticides reduces their reproductive ability.

6.4 Development disorders

Study of development disorders caused by insecticides is to emphasize the links between the concentrations of toxins and dysfunction in normal development from embryonic to puberty periods. So, impairment in the normal development and the growth may reduce the fish's survival chance.

Embryos and larvae may be directly exposed to insecticides, through the yolk or via parental transfer in viviparous fish (Viant *et al.*, 2006). Spinal deformities, mostly scoliosis and lordosis, and morphological abnormalities were among the more adverse effects registered for insecticides toxicity. Other alterations in the embryo of fish exposed to insecticides also consist of yolk sac edema, pericardial edema and crooked body of larvae (Xu *et al.*, 2010). Teratogenic effects of carbaryl insecticides on the embryo of fish have been proved (Todd and Leeuwen, 2002). Similar results were reported in silversides *Menidia beryllina* exposed to tebufos during embryogenesis (Middaugh *et al.*, 1990; Hemmer *et al.*, 1990).

The most important factors decreasing fish growth consist of disorder in feeding behaviors, decrease in feeding rate, dysfunction in metabolism process and waste of energy to overcome the stress caused by insecticide exposure (Tripathi *et al.*, 2003). For example, disorder in the metabolism of carbohydrates, proteins and lipids in various tissues, particularly liver of fish exposed to insecticides, may reduce their growth rates. Begum (2004) found out that protein and carbohydrate metabolism in the liver and muscle tissue is disrupted on the exposure to a carbofuran insecticide. In addition, exposure during embryonic or larval stage can result in behavioral abnormalities, such as decreased ability to capture prey after hatching, functional deficiencies or slowing of growth and finally death (Kuster, 2005; Viant *et al.*, 2006; Arufe *et al.*, 2007). These changes were observed in larvae and embryo of zebra fish (*Danio rerio*) in contact with endosulfan (Velasco-Santamaria *et al.*, 2011), beta-cyprmethrin (Xu *et al.*, 2010); paraoxon-methyl (Küster, 2005) and sevin (carbaryl insecticide) (Todd and Leeuwen, 2002).

6.5 Neurotoxicity

The primary mechanism of organophosphate and carbamate insecticides toxicity is well known – they function as inhibitors of acetycholinestrase enzyme (AChE) or and butyrylcholinesterase (BChE), as well as disturbing the metabolism of other neurotransmitters such as γ-aminobutyrate (GABA). The synthetic pyrethroids change normal neuronal function by interfering in the function of ion channels in the nerve cell membrane, alterations in intracellular calcium ion concentrations and possibly by blocking GABA receptors. Organochlorine insecticides act primarily by changing the transport of ions across the nerve cell membranes, thus altering the ability of nerve to stimulate.

Fish exposure to these insecticides is frequently assessed by determining the alterations in AChE in brain, muscle, plasma and other tissues or probably GABA activity in brain (Banaee, 2010). AChE is an enzyme responsible for inactivating the neurotransmitter acetylcholine (Fulton & Key, 2001). AChE inactivation results in the accumulation of the neurotransmitter acetylcholine in cholinergic synapses space, leading to synaptic blockage and disruption of signal transmission (Ferrari *et al.*, 2004; 2007a, b). Inhibition of AChE induces alteration in the swimming behavior, shaking palsy, spasms and other undesirable effects (Sharbide *et al.*, 2011). Disturbances in AChE activity can also impair feeding, identification and avoidance and escaping from predators, spatial orientation of the species, and reproductive behavior (Bretaud *et al.*, 2000). Thus, AChE inhibition is considered to be a specific biomarker of exposure to organophosphorus and carbamate insecticides like diazinon, chlorpyrifos, propoxur, isoprocarb, (Üner *et al.*, 2006; Cong *et al.*, 2008; 2009; Wang *et al.*, 2009; Banaee *et al.*, 2011;). Similar results have been observed for pyrethroids insecticide toxicity (Koprucu *et al.*, 2006). Disorder in γ-aminobutyrate (GABA) system in brain of rainbow trout exposed to sub-lethal lindane was reported by Aldegunde *et al.*, (1999). GABA receptors inhibit the transmission of nerve impulses; thus disturbances in this receptor would also lead to an over stimulation of the nerves.

In addition, nervous tissue has weaker antioxidant defense system than other kinds of tissue. On the other hand, the brain as center of the nervous system in fish contains low levels of enzymatic and non-enzymatic antioxidant and higher levels of oxidizable unsaturated lipids and catecholamines. So, nerve tissue is very sensitive to oxidative stress damage induced insecticide toxicity compared with other tissues (Üner *et al.*, 2006; Li *et al.*, 2010). These results have been reported by other scientists (Senger *et al.*, 2005).

6.6 Behavioral alterations

Behavioral changes are the most sensitive indicators of potential toxic effects. The behavioral and the swimming patterns of the fish exposed to different insecticides include changes in swimming behavior, feeding activities, predation, competition, reproduction and species-species social interactions such as aggression (Cong *et al.*, 2008; 2009; Werner and Oram, 2008). Banaee *et al.* (2008; 2011) reported similar behavioral responses in common carp and rainbow trout exposed to sub-lethal levels of diazinon. In fact, most insecticides influence the behavior patterns of fish by interfering with the nervous systems and sensory receptors (Keizer *et al.*, 1995; Pan & Dutta, 1998; Cong *et al.*, 2008; 2009); and this incident may impair the identification of situation and development of appropriate response by the fish exposed to insecticide. The effect of certain insecticides on the activity of acetylcholinestrase may lead to a decreased mobility in fish (Bretaud *et al.*, 2000). Researchers have reported the same alterations in *Oryzias latipes, Cyprinus carpio, Labeo rohita, Oncorhynchus tshawytscha, O.latipes, Cirrhinus mrigala, Oreochromis niloticus, Clarias gariepinus* treated with chlorpyrifos (Rice *et al.*, 1997; Halappa & David, 2009), malathion (Patil & David, 2008), diazinon (Scholz *et al.*, 2000), endosulfan (Gormley & Teather, 2003), Fenvalerate (Mushigeri & David, 2005), fenitrothion (Benli & Özkul, 2010), dimethoate (Auta *et al.*, 2002), respectively.

6.7 Genotoxicity

In genetic-toxicology any heritable damage or DNA inactivation resulting from the animal's exposure to xenobiotics is studied. Genotoxic chemicals such as insecticides have common chemicals and physical properties that enable them interact with genetic materials (Campana *et al.*, 1999; Çava & Ergene-Gözükara, 2003; Candioti *et al.*, 2010; Dogan et al., 2011). The mutation that may result from an interaction between a chemical and the genetic material is a heritable change in the cell genotype, and thus the error may be transferred to the daughter cell or the next generation. Carcinogenesis and the formation of some tumors in different tissues of fish exposed to insecticides may also be caused by genotoxic properties of these xenobiotics. One of the ill effects of insecticides' arrival into surface waters may be an induction of chromosomal damage in eggs and larvae of fishes in different stages of development.

Some insecticides that behave as endocrine active compounds can change the expression of vital genes resulting in unusual concentrations of plasma steroid hormones and reproductive dysfunction or immuosuppression (Jin *et al.*, 2010).

6.8 Immunosuppression

The immune system of fish is important for defense against a variety of pathogens. The system is very sensitive to homeostatic adjustments via endocrine regulation and is influenced by the biochemical status of the nervous system. Thus, any impairment in the nervous system and disturbance in the biochemical homeostasis can weaken the immune system of fish. On the other hand, insecticides may alter the function of the immune system and result in immunodepression, uncontrolled cell proliferation, and alterations of the host defense mechanisms including innate immunity and acquire immunity against pathogens.

Different insecticides at sub-lethal levels have been recognized as stressors causing immune-suppression in fish (Werner and Oram, 2008). In addition, some insecticides may exert immunotoxic effects by altering the transcription of important mediators of the fish's immune system (Eder *et al.*, 2009). Effects of insecticides like P,P'-DDE, lindane, cypermethrin, chlorpyrifos, diazinon on the immune factors of fish such as Interleukin-1β (IL-1β), IL-1β receptor (IL-1R1), Interferon gamma (IFN-γ2b), TNFα, MHCIα, MHCIIα, Mx, TLR9, IγML and C- reactive protein (CRP), TCRα in head- kidney leucocytes, Lysozyme activity, chemiluminuscence (CL) response and immunocompetent cells population size, IgM levels, value of white blood cells (WBC) and respiratory burst activity, head kidney phagocytes and peripheral blood leucocytes, etc., have been reported by scholars (Betoulle *et al.*, 2000; Khoshbavar-Rostami *et al.*, 2006; Banaee *et al.*, 2008; Cuesta *et al.*, 2008; Girón-Pérez *et al.*, 2009; Shelley *et al.*, 2009; Ahmadi *et al.*, 2011; Jin *et al.*, 2011, Wang *et al.*, 2011). The exposure to sub-lethal concentrations of insecticides is what probably makes fish vulnerable to infectious diseases because of their immune-depressive effects (Zelikoff *et al.*, 2000). For example, the susceptibility of juvenile chinoock salmon (*O.tshawytscha*) to infectious hematopoietic necrosis virus was significantly increased in fish exposed to sub-lethal concentrations of esfenvalerate (Clifford *et al.*, 2005). Similar results were reported in goldfish and common carp that were exposed to carbaryl and lindane respectively (Shea, 1983; Shea & Berry, 1984; Cossarini-dunier & Hattenberger, 1988).

7. References

Agrahari, S. Pandey, K.C. & Gopal, K. (2007). Biochemical alteration induced by monocrotophos in the blood plasma of fish, *Channa punctatus* (Bloch). Pesticide Biochemistry and Physiology 88 268–272.

Aldegunde, M. Soengas, J.L. Ruibal, C. & Andrés, M.D. (1999). Effects of chronic exposure to γ-HCH (lindane) on brain serotonergic and gabaergic systems, and serum cortisol and thyroxine levels of rainbow trout, *Oncorhynchus mykiss*. Fish Physiology and Biochemistry, 20: 325-330.

Al-Kahtani, M.A. (2011). Effect of an Insecticide Abamectin on Some Biochemical Characteristics of Tilapia Fish (*Oreochromis Niloticus*). American Journal of Agricultural and Biological Sciences, 6 (1): 62-68.

Arjmandi, R.Tavakol, M. & Shayeghi, M. (2010). Determination of organophosphorus insecticide residues in the rice paddies. International Journal Environmental Science Technology, 7 (1): 175-182.

Arufe, M.I. Arellano, J.M. García, L. Albendín, G. & Sarasquete, C. (2007). Cholinesterase activity in gilthead seabream (*Sparus aurata*) larvae: Characterization and sensitivity to the organophosphate azinphosmethyl. Aquatic Toxicology, 84: 328-336.

Arukwe, A. 2001. Cellular and molecular responses to endocrine-modulators and the impact on fish reproduction. Marine Pollution Bulletin, 42(8): 643-655.

Auta, J. Balogun, J.K. Lawal, F.A. & Ipinjolu, J.K. (2002). Short-term effect of dimethoate on behavior of juveniles of *Oreochromis niloticus* (Trewavas) and *Clarias gariepinus* (Teugels). Journal of Tropical Biosciences, 2(1): 55-59.

Bagheri, F. (2007) Study of pesticide residues (Diazinon, Azinphosmethyl) in the rivers of Golestan province (GorganRoud and Gharehsou), M.Sc. Thesis, Tehran University of Medical Science. Tehran, Iran, 1-125.

Ballesteros, M.L. Gonzalez, M.Wunderlin, D.A. Bistoni, M.A. & Miglioranza, K.S.B. (2011). Uptake, tissue distribution and metabolism of the insecticide endosulfan in *Jenynsia multidentata* (Anablepidae, Cyprinodontiformes). Environmental Pollution, 159: 1709-1714.

Banaee, M. (2010). Influence of silymarin in decline of sub-lethal diazinon-induced oxidative stress in rainbow trout (*Oncorhynchus mykiss*). Ph.D. Thesis, Aquaculture Department, Natural Resource Faculty, University of Tehran, Iran, pp. 149.

Banaee, M. Mirvagefei, A. R. Rafei, G. R. & Majazi Amiri, B. (2008). Effect of sub-lethal Diazinon Concentrations on Blood Plasma Biochemistry. Int. J. Environ. Res., 2(2): 189-198.

Banaee, M. Mirvaghefei, A. R. Majazi Amiri, B. & Rafei, G.R. (2012). Biochemical Blood and Histopathological Study of Experimental Diazinon Poisoning in common carp fish (*Cyprinus carpio*). Journal of Fisheries (Iranian Journal of Natural Resources), (Article in press).

Banaee, M. Mirvaghefi, A. R. Majazi Amiri, B. Rafei, G.R. & Nematdost, B. (2011). Hematological and Histopathological Study of Experimental Diazinon Poisoning in common carp fish (*Cyprinus carpio*). Journal of Fisheries (Iranian Journal of Natural Resources), 64(1): 1-14.

Banaee, M. Mirvaghefi, A.R. Ahmadi, K. & Ashori, R. (2009). The effect of diazinon on histophatological changes of testis and ovaries of common carp (*Cyprinus carpio*). Scientific Journal of Marine Biology, 1(2): 25-35.

Banaee, M. Mirvaghefi, A.R. Ahmadi, K. & Banaee, S. (2008). determination of LC$_{50}$ and investigation of acute toxicity effects of diazinon on hematology and serology indices of common carp (*Cyprinus carpio*). Journal of Marine Science and Technology Research, 3(2): 1-10.

Banaee, M. Sureda, A. Mirvaghefi, A.R. & Ahmadi, K. (2011). Effects of diazinon on biochemical parameters of blood in rainbow trout (Oncorhynchus mykiss). Pesticide Biochemistry and Physiology, 99: 1-6.

Begum, G. (2004). Carbofuran insecticide induced biochemical alterations in liver and muscle tissues of the fish *Clarias batrachus* (linn) and recovery response. Aquatic Toxicology, 66: 83–92.

Begum, G. (2005). In vivo biochemical changes in liver and gill of *Clarias batrachus* during cypermethrin exposure and following cessation of exposure. Pesticide Biochemistry and Physiology, 82: 185–196.

Behrens, A. & Segner, H. (2001). Hepatic biotransformation enzymes of fish exposed to non-point source pollution in small streams. Journal of Aquatic Ecosystem Stress and Recovery 8: 281–297.

Benli, A.Ç.K. & Özkul, A. (2010). Acute toxicity and histopathological effects of sublethal fenitrothion on Nile tilapia, Oreochromis niloticus. Pesticide Biochemistry and Physiology, 97: 32–35.

Betoulle, S. Duchiron, C. & Deschaux, P. (2000). Lindane differently modulates intracellular calcium levels in two populations of rainbow trout (Oncorhynchus mykiss) immune cells: head kidney phagocytes and peripheral blood leucocytes. Toxicology, 145: 203–215.

Bowman, J. (1988a) Acute Flow-Through Toxicity of Chlorpyrifos Technical to Bluegill (Lepomis macrochirus): Project ID:37189. Unpublished study prepared by Analytical Bio-Chemistry Laboratories, Inc. 188 p.

Bowman, J. (1988b) Acute Flow-Through Toxicity of Chlorpyrifos to Rainbow Trout (Salmo gairdneri): Project ID: 37188. Unpublished study prepared by Analytical Bio-Chemistry Laboratories, Inc. 174 p.

Bretaud, S. Toutant, J.P. & Saglio, P. (2000). Effects of carbofuran, diuron, and nicosulfuron on acetylcholinesterase activity in goldfish (Carassius auratus). Ecotoxicol. Environ. Saf., 47: 117–124.

Calmbacher, C. (1978) Acute Toxicity of San 3261 Lot #7801 to Bluegill, Rafinesque: UCES Project # 11506-16-03. (Unpublished study received Mar 1, 1979 under 11273-EX-15. prepared by Union Carbide Corp., submitted by Sandoz, Inc., Crop Protection, San Diego, CA. CDL:097841-AD).

Campana, M.A. Panzeri, A.M. Moreno, F V.J. & Dulout, F.N. (1999). Genotoxic evaluation of the pyrethroid lambda-cyhalothrin using themicronucleus test in erythrocytes of the fish Cheirodon interruptus interruptus, Mutat. Res. 438: 155–161.

Candioti, J.V. Soloneski, S. & Larramendy, M.L. (2010). Genotoxic and cytotoxic effects of the formulated insecticide Aficida on Cnesterodon decemmaculatus (Jenyns, 1842) (Pisces: Poeciliidae). Mutation Research, 703: 180–186.

Capkin, E. Altinok, I. & Karahan, S. (2006). Water quality and fish size affect toxicity of endosulfan, an organochlorine pesticide, to rainbow trout. Chemosphere, 64:1793–1800.

Çava, T. & Ergene-Gözükara, S. (2003). Evaluation of the genotoxic potential of lambda-cyhalothrin using nuclear and nucleolar biomarkers on fish cells. Mutation Research, 534: 93–99.

Cengiz, E.I. & Unlu, E. (2006). Sublethal effects of commercial deltamethrin on the structure of the gill, liver and gut tissues of mosquitofish, Gambusia affinis: A microscopic study. Environmental Toxicology and Pharmacology, 21: 246–253.

Cengiz, E.I. (2006). Gill and kidney histopathology in the freshwater fish Cyprinus carpio after acute exposure to deltamethrin. Environmental Toxicology and Pharmacology, 22: 200–204.

Clifford, M.A. Eder, K.J. Werner, I. & Hedrick, R.P. (2005). Synergistic effects of esfenvalerate and infectious hematopoietic necrosis virus on juvenile Chinook salmon mortality. Environ. Toxicol. Chem., 24 (7): 1766–1772.

Cong, N.V. Phuong, N.T. & Bayley, M. (2008). Brain cholinesterase response in the snakehead fish (*Channa striata*) after field exposure to diazinon. Ecotoxicology and Environmental Safety, 71: 314– 318.

Cong, N.V. Phuong, N.T. & Bayley, M. (2009). Effects of repeated exposure of diazinon on cholinesterase activity and growth in snakehead fish (*Channa striata*). Ecotoxicology and Environmental Safety, 72: 699–703.

Cossarini-dunier, M. & Hattenberger, A.M. (1988). Effect of pesticides (atrazine and lindane) on the replication of spring viremia of carp virus *in vitro*. Ann. Rech. Vet., 19: 209–211.

Cuesta, A. Meseguer, J. & Esteban, M.Á. (2008). Effects of the organochlorines p, p'-DDE and lindane on gilthead seabream leucocyte immune parameters and gene expression. Fish Shellfish Immunol.25:682-8.

Davey, R.B. Meisch, M.V. & Carter, F.L. (1976). Toxicity of five rice field pesticides to the mosquitofish, *Gambusia affinis*, and green sunfish, *Lepomis cyanellus*, under laboratory and field conditions in Arkansas. Environmental Entomology 5(6):1053-1056. (Also In unpublished submission received Mar 30, 1977 under 279-2712. submitted by FMC Corp., Philadelphia, Pa. CDL:229241-Q).

Dobšíková, R. (2003). Acute Toxicity of Carbofuran to Selected Species of Aquatic and Terrestrial Organisms. Plant Protect. Sci. 39(3): 103–108.

Dogan, D. Can, C. Kocyigit, A. Dikilitas, M. Taskin, A. & Bilinc, H. (2011). Dimethoate-induced oxidative stress and DNA damage in *Oncorhynchus mykiss*. Chemosphere (Article in press).

Duttaa, H.M. & Meijer, H.J.M. (2003). Sublethal effects of diazinon on the structure of the testis of bluegill, Lepomis macrochirus: a microscopic analysis. Environmental Pollution 125: 355–360.

Eder, K.J. Leutenegger, C.M. Köhler, H.R. & Werner, I. (2009). Effects of neurotoxic insecticides on heat-shock proteins and cytokine transcription in Chinook salmon (*Oncorhynchus tshawytscha*). Ecotoxicology and Environmental Safety 72: 182– 190.

Fadakar Masouleh, F. Mojazi Amiri, B. Mirvaghefi, A.R. & Nematollahi, M.A. (2011). *In vitro* effects of diazinon on male reproductive tissue and sperm motility of Caspian Kutum (*Rutilus frisii kutum*). Research Journal of Environmental Toxicology, 5(2): 108-116.

Fanta, E. Sant'Anna Rios, F. Romao, S. Vianna, A.C.C. & Freiberger, S. (2003). Histopathology of the fish *Corydoras paleatus* contaminated with sublethal levels of organophosphorus in water and food. Ecotoxicology and Environmental Safety, 54: 119–130.

Favari, L. Lopez, E. Martınez-Tabche, L. & Dıaz-Pardo, E. (2002). Effect of Insecticides on Plankton and Fish of Ignacio Ramirez Reservoir (Mexico): A Biochemical and Biomagnification Study. Ecotoxicology and Environmental Safety, 51: 177–186.

Ferrari, A. Venturino, A .& de D'Angelo, A.M.P. (2007a). Effects of carbaryl and azinphos methyl on juvenile rainbow trout (*Onchorhynchus mykiss*) detoxifying enzymes. Insecticide Biochemistry and Physiology, 88: 134-142.

Ferrari, A. Venturino, A. & de D'Angelo, A.M.P. (2004). Time course of brain cholinesterase inhibition and recovery following acute and subacute azinphosmethyl, parathion and carbaryl exposure in the goldfish (*Carassius auratus*). Ecotoxicology and Environmental Safety, 57: 420–425.

Ferrari, A. Venturino, A. & de D'Angelo, A.M.P. (2007b). Muscular and brain cholinesterase sensitivities to azinphos methyl and carbaryl in the juvenile rainbow trout *Oncorhynchus mykiss*. Comparative Biochemistry and Physiology, Part C 146: 308–313.

Fulton, M.H. & Key, P.B. (2001). Acetylcholinesterase inhibition in estuarine fish and invertebrates as an indicator of organophosphorus insecticide exposure and effects. Environmental Toxicology & Chemistry 20 37-45.

Ganai, F.A. Malik, M. & Nisar, Z. (2011). Genotoxic effects of organophosphate pesticide phorate in some exotic fishes of Kashmir. Journal of American Science, 7(4): 46-50.

Girón-Pérez, M.I. Velázquez-Fernández, J. Díaz-Resendiz, K. Díaz-Salas, F.Canto-Montero, C. Medina-Dííaz, I. Robledo-Marenco, M. Rojas-García, A. & Zaitseva, G. (2009). Immunologic parameters evaluations in Nile tilapia (*Oreochromis niloticus*) exposed to sublethal concentrations of diazinon. Fish & Shellfish Immunology 27 383-385.

Gormley, K.L. & Teather, K.L. (2003). Developmental, behavioral, and reproductive effects experienced by Japanese medaka (*Oryzias latipes*) in response to short-term exposure to endosulfan. Ecotoxicology and Environmental Safety, 54: 330–338.

Gül, A. (2005). Investigation of acute toxicity of chlorpyrifos-methyl on Nile tilapia (*Oreochromis niloticus* L.) larvae. Chemosphere, 59: 163–166.

Haider, S. & Upadhyaya, N. (1985). Effect of commercial formulation of four organophosphate insecticides on the ovaries of a freshwater teleost, *Mystus vitattus* (Bloch): a histological and histochemical study. J. Environ. Sci. Health. B20: 321-340.

Halappa, R. & David, M. (2009). Behavioural responses of the fresh water fish, *Cyprinus carpio* (Limmaeus) following sublethal exposure to chlorpyrifos. Turkish Journal of Fisheries and Aquatic Sciences, 9: 233-238.

Hemmer, M.J. Middaugh, D.P. & Moore, J.C. (1990). Effects of temperature and salinity on *Menidia berllina* embryos exposed to terbufos. Diseases Aquatic Organisms, 8: 127-136.

Holcombe, G. Phipps, G. & Tanner, D. (1982) The acute toxicity of kelthane, dursban, disulfoton, pydrin, and permethrin to fathead minnows Pimephales promelas and rainbow trout Salmo gairdneri. Environ. Pollution 29:167-178.

Honarpajouh, K. (2003). Study and Identification of OP pesticides residues (Azinphosmethyl and Diazinon) in the Mahabad and Siminerood Rivers, M.Sc. Thesis, Tehran University of Medical Science. Tehran, Iran, pp. 95.

Jaensson, A. Scott, A.P. Moore, A. Kylin, H. & Olsén, K.H. (2007). Effects of a pyrethroid pesticide on endocrine responses to female odours and reproductive behaviour in male parr of brown trout (*Salmo trutta* L.). Aquatic Toxicology, 81: 1-9.

Jin, Y.X Chen, R.J. Liu, W.P. & Fu, Z.W. (2010). Effect of endocrine disrupting chemicals on the transcription of genes related to the innate immune system in the early developmental stage of zebrafish (*Danio Rerio*). Fish Shellfish Immunol. 28:854-61.

Jin, Y.X. Zheng, S. & Fu, Z. (2011b). Embryonic exposure to cypermethrin induces apoptosis and immunotoxicity in zebrafish (Danio rerio). Fish & Shellfish Immunology, 30: 1049-1054.

Jin, Y.X. Zheng, S.S. Pu, Y. Shu, L.J. Sun, L.W. & Liu, W.P. (2011). Cypermethrin has the potential to induce hepatic oxidative stress, DNA damage and apoptosis in adult zebrafish (*Danio rerio*). Chemosphere 82:398-404.

John, P.J. (2007). Alteration of certain blood parameters of freshwater teleost *Mystus vittatus* after chronic exposure to metasystox and sevin. Fish Physiology Biochemistry, 33: 15-20.

Johnson, W. & Finley, M. (1980). Handbook of Acute Toxicity of Chemicals to Fish and Aquatic Invertebrates: resource Publication 137. USDI, Fish and wildlife service, Washington, D.C. 98 p.

Jones, F. & Davis, J. (1994). DDVP Technical Grade: Acute Toxicity to Sheepshead Minnow (*Cyprinodon variegatus*) Under Flow-through Test Conditions: Lab Project Numbers: J9403007F: J9403007B. Unpublished study prepared by Toxikon Environmental Sciences, 59 p.

Kapur, K & Toor, H.S. (1978). The effect of fenitrothion on reproduction of a teleost fish, *Cyprinus carpio communis* Lim: a biochemical study. Bull. Environ. Contam. Toxicol. 20: 438-442.

Kavitha, P. & Rao, J. V. (2007). Oxidative stress and locomotor behaviour response as biomarkers for assessing recovery status of mosquito fish, *Gambusia affinis* after lethal effect of an organophosphate pesticide, monocrotophos. Pesticide Biochemistry and Physiology, 87: 182–188.

Kavitha, P. & Rao, V.J. (2009). Sub-lethal effects of profenofos on tissue-specific antioxidative responses in a Euryhyaline fish, *Oreochromis mossambicus*. Ecotoxicology and Environmental Safety 72: 1727–1733.

Keizer, J. D'Agostino, G. Nagel, R. Volpe, T. Gnemid, P. & Vittozzi, L. (1995). Enzymological differences of AChE and diazinon hepatic metabolism: correlation of *in vitro* data with the selective toxicity of diazinon to fish species. The Science of the Total Environment 171: 213-220.

Khoshbavar-Rostami, H.A. Soltani, M. & Hassan, H.M.D. (2006). Immune response of great sturgeon (*Huso huso*) subjected to long-term exposure to sublethal concentration of the organophosphate, diazinon. Aquaculture, 256: 88–94.

Kitamura, S. Kadota, T. Yoshida, M. Jinno, N. & Ohta, S. (2000). Whole-body metabolism of the organophosphorus pesticide, fenthion, in goldfish, Carassius auratus. Comp. Biochem. Physiol. C 126: 259-266.

Koprucu, S.S. Koprucu, K. & Urail, M.S. (2006). Acute toxicology of synthetic pyrethroid deltamethrin to fingerling European catfish (*Silirus glanis* L.). Bulletin of Environmental Contamination and Toxicology, 76: 59-65.

Kumar, K. & Ansari, B.A. (1984). Malathion toxicity: skelet deformities in zebrafish (*Brachydanio rerio*, Cyprinidae). Pestic. Sci. 15: 107-111.

Küster, E. (2005). Cholin- and carboxylesterase activities in developing zebrafish embryos (*Danio rerio*) and their potential use for insecticide hazard assessment. Aquatic Toxicology 75: 76–85.

Lawson, E.O. Ndimele, P.E. Jimoh, A.A. & Whenu, O.O. (2011). Acute Toxicity of Lindane (Gamma Hexachloro-Cyclohexane) to African Catfish (*Clarias gariepinus*, Burchell, 1822). International Journal of Animal and Veterinary Advances, 3(2): 63-68.

Li, Z.H. Zlabek, V. Velisek, J. Grabic, R. Machova, J. & Randak, T. (2010). Modulation of antioxidant defence system in brain of rainbow trout (*Oncorhynchus mykiss*) after chronic carbamazepine treatment. Comparative Biochemistry and Physiology, Part C, 151: 137-141.

Lopes, R.B. Paraiba, L.C. Ceccarelli, P.S. & Tornisielo, V.L. (2006). Bioconcentration of trichlorfon insecticide in pacu (*Piaractus mesopotamicus*). Chemosphere, 64: 56-62.

Ma, Y. Chen, L. Lu, X. Chu, H. Xu, C. & Liu, W. (2009). Enantioselectivity in aquatic toxicity of synthetic pyrethroid insecticide fenvalerate. Ecotoxicology and Environmental Safety, 72: 1913-1918.

Magesh, S. & Kumaraguru, A.K. (2006). Acute toxicity of endosulfan to the milk fish, *Chanos chanos*, of the southeast coast of India. Bull. Environ. Contam. Toxicol., 76: 622-628.

Matos, P. Fontainhas-Fernandes, A. Peixoto, F. Carrola, J. & Rocha, E. (2007). Biochemical and histological hepatic changes of Nile tilapia *Oreochromis niloticus* exposed to carbaryl. Pesticide Biochemistry and Physiology, 89: 73-80.

Maxwell, L.B. & Dutta, H.M. (2005). Diazinon-induced endocrine disruption in bluegill sunfish, *Lepomis macrochirus*. Ecotoxicology and Environmental Safety, 60: 21-27.

Mayer, F. L. & Ellersieck, M. R. (1986). Manual of acute toxicity: interpretation and data base for 410 chemicals and 66 species of freshwater animals. U.S. Department of the Interior, Fish and Wildlife Service, Resource Publication 160. 579 p. Author: Mayer, F. Ellersieck, M. Wetzel, J. (1986) Static Acute 96-hour LC50 of Linuron (INZ-326118) to Rainbow Trout (*Salmo gairdneri*): Rept. No. HLR 525-86.

McAllister, W. Swigert, J. & Bowman, J. (1986a). Acute Toxicity of Betasan Technical to Bluegill (*Lepomis macrochirus*): Static Acute Toxicity Report #34027: T-12394. Unpublished study prepared by Analytical Bio-Chemistry Laboratories, Inc. 47 p.

McAllister, W. Swigert, J. & Bowman, J. (1986b). Acute Toxicity of Betasan Technical to Rainbow Trout (Salmo gairdneri): Static Acute Toxicity Report #34028: T-12395. Unpublished study prepared by Analytical Bio-Chemistry Laboratories, Inc. 51 p.

McKim, J.M. & Lein, G.L. (2001). Toxic responses of the skin. In: Target Organ Toxicity in Marine and Freshwater Teleosts, edited by D. Schlenk and W.H. Benson, London, UK, Taylor & Francis, pp. 151-223.

Middaugh, D.P. Fournie, J.W. & Hemmer, M.J. (1990). Vertebral abnormalities in Juvenile inland silversides *Menidia beryllina* exposed to tebufos during embryogenesis. Diseases Aquatic Organisms, 9: 109-116.

Mishra, D. Srivastav, S.K. & Srivastav, A.K. (2005). Effects of the insecticide cypermethrin on plasma calcium and ultimobranchial gland of a teleost, *Heteropneustes fossilis*. Ecotoxicology and Environmental Safety, 60: 193-197.

Modesto, K.A. & Martinez, C.B.R. (2010). Roundup causes oxidative stress in liver and inhibits acetylcholinesterase in muscle and brain of the fish *Prochilodus lineatus*. Chemosphere 78: 294-299.

Monteiro, D.A. Alves de Almeida, J. Rantin, F.T. & Kalinin, A.L. (2006). Oxidative stress biomarkers in the freshwater characid fish, *Brycon cephalus*, exposed to organophosphorus insecticide Folisuper 600 (methyl parathion). Comparative Biochemistry and Physiology, Part C 143: 141–149.

Monteiro, D.A. Rantin, F.T. & Kalinin, A.L. (2009). The effects of selenium on oxidative stress biomarkers in the freshwater characid fish matrinxã, *Brycon cephalus* (Günther, 1869) exposed to organophosphate insecticide Folisuper 600 BR® (methyl parathion). Comparative Biochemistry and Physiology, Part C 149: 40–49.

Moore, A. & waring, C.P. (1996). Sublethal effect of the pesticide diazinon on olfactory function in mature atlantic salmon parr. J. Fish. Biol. 48: 758-775.

Moore, A. & Waring, C.P. (2001). The effects of a synthetic pyrethroid pesticide on some aspects of reproduction in Atlantic salmon (*Salmo salar* L.). Aquatic Toxicology, 52: 1–12.

Mushigeri, S.B. & David, M. (2005). Fenvalerate induced changes in the Ach and associated AchE activity in different tissues of fish *Cirrhinus mrigala* (Hamilton) under lethal and sub-lethal exposure period. Environmental Toxicology and Pharmacology, 20: 65–72.

Nabb, D.L. Mingoia, R.T. Yang, Ch. & Han, X. (2006). Comparison of basal level metabolic enzyme activities of freshly isolated hepatocytes from rainbow trout (*Oncorhynchus mykiss*) and rat. Aquat. Toxicol., 80: 52-59.

Nebbia, C. (2001). Biotransformation Enzymes as Determinants of Xenobiotic Toxicity in Domestic Animals. The Veterinary Journal 2001, 161, 238-252

Pan, G. & Dutta, H.M. (1998). The inhibition of brain acetylcholinesterase activity of juvenile largemouth bass, *Micropterus salmoides* by sub-lethal concentrations of diazinon. Environmental Research Section A, 79: 133-137.

Pant, J. Tewari, H. & Gill, T.S. (1987). Effects of aldicarb on the blood and tissues of a freshwater fish. Bull. Environ. Contam. Toxicol. 38: 36-41.

Patil, V.K. & David, M. (2008). Behaviour and respiratory dysfunction as an index of malathion toxicity in the freshwater fish, *Labeo rohita* (Hamilton). Turkish Journal of Fisheries and Aquatic Sciences, 8: 233-237.

Patnaik, L. (2010). Biochemical Alterations Induced by Sevin in *Clarias batrachus*. ASIAN J. EXP. BIOL. SCI., 1(1):124 - 127.

Peddie, S. Zou, J. Cunningham, C. & Secombes, C.J. (2001). Rainbow trout (*Oncorhynchus mykiss*) recombinant IL-1beta and derived peptides induce migration of head-kidney leucocytes in vitro. Fish Shellfish Immun. 11, 697–709.

Rand, G.M. (2004). Fate and effects of the insecticide–miticide chlorfenapyr in outdoor aquatic microcosms. Ecotoxicology and Environmental Safety, 58: 50–60.

Rao, J.V. (2006a). Sublethal effects of an organophosphorus insecticide (RPR-II) on biochemical parameters of tilapia, *Oreochromis mossambicus*. Comparative Biochemistry and Physiology, Part C 143: 492–498.

Rao, J.V. (2006b). Biochemical alterations in euryhaline fish, *Oreochromis mossambicus* exposed to sub-lethal concentrations of an organophosphorus insecticide, monocrotophos. Chemosphere, 65: 1814–1820.

Rao, J.V. (2006c). Toxic effects of novel organophosphorus insecticide (RPR-V) on certain biochemical parameters of euryhaline fish, *Oreochromis mossambicus*. Pesticide Biochemistry and Physiology, 86: 78–84.

Rice, J.P. Drews, C.D. Klubertans, T.M. Bradbury, S.P. & Coats, J.R. (1997). Acute toxicity and behavioural effects of chlorypyrifos, permethrin, phenol, strychnine and 2,4-dinitrophenol to 30 day old Japanse medaka (*Oryzias latipes*). Environmental Toxicology and Chemistry, 16: 696-704.

Saha, S. & Kaviraj, A. (2009). Effects of cypermethrin on some biochemical parameters and its amelioration through dietary supplementation of ascorbic acid in freshwater catfish *Heteropneustes fossilis*. Chemosphere, 74: 1254–1259.

Sayeed, I. Parvez, S. Pandey, S. Bin-Hafeez, B. Haque, R. & Raisuddin, S. (2003). Oxidative stress biomarkers of exposure to deltamethrin in freshwater fish, *Channa punctatus* Bloch. Ecotoxicology and Environmental Safety 56: 295–301.

Scholz, N.L. Truelove, N.K. French, B.L. Berejikian, B.A. Quinn, T.P. Casillas, E. Collier, T.K. (2000). Diazinon disrupts antipredator and homing behaviors in chinook salmon (*Oncorhynchus tshawytscha*). Can. J. Fish. Aquat. Sci. 57: 1911–1918.

Senger, M.R. Rico, E.P. de Bem Arizi, M. Rosemberg, D.B. Dias, R.D. Bogo, M.R. & Bonan, C.D. (2005). Carbofuran and malathion inhibit nucleotide hydrolysis in zebrafish (*Danio rerio*) brain membranes. Toxicology, 212: 107–115.

Sharbide, A.A. Metkari, V. & Patode, P. (2011). Effect of diazinon on acetylcholinestrase activity and lipid peroxidation of *Poecilia reticulate*. Research Journal of Environmental Toxicology, 5(2): 152-161.

Shayeghi, M. Darabi, H. Abtahi, H. Sadeghi, M. Pakbaz, F. & Golestaneh., S.R. (2007). Assessment of persistence and residue of diazinon and malathion in three Rivers (Mond, Shahpour and Dalaky) of Bushehr province in 2004-2005 years. Iranian South Medical Journals 10(1) 54-60.

Shea, T.b. & Berry, E.S. (1984). Suppression of interferon synthesis by the pesticide carbaryl as a mechanism for enhancement of goldfish virus-2 replication. Appl. Environ. Microbiol., 47: 250-252.

Shea, T.B. (1983). Enhancement of goldfish virus-2 *in vitro* replication by the pesticides carbaryl and toxaphane. Appl. Environ. Microbiol., 45: 1859-1864.

Shelley, L.K. Balfry, S.K. Ross, P.S. & Kennedya, C.J. (2009). Immunotoxicological effects of a sub-chronic exposure to selected current-use pesticides in rainbow trout (*Oncorhynchus mykiss*). Aquatic Toxicology, 92: 95–103.

Siang, H.Y. Yee, L.M. & Seng, C.T. (2007). Acute toxicity of organochlorine insecticide endosulfan and its effect on behaviour and some hematological parameters of Asian swamp eel (*Monopterus albus*, Zuiew). Pesticide Biochemistry and Physiology, 89: 46–53.

Singh, H. (1989). Interaction of xenobiotics with reproductive endocrine functions in a protogynous teleost, *Monopterus albus*. Mar. Environ. Res., 28: 285-289.

Singh, S. & Singh, T.P. (1987). Evaluation of toxicity limit and sex hormone production in response to Cythion and BHC in the vitellogenic catfish *clarias batrachus*. Environ. Res., 42: 428-488.

Singh, S.K. Tripathi, P.K. Yadav, R.P. Singh, D. & Singh, A. (2004). Toxicity of malathion and carbaryl pesticides: effects on some biochemical profiles of the freshwater fish *Colisa fasciatus*. Bull. Environ. Conram. Toxicol. 72: 592-599.

Skandhan, K.P. Sahab Khan, P. & Sumangala, B. (2001). DDT and male reproductive system. Research Journal of Environmental Toxicology, 5(2): 76-80.

Stegeman, J.J. & Hahn, M.E. (1994). Biochemistry and molecular biology of monooxygenases: current perspectives on forms, functions and regulation of cytochrome P450 in aquatic species, in: Malins, D.C. Ostrander, G.K. (Eds.), Aquatic Toxicology: Molecular, Biochemical Cellular Perspectives, CRC Press, Boca Raton, Florida, pp. 87-206.

Straus, D.L. Schlenk, D. & Chambers, J.E. (2000). Hepatic microsomal desulfuration and dearylation of chlorpyrifos and parathion in fingerling channel catfish: lack of effect from Aroclor 1254. Aquat. Toxicol. 50: 141-149.

Talebi, K. (1998). Diazinon Residues in the Basins of Anzali Lagoon, Iran. Bulletin Environmental Contamination Toxicology, 61: 477-483.

Tarahi Tabrizi, S. (2001). Study of pesticide residues (diazinon, malathion, metasytoux) in the Tabriz Nahand River, M.Sc. Thesis, Tehran University of Medical Science, Tehran, Iran, 1-88.

Todd, N.E. & Leeuwen, M.V. (2002). Effects of Sevin (Carbaryl Insecticide) on Early Life Stages of Zebrafish (*Danio rerio*). Ecotoxicology and Environmental Safety, 53: 267-272.

Tripathi, P.K. Srivastava, V.K. & Singh, A. (2003). Toxic effects of dimethoate (organophosphate) on metabolism and enzyme system of freshwater teleost fish *Channa punctatus*. Asian Fisheries Science, 16: 349-359.

Üner, N. Oruç, E.Ö. Sevgiler, Y. Şahin, N. Durmaz, H. & Usta, D. (2006). Effects of diazinon on acetylcholinesterase activity and lipid peroxidation in the brain of *Oreochromis niloticus*. Environmental Toxicology and Pharmacology, 21: 241-245.

Velasco-Santamaría, Y.M. Handy, R.D. & Sloman, K.A. (2011). Endosulfan affects health variables in adult zebrafish (*Danio rerio*) and induces alterations in larvae development. Comparative Biochemistry and Physiology, Part C 153: 372-380.

Velisek, J. Svobodova, Z. & Piackova, V. (2009). Effects of acute exposure to bifenthrin on some haematological, biochemical and histopathological parameters of rainbow trout (*Oncorhynchus mykiss*). Veterinarni Medicina, 54, (3): 131-137.

Velmurugan, B. Selvanayagam, M. Cengiz, E.I.& Unlu, E. (2009). Histopathological Changes in the Gill and Liver Tissues of Freshwater Fish, *Cirrhinus mrigala* Exposed to Dichlorvos. An International Journal Brazilian Archives of Biology and Technology, 52(5): 1291-1296.

Viant, M.R. Pincetich, C.A. & Tjeerdema, R.S. (2006). Metabolic effects of dinoseb, diazinon and esfenvalerate in eyed eggs and alevins of Chinook salmon (*Oncorhynchus tshawytscha*) determined by H NMR metabolomics. Aquatic Toxicology, 77: 359-371.

Viran, R. Erkoc, F.U. Polat, H. & Kocak, O. (2003). Investigation of acute toxicity of deltamethrin on guppies (*Poecilia reticulata*). Ecotoxicology and Environmental Safety, 55: 82–85.

Vryzas, Z. Vassiliou, G. Alexoudis, C. & Papadopoulou-Mourkidou, E. (2009). Spatial and temporal distribution of pesticide residues in surface waters in northeastern Greece. Water Research, 43: 1-10.

Wang, C. Lu, G. Gui, J. & Wang, P. (2009). Sublethal effects of pesticide mixtures on selected biomarkers of *Carassius auratus*. Environmental Toxicology and Pharmacology, 28: 414-419.

Wang, X. Xing, H. Li, X. Xu, S. & Wang, X. (2011). Effects of atrazine and chlorpyrifos on the mRNA levels of IL-1β and IFN- γ2b in immune organs of common carp. Fish & Shellfish Immunology, 31: 126-133.

Werimo, K. Bergwerff, A.A. & Seinen, W. (2009). Residue levels of organochlorines and organophosphates in water, fish and sediments from Lake Victoria-Kenyan portion. Aquatic Ecosystem Health & Management 12 337-341.

Wetzel, J. (1986) Static Acute 96-hour LC50 of Linuron (IN Z-326118) to Bluegill (Lepomis macrochirus): Rept. No. 52786. Unpublished study prepared by E.I. du Pont de Nemours and Co., Inc. 14 p.

Woodward, D. & Mauck, W. (1980) Toxicity of five forest insecticides to cutthroat trout and two species of aquatic invertebrates. Bull. Environm. Contam. Toxicol. 25:846-854. (Submitter 69597. also in unpublished submission received Dec 9, 1982 under 3125327. submitted by Mobay Chemical Corp., Kansas City, MO. CDL: 248989-O).

Xu, C. Tu, W. Lou, C. Hong, Y. & Zhao, M. (2010). Enantioselective separation and zebrafish embryo toxicity of insecticide beta-cypermethrin. Journal of Environmental Sciences, 22(5): 738–743.

Xu, W.N. Liu, W.B. & Liu, Z.P. (2009). Trichlorfon-induced apoptosis in hepatocyte primary cultures of *Carassius auratus gibelio*. Chemosphere, 77:895-901.

Yaji, A.J. Auta, J. Oniye, S.J. Adakole, A.J. & Usman, J.I. (2011). Effects of cypermethrin on behavior and biochemical indices of fresh water fish *Oreochromis niloticus*. EJEAFChe, 10(2): 1927-1934.

Yi, M.Q. Liu, H.X. Shi, X.Y. Liang, P. & Gao, X.W. (2006). Inhibitory effects of four carbamate insecticides on acetylcholinesterase of male and female *Carassius auratus in vitro*. Comparative Biochemistry and Physiology, Part C, 143: 113–116.

Zaheer Khan, M. & Law, F.C.P. (2005). Adverse effects of pesticides and related chemicals on enzyme and hormone systems of fish, amphibians and reptiles: a review. Proc. Pakistan Acad. Sci. 42(4): 315-323.

Zelikoff, J.T. Raymond, A. Carlson, E. Li, Y. Beaman, J.R. & Anderson, M. (2000). Biomarkers of immunotoxicity in fish: from the lab to the ocean. Toxicol. Lett. (Amst.), 325–331.

6

Development of a Prophylactic Butyrylcholinesterase Bioscavenger to Protect Against Insecticide Toxicity Using a Homologous Macaque Model

Yvonne Rosenberg, Xiaoming Jiang, Lingjun Mao,
Segundo Hernandez Abanto, Keunmyoung Lee
PlantVax Inc.
USA

1. Introduction

Organophosphorus (OP) and carbamate pesticides are extensively used to control agricultural, household and structural pests. Each year approximately 5.6 billion pounds of pesticides are used worldwide potentially exposing ~1.8 billion people who use pesticides to protect the food and commercial products that they produce (Alavanja, 2009). Although unintentional occupational poisonings represent only a small number, estimated to be ~10% (Litchfield, 2005) or 25 million agricultural workers globally (Jeyaratnam, 1990), large scale exposure of both civilian and military personnel has become an ever increasing threat, as a result of deliberate insecticide contamination of the environment and critical water supplies by terrorists. In this context, pesticide use is one of only two exposures consistently identified by Gulf War epidemiologic studies to be significantly associated with the multisymptom illness profiles described as Gulf War illness (Cao et al., 2011). Pesticide use has also been associated with neurocognitive deficits and neuroendocrine alterations in Gulf War veterans in clinical studies conducted following the end of the war.

While OP nerve agents and WHO Class I and Class II OP pesticides constitute a diverse group of chemical structures, all potentially exhibit a common mechanism of toxicity, that is, active site phosphorylation of acetylcholine (AChE) resulting in AChE inhibition and accumulation of acetylcholine, overstimulation of cholinergic receptors, and consequent clinical signs of cholinergic toxicity such as seizures, brain damage and cognitive and behavioural defects (Millard et al., 1999; Rosenberry et al., 1999; Colosio et al., 2009). The relationship between AChE inhibition and symptoms showed that prevalence ratios were significantly >1 for respiratory, eye and central nervous system symptoms for workers with >30% inhibition (Ohayo-Mitoko et al., 2000). More recent studies indicate that insecticide exposure to DFP (diisopropyl fluorophosphate) causes a prolonged increased in hippocampal neuronal Ca++ plateau which may underlie morbidity and mortality (Deshpande et al., 2010). These findings are consistent with those indicating persistent changes in locus coeruleus noradrenergic neuronal activity and lasting changes in this brain area after removal of the insecticide chlorpyrifos oxon; reminiscent of the lasting cognitive symptoms of Gulf War illness in soldiers exposed to these compounds (US DOD, Pesticides-Final Report, 2003).

Currently, the standard (approved) treatment for acute OP pesticide poisoning involves administration of intravenous (iv) atropine and an oxime e.g. obidoxime, pralidoxime to reactivate inhibited AChE (Worek et al., 2010). However, the effectiveness and safety of oxime administration in acute OP pesticide-poisoned patients has been challenged and a recent clinical trial showed no clinical benefits and a trend towards harm in all sub-groups, despite clear evidence that these doses reactivated AChE in the blood (Buckley et al., 2011).

An efficacious prophylactic therapeutic treatment for preventing insecticide poisoning that can bind and scavenge the OP before it reaches and targets AChE in neuromuscular junctions is therefore a high priority. The leading candidate of this type is native (plasma) butyrylcholinesterase (BChE) whose potent OP bioscavenging ability has been demonstrated in many animal models and against varied OP neurotoxins (Doctor et al., 2001; Lenz et al., 2001). While several new catalytic and other stoichiometric enzymes also exhibit this ability (Lenz et al., 2007), based on availability, broad spectrum efficacy, stability and safety (Sun et al., 2005), BChE is the most advanced in terms of development of a human treatment. In Turkey, frozen plasma (BChE levels of 3,000 - 5,700 units) given as an alternative or adjunctive treatment with atropine and oximes, has been shown to prevent mortality and intermediate syndrome in acutely insecticide-exposed and hospitalized individuals (Güven et al., 2004). Currently, BChE also finds use as a treatment of cocaine overdose and for the alleviation of succinylcholine-induced apnea.

Structurally, BChE (also known as pseudocholinesterase or non-specific cholinesterase) is a serine esterase (MW=345,000) comprised of four identical subunits each containing 574 amino acids, held together by non-covalent bonds, with 36 carbohyrdrate chains (23.9% by weight) (Lockridge, 1990; Nachon et al., 2002). BChE is found in all species at levels of 1-20 ug/ml in plasma (Rosenberg, unp. data) and is also abundant in liver, intestine and lung. Recombinant (r) human butyryl-cholinesterase (HuBChE), like the native form, is also a potent bioscavenger of OP neurotoxins (Doctor et al., 2001; Lenz et al., 2001; Raveh et al., 1997) but its development as a human treatment for pesticide exposure has been disadvantaged by: (i) poor in vivo stability (bioavailability) of the unmodified forms and the presence of potentially immunogenic glycans using certain expression systems (ii) a 1:1 stoichiometry between the enzyme and OP (Raveh et al., 1997) and (iii) the high LD_{50} of insecticides (ug-mg/kg levels). This necessitates the delivery of large, costly, rBChE doses to detoxify exposed individuals which is problematic when intramuscular (im) or subcutaneous (sc) injections are the chosen routes of delivery. In this chapter, we shall describe our experience of how the chemistry, glycosylation, chemical modification, animal model and route of administration may reduce or enhance the potential of BChE bioscavengers as prophylactic therapeutic human antidotes for OP insecticide exposure.

2. Production of tetrameric and monomeric forms of rMaBChE and rHuBChE

Macaque (Ma) and human (Hu) BChE molecules are very similar molecules differing by only 22 amino acids and sharing ~96% DNA sequence identity, critical glycosylation sites, cysteines and disulfide bridging (Boeck et al., 2002; Rosenberg et al., 2010). Thus, most anti-BChE antisera react with both molecules. Native HuBChE and MaBChE in plasma are composed predominantly of tetramers (98%) with the tetramerization domain being located within the last 40 C-terminal residues of each monomeric subunit (534-574) (Blong et al.,

1997). In human serum, the association of lamellipodin proline rich peptides with the monomeric chains results in the formation of BChE tetramers (Li et al., 2008). Recombinant BChE produced in mammalian cells, in contrast, has only 10-20% tetrameric forms and therefore optimal tetramerization requires the addition of either poly(L-proline) to the culture medium or co-expression of the full length BChE monomers with the proline-rich attachment domain (PRAD) of ColQ gene (Altamirano & Lockridge, 1999).

To date, rHuBChE and rMaBChE molecules have been produced in transgenic mammalian cells (Chilukuri et al., 2008; Rosenberg et al, 2010), goat milk (Huang et al., 2007) and in plants (Geyer et al., 2010; Jiang, unpub. data). Our approach has been to utilize two expression systems for the production of rMaBChE and rHuBChE. Initially, Chinese hamster ovary cells (CHO) were used because of their human-like glycosylation. More recently, a transient plant expression platform was adopted to increase the yield and reduce the time and cost of producing rBChE. Although CHO cells and plants are able to produce significant levels of tetrameric BChE molecules (Li et al., 2008; Geyer et al., 2010), in the present studies, co-transfection of the BChE and PRAD genes has been shown to increase both levels of tetramerization and yields in each expression system. While the CHO cell expression of recombinant proteins is very well established, recent innovations in transient plant expression systems e.g. Bayer's Magnifection system (Gleba et al., 2005) and the Cow Pea Mosaic Virus Hyper-translatable Protein Expression System (PBL Technology) (Sainsbury et al., 2008) have been shown to be some of the most rapid, cost effective and productive expression systems in existence; capable of producing grams of recombinant proteins in weeks (Goodin et al., 2008).

CHO-derived (Stable Transfection)*				Plant-derived (Transient Transfection)*		
rMaBChE[#+]		rHuBChE		rMaBChE[#]		
					N. tobacum	N. benthamiana
Monomeric	Tetrameric	Monomeric	Tetrameric	Monomeric	Tetrameric	Tetrameric
8U/ml (9mg/L)	25U/ml (28mg/L)	16 U/ml (22.9mg/L)	45 U/ml (64.3 mg/L)	60 U/gm (66.6 mg/kg)	140 U/gm (155.5 mg/kg)	400 U/gm (444 mg/kg)

*All tobacco plants and leaves from *Nicotiana tobacum* and *N. benthamiana* were transfected using Agrobacterium-mediated infiltration

[#] CHO supernatants and whole leaf extracts are prepared for purification.

[+] BChE activity is determined spectrophotometrically (Grunwald at al., 1997), using butyrylthiocholine (BTC) (0.5 mM each) as substrate. One unit of enzyme activity is the amount required to hydrolyze 1 umol substrate/min. One mg MaBChE has 900 units of activity and one mg HuBChE has 700units.

Table 1. Expression levels of different forms of rBChE using CHO-and plant-based expression systems.

In addition to the tetrameric forms, a truncated monomeric form of rBChE (MW=~81KDa) that is incapable of oligomerization has also been produced by the insertion of a stop codon at G534 resulting in a monomeric form lacking 41 C-terminal residues (Blong et al., 1997). The smaller monomeric molecules may more rapidly gain access to the blood from muscle or lungs (depending on the route of delivery) with transiently higher bioavailablity in the plasma, which would be advantageous in emergency situations that require real time responses and rapid treatment or booster administrations.

3. In vitro biological properties of rMaBChE

To test the chemical properties of CHO- and tobacco-derived rMaBChE, inhibition and reactivation assays using   diisopropyl fluorophosphate (DFP) and paraoxon (diethyl 4-nitrophenyl phosphate) have been performed with and without the oxime 2-PAM (pyridine-2-aldoxime methochloride)(Luo et al., 2008). DFP is an OP compound that has been used as an experimental insecticide agent in neuroscience because of its ability to inhibit cholinesterase and cause delayed peripheral neuropathy. Paraoxon is an insecticide and will be described in detail in a later section. Following purification of the CHO supernatant and the plant leaf extract using procainamide sepharose, rMaBChE molecules conjugated with polytheleneglycol (PEG) using succinimidyl-propionate-activated methoxy-PEG-20K (SPA-PEG-20K; Nektar Inc., Birmingham, AL) or Sunbright ME-200HS 20K PEG (NOF, Tokyo, Japan) (Chilukuri et al., 2008a; Cohen et al., 2001) to test the effects of PEGylation on enzyme plasma stability. Initially, the biochemical properties of both the unmodified and PEGylated forms of both monomeric and tetrameric rMaBChE were examined using DFP inhibition; bimolecular rate constants (k_i ($\times$ 10^7) M^{-1} min^{-1}) for inhibition of all the recombinants forms ranging from 2.58 - 2.23 ($\times$ 10^7) M^{-1} min^{-1} which were indistinguishable from the well characterized native HuBChE (2.29 +/- 0.1) and native MaBChE (2.22 +/- 0.1) (data not shown).

3.1 Inhibition and reactivation of plant derived CHO-derived and plant-derived rBChE by paraoxon

The kinetics of inhibition of both plant-derived and CHO-derived rMaBChE by paraoxon were further examined as shown in Fig. 1A. At low paraoxon concentrations (0.01 and 0.02uM), the reciprocal value of $E_t/E_{t,0}$ was highly correlated with the reaction time; the reaction rate constant of plant-derived rMaBChE at 0.01uM paraoxon being slightly faster than that of CHO-derived MaBChE (0.035 $M^{-1}min^{-1}$ vs 0.022 $M^{-1}min^{-1}$ respectively). These values follow the simple 2nd-order (reciprocal) model.

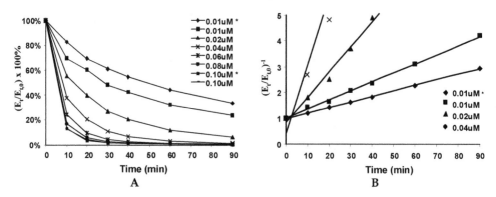

Fig. 1. Inhibition kinetics of plant- and CHO-derived* rMaBChE by different concentrations of paraoxon (0.01uM - 0.10uM) A: Percent inhibition of BChE by paraoxon. (Percent BChE activity was obtained by dividing the BChE activity with paraoxon at each time point with control BChE activity at the same time point. B: Reciprocal plot of BChE inhibition by paraoxon.

3.2 Reactivation of paraoxon-inhibited plant-derived rMaBChE by 2-PAM

Since a 1 hour incubation of 0.016 uM plant-derived MaBChE (1.2U/ml) with 0.02uM paraoxon resulted in 80-90% inhibition of the enyme (Fig. 1A), the same conditions (incubation of paraoxon with rMaBChE at a final enzyme concentration of 0.04-0.05uM), was used to prepare inhibited rMaBChE. Reactivation of inhibited rMaBChE was then initiated by adding different concentrations of 2-PAM (0.4mM-6.4mM) for various times (Fig.2). The kinetics of reactivation of paraoxon-inhibited CHO- and plant-derived rMaBChE were found to follow the simple first-order (mono-exponential) model.

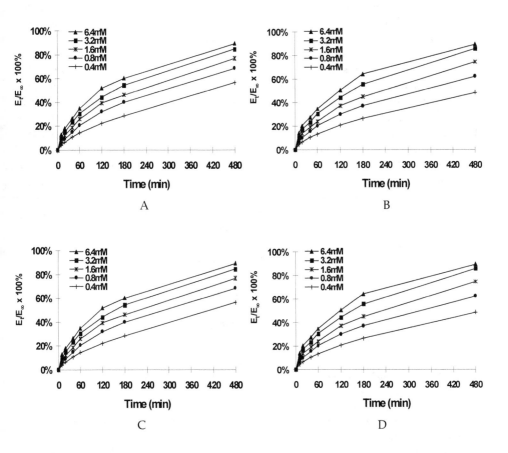

Fig. 2. Reactivation kinetics of paraoxon inhibited plant- and CHO-derived MaBChE by 2-PAM. A, C: Plant-derived MaBChE; B, D: CHO-derived MaBChE; A and B: Direct plot of the time course vs % reactivation; C and D: Semi-logarithmic plot of time course of reactivation. For inhibition controls, inhibited BChE was incubated with reaction buffer without 2-PAM. Triplicate BChE assays were performed at the times indicated.

The results indicate that both paraoxon-inhibited plant- and CHO-derived rMaBChE showed very similar patterns of reactivation by different concentrations of 2-PAM (Fig. 2A, 2B) with nearly 100 % reactivation of each rMaBChE form being achieved by 24 hours at >1.60 mM 2-PAM; the k_{app} values of CHO-rMaBChE ranging from 0.0014 to 0.004 min^{-1} and plant-rMaBChE from 0.0013 to 0.0051 min^{-1} (Fig. 2C, 2D). The reactivation k_{app}s at each 2-PAM concentration was linear when plotted against 2-PAM concentration (mM) expressed logrithmically.

4. In vivo testing of rBChE

In the area of insecticide exposure/contamination, there is a high likelihood that agricultural workers or military personnel will be exposed multiple times during their lives and thus multiple prophylactic treatments must be considered a possibly. This is often problematic since administration of heterologous HuBChE into macaques or other species eg mice has been shown to generate anti-BChE antibody responses and rapidly eliminate enzyme on repeated injections (Matzke et al., 1999; Chiluluri et al., 2008b; Sun et al., 2009). Thus, in vivo retention times of exogenously administered recombinant proteins can only be accurately assessed using homologous systems (rMaBChE → macaques and rHuBChE → humans) in which antibodies or other immune responses are not induced. In this context, homologous BChE enzyme has been shown to have a long half-life (8-12 days) with no adverse effects and no immunogenicity following either (i) transfusions of human plasma into humans (ii) daily administrations of partially purified native HuBChE into humans for several weeks (Jenkins et al., 1967; Cascio et al., 1988) or (iii) injection of purified native MaBChhE or PEG-rMaBChE into macaques (MRT= 200-300 h)(Rosenberg et al., 2002, 2010). These data are in contrast to exogenously administered heterologous HuBChE which displayed a rapid clearance in macaques (MRT = 33.7 h) (Raveh et al., 1989). While the choice of the animal model for PK, immunogenicity and efficacy testing is always important, the animal species utilized for the evaluation of an efficacious human cholinesterase bioscavenger is critical, since potential treatments against OP toxicity cannot be tested in humans and will require extensive testing in animal models and the Animal Rule (CFR 601.90 for biologics) for regulatory approval.

4.1 Pharmacokinetics of clearance in rodent and macaque models

Pharmacokinetic profiles following administration of biologics in many rodent and primate species are used to indicate the periods after administration that such biologics are likely to exhibit optimal benefit or protection. An efficacious therapeutic for preventing OP poisoning is a molecule that: (i) can bind and scavenge the OP before it reaches the targeted AChE in neuromuscular junctions and (ii) has the ability to remain at therapeutic levels in the blood for prolonged periods to counteract a known or impending OP exposure. The in vivo parameters generally used to assess PK performance after administration are mean retention time (MRT), maximal concentration (Cmax), time to reach maximal concentration (Tmax), elimination half life (T1/2) and area under the plasma concentration curve extrapolated to infinity (AUC).

Generally pharmacokinetics of recombinant molecules differs considerably depending on the structure, glycoslyation, size, route of administration, immunogenicity, and animal model utilized. Interestingly, despite protein sequence identity, rBChE proteins, similar to many other recombinant biologics, have been shown to be rapidly cleared following injection (Saxena et al., 1998; Cohen et al., 2006) in contrast to the good plasma stability of native BChE. Thus, rBChE molecules require post-translational modification to provide protection as therapeutic scavengers. A common means of increasing the radius of the target molecule permitting slower renal clearance and prolonging plasma retention is by PEG conjugation. This has been successfully used with proteins, peptides, oliogonucleotides and antibody fragments to improve pharmacokinetic and immunological profiles (Kang et al., 2009). Accordingly, both monomeric and tetrameric forms of rMaBChE have been conjugated with 20KD PEG (without interference of in vitro biological properties) and the pharmacokinetic profiles of the unmodified and PEG-conjugated rMaBChE forms compared in monkeys and mice (Rosenberg et al., 2010).

Figure 3 shows the PK profiles in 24 monkeys following iv injection of 1.2 -3 mg/kg of unmodified or PEG-rMaBChE and illustrates several aspects of BChE clearance: (i) PEG-rMaBChE exhibits good stability in the lower range of the native form; the hierarchy of clearance being native BChE ~ PEG-rMaBChE >>> unmodified monomeric rMaBChE > unmodified tetrameric rMaBChE. (ii) Surprisingly, five of the monkeys demonstrated unexpected dramatic decreases in BChE levels (shown in bold between days 150 and 230 days post injection). In each case, these decreases always occurred immediately after the weekend treatment of the grass surrounding the animal facility and presumably resulted from exposure of the housed monkeys to insecticide; highlighting the unintentional consequences of routine insecticide use on plasma BChE activity and (iii) despite very poor retention of the unmodified monomeric rBChE, administration of the PEGylated monomeric rMaBChE showed overlapping pharmacokinetic parameters with the larger PEG-rMaBChE tetrameric form despite lack of oligomerization.

Importantly, the extended circulatory retention afforded by PEG conjugation of rMaBChE in monkeys (injected iv) was not observed in mice (injected ip) where unmodified and modified monomeric and tetrameric rMaBChE all exhibited the same high MRT and T1/2 (Rosenberg et al., 2010). This indicates that, depending on the parameter measured, the mouse model does not accurately predict the outcome in monkeys with MRT and T1/2 values appearing to be less predictive indicators of circulatory stability in macaques than parameters such as AUC and Cmax. Similar differential pharmacokinetic behaviour was observed following the administration of recombinant rhesus (Rh) and HuAChE in mice and monkeys (Cohen et al., 2004).

These studies highlight the potential problems inherent in choosing an animal model to test human biologics. Notwithstanding the differences in pharmacokinetic behaviour of the same protein in different species and the high potential for immunogenicity in rodents due to the evolutionary distance between rodents and humans, other influences may also play a role in the circulatory stability of proteins following even the first injections into heterologous species. Table 2 shows the pharmacokinetic parameters (MRT, Cmax, Tmax, T1/2 and AUC) following injection of different forms of BChE into several different animal species determined from the time course curve of blood BChE concentrations and using a Windows-based program for non-compartmental analysis. Several conclusions can be made.

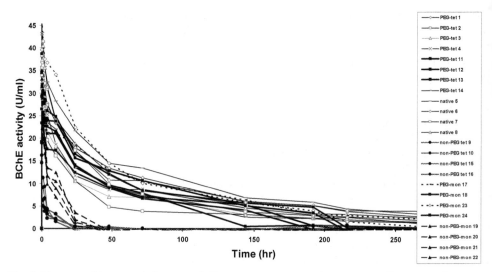

Fig. 3. Pharmacokinetics of clearance following iv injection of 1.2 - 3.0 mg/kg rMaBChE into 24 monkeys. Each line represents a single monkey. Different forms of rMaBChE were used except for 4 macaques receiving native BChE.

For example, while the Cmax following first injections appear to be similar in any animal model at comparable doses, the AUC, MRT and T1/2 are often significantly higher in homologous systems (e.g. PEG-rMaBChE into macaques and native mouse (Mo) BChE into mice) than heterologous injections (native HuBChE into monkeys or mice or PEG-rHuBChE into monkeys). This indicates that heterologous proteins, even when PEGylated and given at a time when anti-BChE titers are absent or low, appear to be eliminated faster than homologous proteins suggesting that pharmacokinetic parameters are less than optimal in all heterologous systems.

It should also be noted, that while PEG conjugation markedly improves the pharmacokinetic profile of therapeutic rMaBChE and other biologics, effects relating to immunogenicity have been mixed. Thus, reduced immunogenicity has been observed following PEGylation of enzymes, cytokines and hormones, while administration of PEGylated interferon-β1a to monkeys actually resulted in increased immunogenicity (Pepinsky et al., 2001). In the case of rHuBChE produced in HEK-293 cells, PEGylation failed to eliminate immunogenicity in mice as demonstrated by the rapid clearance of a repeat 100U injection of (heterologous) PEG-rHuBChE, coincident with induction of high levels of serum anti-BChE antibody (Sun et al., 2009). Likewise, when tested in a sandwich ELISA, the presence of 4–7 PEG molecules per rMaBChE monomer did not prevent the binding of BChE epitopes to either an anti-BChE MAb or a polyclonal rabbit anti-BChE antibody when antigen concentrations were increased to as little as 4–8 U/ml (Rosenberg et al., 2010) which, as mentioned above, is in the range of BChE in normal plasma. These studies raise the question whether chemical modification by PEG will be able to mask any "foreign" rBChE epitopes, such as non-human glycans, sufficient to prevent humoral immune responses and also highlights the importance of using homologous animal models to perform in vivo PK, immunogenicity and efficacy testing.

Human and Mouse BChE								
BChE	**Dose** [Units, mg, mg/kg]	**Animal**	**Route**	**MRT** (hr)	**AUC** (U/ml.h)	**Cmax** (U/ml)	**Tmax** (hr)	**T1/2** (hr)
natHuBChE (Raveh,1997)	11.5 mg (8,000 U)	Monkey	iv	33	710			
natHuBChE	11.5 mg (8,000 U)	Monkey	im		582	16.2	9.5	
natHuBChE (Lenz,2005)	5.25 mg/kg (12,000 U)	Monkey	im		2576	21	9.27	79.3
	8.75 mg/kg (20,000 U)	Monkey	im		3822	33	10.3	73.5
natHuBChE (Sun, 2005)	34 mg/kg (30,000 U)	Monkey	iv	73	16,538	222	0	37
natHuBChE (Sun, 2009)	100 U	Mouse	im	48	1,300	19		21
natMaBChE *	100 U	Mouse	im	73	2,500	25		24
Monkey BChE								
BChE	**Dose** [Units, mg, mg/kg]	**Animal**	**Route**	**MRT** (hr)	**AUC** (U/ml.h)	**Cmax** (U/ml)	**Tmax** (hr)	**T1/2** (hr)
natMaBChE *	3 -5 mg/kg (7,000 U)	Monkey	iv	191				
(Rosenbberg, 2002)	1.3 - 1.65 mg/kg (3,000 U)	Monkey	iv	50				
natMaBChE (unpub)*	1.8 mg/kg	Monkey	iv	142	2950	27		
	1.8 mg/kg	Monkey	iv	142	4010	37		
natMaBChE*	2.9 mg/kg	Monkey	iv	224	4431	38		143
(Rosenberg, 2010)	2.9 mg/kg	Monkey	iv	307	4299	40		126
	1.9 mg/kg	Monkey	iv	200	2097	26		157
PEG-rMaBChE*	2.9 mg/kg	Monkey	iv	168	2141	33		112
(Rosenberg, 2010)	2.9 mg/kg	Monkey	iv	223	3312	39		85
	1.9 mg/kg	Monkey	iv	134	1724	24		97
PEG-rMaBChE (unpub)*	3.0 mg/kg	Monkey	iv		4359	51		
PEG-rHuBChE (unpub)*	3.0 mg/kg	Monkey	iv		1101	40		

MRT: mean retention time, Cmax: maximal concentration, Tmax: time to reach maximal concentration, T1/2: elimination half life, AUC: area under the plasma concentration curve extrapolated to infinity. nat: native, Mon: monomeric, Tet: tetramer.

Table 2. Pharmacokinetic parameters of different forms of BChE in homologous* and heterologous systems.

4.2 The role of glycosylation and oligomerization on pharmacokinetics

The BChE molecule is a soluble protein, protected from proteolysis by a heavy sugar coating from nine N-linked glycans (Li et al., 2008). N-glycosylation is one of the major post-translational modifications of proteins and can be critical to their bioavailability. Importantly, while the first steps in the N-glycosylation pathway, leading to the formation of oligomannosidic structures, are conserved in plants and animals, the final steps in the formation of complex N-glycans may differ with the expression system used. Thus, in contrast to native HuBChE molecules which have highly sialylated bi- and triantennary type glycans (Saxena et al., 1998; Kolarich et al. 2008) containing the N-acetyl neuraminic acid (NANA, NeuAc) form of sialic acid (Varki, 2001), rHuBChE molecules may exhibit under-sialyated or immunogenic non-human glycan structures that accelerate in vivo clearance due to rapid uptake by asialoglycoprotein and mannose receptors in the liver or by antibody-mediated mechanisms (Park et al., 2005). For example, CHO cells produce recombinant proteins which contain human-like glycans that may be undersialyted, compared to those produced in livestock systems which append the non-human galactose-α-1,3-galactose and the N-glycolyl neuraminic (NGNA, NeuGc) form of sialic acid (Chung et al., 2008; Diaz et al., 2009) and those produced in plants which are non-sialylated and append the non-human β-1,2 xylose and α-1,3 fucose containing glycans (Altmann, 2007).

The relationship between sialic acid levels and oligomerization of recombinant molecules with their circulatory longevity has been extensively studied. For example, administration to mice of recombinant bovine and rhesus acetylcholinesterase (rBoAChE, rRhAChE) as well as plant-derived rHuBChE have supported the idea that pharmacokinetic behaviour is governed by hierarchical rules (Kronman et al., 2000); efficient enzyme tetramerization and high sialic acid occupancy both being required for optimal plasma retention. However, other data from monkey and mice studies do not closely obey these classical rules for circulatory retention. For example: (i) the requirement for tetramerization of rAChE molecules was less important when performed in macaques rather than mice (Cohen et al., 2004) (ii) CHO-derived monomeric PEG-rMaBChE resulted in high MRT when injected into in monkeys (Fig.3, Rosenberg et al., 2010) and (iii) the MRT and T1/2 of unmodified and PEG-modified monomeric rMaBChE were both unexpectedly high following injection into mice; PEG-conjugation offering no significant advantages.

While the short lived circulatory retention of asialylated BChE attests to the importance of sialylation in retention/clearance, the degree to which silaic acid occupancy is required does not always seem straight forward. Thus, although the rapid clearance of monomeric (13% non-silayted) and tetrameric (25% nonsialyted) rMaBChE in monkeys, compared to the native or PEGylated forms, has been thought to result from undersialylation, glycan analysis by MALDI-TOF of the highly stable native HuBChE and MaBChE proteins indicates that these also contain a significant percentage of nonsialyted or undersilayted proteins. For example, native HuBChE contains 23% monosialyted glycans (99.9% NANA) and a significant percentage of non-sialyted glycans (Kolarich et al., 2008) while native MaBChE is comprised of 21.3% non-sialayted glycans and 21.8% monosialylated glycans (99.9% NGNA) (Rosenberg, unp. data). This means that heterologous animal models invariably involve the administration of native or CHO-derived human proteins containing NANA into animals containing the NGNA form of sialic acid (monkeys, rodents). These findings showing either high percentages of undersialylated glycans in the stable native proteins and those showing lower pharmacokinetic parameters following heterologous injections, raise the interesting question as to whether the type of sialic acid type as well as the degree of sialic acid occupancy may determine the rate of clearance of recombinant glycoproteins.

It is also important to note that recent engineering of different expression systems is now permitting the production of glycoproteins with human-like glycans. For example, while the inability to perform appropriate N-glycosylation has been a major limitation of plants as expression systems, these are being overcome by new approaches involving the generation of knockout or knockdown plants that: (i) completely lack xylosyl transferase (XylT) and fucosyl transferase (FucT) activity (Strasser et al., 2004) and accumulate high amounts of human-like N-glycan structures that contain no 1,2-xylose or core a1,3-fucose (ii) lack complex N-glycans resulting from the inactivity of N-acetlyglucosaminyltransferase 1 (GnT1) (Strasser et al, 2005; Wenderoth & von Schaewen, 2000) and (iii) contain glycans terminating in sialic acid (Paccalet et al., 2007; Castilho et al., 2010). In addition, different glycoforms of plant derived proteins can be generated by protein targeting to different compartments (i) cytosol (aglycosylated) (ii) ER (high mannose) or (iii) secreted into the apoplast (complex) (Stoger et al., 2005)

4.3 Effects of the route of administration on pharmacokinetics

As mentioned, delivery of PEG-rBChE as a pre-exposure modality is disadvantaged by its large size and a 1:1 stoichiometry between the enzyme and OP requiring high doses due to the high LD_{50} of many insecticides (ug-mg/kg levels). The route of systemic delivery of high doses of native BChE (MW~350KDa) and tetrameric PEG-rMaBChE (MW>800KDa) will determine the pharmacokinetics (PK) of clearance and is critical to efficacy and safety. Currently very little monkey data exists on the delivery of a stoichiometrically equivalent dose of PEG-rBChE calculated to protect against a known LD_{50} of a toxic OP insecticide. Although immediate release requiring intravenous (iv) injection may be necessary in certain high threat situations, these are usually impractical in the field. Needleless cutaneous delivery via the dermis and epidermis (chemical mediators, electroporation) appear quite promising, but are unlikely to deliver high doses. Thus, self-administered transdermal injections through the skin either by subcutaneous (sc) or intramuscular (im) routes have been the approaches most commonly used; virtually all human vaccines currently on the market being administered via these routes. Traditionally, autoinjectors, devices for im delivery of a self administered single dose of a drug are used in the military to protect personnel from chemical warfare agents and are currently used to deliver morphine for pain and atropine, diazepam and 2-PAM-Cl for first-aid against nerve agents. For this reason, most animal protection studies with OP bioscavengers have routinely been delivered im to rodents (Lenz et al., 2005; Mumford et al, 2010; Saxena, et al., 2011).

Despite all the pharmacokinetics data generated using im and sc routes of delivery of many drugs and biologics, little is known about the factors that govern the rate and extent of protein absorption from the injection site and the role of the lymphatic system in the transport of large molecules to the systemic circulation. With smaller molecules, the time to maximal concentration is usually shorter following im injections compared to sc injections where absorption is slow and prolonged and accounts for the lag in entering the blood. However with larger therapeutic molecules (MW>16KDa), the lymphatics are thought by some groups to be the primary route of absorption from sc (and im) injection sites. Large molecules are thought to exit the interstitium via cleft like openings into the lymph and enter the systemic circulation via the thoracic duct (Supersaxo et al., 1990; Porter et al., 2001; McLennan et L., 2006). To assess the effects of different routes of delivery, pharmacokinetic behaviour using different doses of PEG-rMaBChE tetrameric molecules was compared in monkeys following im and sc injections.

4.3.1 Intramuscular delivery of PEG-rMaBChE

Four monkeys each received an im injection of either 2.5 or 3 mg/kg of PEG-rMaBChE. As shown in Fig. 4, unlike the delivery of the smaller native HuBChE which appear to behave uniformly following im injection (Lenz et al., 2005), the much larger PEG-conjugated form exhibits very variable results when delivered into the muscle with Tmax values in the 4 macaques having values of 8, 24, 48 and 48 hr respectively; the 8-hour peak looking more like an iv injection than an im injection. It is not clear whether this more rapid exit from the muscle injection site into the blood reflects a more vascularised muscle or whether im delivery has more potential to damage blood vessels and promote faster draining. It is clear however that delivery of large doses of a therapeutic such as PEG-rHuBChE will require many im injections to achieve required peak values and will increase the likelihood of targeting a blood vessel. The stoichiometric dose of BChE required to protect humans against 2 LD_{50} of soman has been considered to be 3 mg/kgm (200 mg/70 kg); the antidotal efficacy of BChE being contingent upon both the rate of OP detoxification and its levels in blood (Raveh, 1997; Ashani & Pistinner, 2004). It would be unlikely that Cmax values (20 and 23 U/ml at 3 mg/kg and 17 and 10 U/kg at 2.5 mg/kg) following im administration would be sufficient for protection. In addition, the variable times of peak enzyme make it difficult to choose a time for prophylactic dosing.

4.3.2 Subcutaneous delivery of PEG-rMaBChE

Extensive pharmacokinetics have been performed on many well known biologics in monkeys and humans, either PEGylated or unmodified, using the sc routes of delivery (Boelaert et al., 1989; Ramakrishnan et al., 2003; Heatherington et al., 2001; Radwanski et al., 1987; Mager et al., 2005), although extrapolation from these studies may be problematic because all used considerably smaller molecules than native or PEG-rBChE. Generally, sc injections have been the delivery route of choice for compounds with limited oral bioavailability, as a means of modifying or extending the release profiles of these molecules, or as a means of delivering drugs that require large quantities (Yang, 2003) since larger volumes may be injected. In one case, a highly concentrated form of a therapeutic requiring large doses for its effects has be prepared as a crystalline and successfully delivered sc in a small volume (Yang et al. 2003).

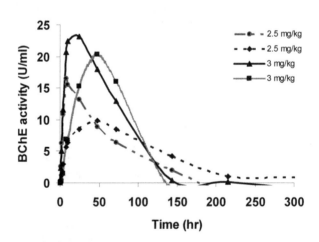

Fig. 4. Pharmacokinetic profiles of PEG-rMaBChE delivered by im injection. Four monkeys were injected into the thigh muscles using a 1-ml syringe.

Figure 5 shows the pharmacokinetic profiles following sc delivery of the tetrameric PEG-rMaBChE at 2.5, 3 and 5 mg/kg. Tmax values were all consistently ~48 hrs, regardless of the dose. However, while Cmax was generally associated with dose, there was a good deal of overlap between the 3 mg/kg and 5mg/kg doses; the larger doses being retained at higher levels in the blood for many days. This once again raises the question as to whether a high dose of very large molecules can leave the site of the sc injection and enter the blood at levels required for protection. By contrast 3 mg/kg delivered iv reaches a peak of >50 U/ml. It is important to note that despite the apparent low bioavailability of sc administered proteins compared to those given intravenously (17-65%), sc delivery often produces equivalent efficacy to iv administration and is assumed to be due to prolonged absorption leading to reduced receptor saturation.

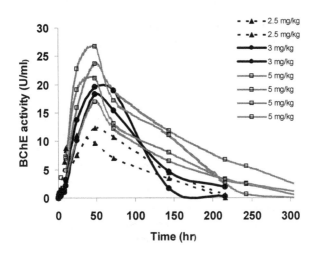

Fig. 5. Pharmacokinetics of PEG-rMaBChE delivered by sc injection. Eight monkeys were injected with the doses indicated in 2-4 ml sc between the shoulder blades.

A direct comparison of the pharmacokinetic parameters following im versus sc injections of 4 monkeys at does of 2.5 mg/kg and 3 mg/kg is shown in Table 3 and indicates that the im and sc values are quite similar. Overall, the results indicate that for a very high MW protein such as PEG-rMaBChE or PEG-rHuBChE, neither im or sc administrarion are optimal to achieve good plasma retention with high PK parameters. For this reason, a different non-parenteral route of delivery via the lung, where the high MW becomes an advantage, is now the choice route of delivery.

Parameters	Subcutaneous injection				Intramuscular injection			
	Four individual monkeys				Four individual monkeys			
MRT (h)	62.23	90.12	110.2	73.4	49.37	60.99	58.6	108.0
T1/2 (h)	25.2	42.3	77.8	37.8	23.3	19.4	24.0	58.7
Cmax (U/ml)	19.6	18.3	12.3	11.0	23.1	20.3	16.5	9.8
AUC (U/ml·h)	1706	1856	1489	1128	1762	1675	1089	1367

Table 3. Comparison of the pharmacokinetics parameters four following sc and im injections performed in parallel.

4.4 Protection studies with PEG-rMaBChE

Many studies have demonstrated efficacy of native HuBChE, both pre-and post–exposure, in rodents and monkeys to protect against OP nerve agents delivered by sc injection, iv injection or vapour. (Lenz et al., 2005; Sun et al., 2008; Saxena et al., 2011; Mumford et al, 2010). Protection has also been shown in mice and guinea pigs using PEG-rBChE produced in goat and plants (Huang et al. 2007, Geyer et al., 2010). However, very few studies have utilized the non-human primate monkey model for assessing insecticide toxicity and none have used respiratory exposure.

Two types of protection studies using different routes of delivery are currently being performed to assess the ability of BChE to protect against toxicity resulting from exposure to the insecticide paraoxon.

1.   Aerosolized PEG-rMaBChE 1 hr prior to aerosolized paraoxon exposure.
2.   Intravenous delivery of PEG-rMaBChE 1 hr prior to sc delivery of paraoxon.

4.4.1 Paraoxon

The majority of OP insecticides are lipophilic, not ionised, and are absorbed rapidly following inhalation or ingestion (Vale, 1998). Dermal absorption is slower and can be prevented by removing clothes and bathing, but severe poisoning may still ensue if exposure is prolonged. Respiratory pesticide exposure by inhalation of powders, airborne droplets or vapours is particularly hazardous because pesticide particles can quickly enter the bloodstream via the lungs and cause serious damage. Under low pressure, droplet size is too large to remain airborne. However, when high pressure, ultra low volume application (ULV) or fogging equipment is used for agricultural purposes, respiratory exposure is increased due to the production of mist- or fog-size particles, which can be carried on air currents for a considerable distance (Armed Forces Pest Management Board Technical Guide No. 13). Small children are highly vulnerable because they breathe in greater volumes of air, relative to their body weight, than adults.

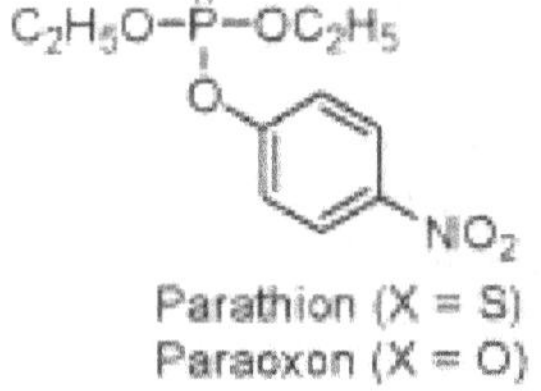

Parathion (X = S)
Paraoxon (X = O)

Fig. 6. Chemical structure of parathion and paraoxon.

Paraoxon is the active metabolite of the inactive parathion (Fig. 6) produced by a sulfur-for-oxygen substitution carried out predominantly in the liver by the mixed-function oxidases (Dauterman, 1971). It was chosen for these studies because it inhibits AChE, BChE and carboxylesterase (Levine, 2006), it has a relatively low LD_{50}, and low volatility and stability in aqueous solution. Parathion has probably been responsible for more cases of accidental poisoning and death than any other OP insecticide (Lotti & Moretto, 2005) and was recently phased out of use in the US. In humans, parathion is absorbed via skin, mucous membranes, and orally and is rapidly metabolized to paraoxon which can result in headaches, convulsions, poor vision, vomiting, abdominal pain, severe diarrhea, unconsciousness, tremor, dyspnea and finally lung-edema as well as respiratory arrest. Symptoms of severe poisoning are known to last for extended periods of time, sometimes months. Additionally, peripheral neuropathy including paralysis is noticed as late sequelae after recovery from acute intoxication (http://extoxnet.orst.edu/pips/parathio.htm). Parathion has been extensively used for committing suicide and potentially for the deliberate killing of people.

4.4.2 Aerosolized PEG-rMaBChE protection against aerosolized paraoxon exposure

As an alternative to delivering high doses of a large molecule into the systemic circulation by sc or im routes, studies are currently being performed using aerosol therapy for delivering rBChE directly to the lung in order to create an effective "pulmonary bioshield" that will detoxify incoming inhaled insecticide in situ and prevent or reduce respiratory toxicity. This takes advantage of the large size of the molecule which will be retained in the lung due to its inability to pass through the lung endothelium into the blood. In this context, inhalation serves as a major means of intoxication because of rapid accesses of the OP to the blood. An efficient pre-exposure pulmonary therapeutic in the form of aerosolized PEG-rBChE could be delivered before a known use/release of insecticides and prevent the lung damage and delayed neuropathy often associated with exposure, while reducing the need for post-exposure atropine and oximes.

Maxwell et al. (2006) have recently shown that for OP compounds (including the insecticides paraoxon, DFP and dichlorvos) the primary mechanism of in vivo toxicity is the inhibition of AChE and the residual unexplained variation in OP toxicity represents <10% of the total variation in toxicity. Almost all of the variation in the LD_{50} of OP compounds in rats was explained by the variation in their in vitro rate constants for inhibition of AChE. Thus, to develop a paraoxon/monkey animal model for aerosolized insecticide exposure and to avoid unnecessary stressing and killing of monkeys in developing the model, the dose of aerosolized paraoxon required to achieve a ~50% inhibition of RBC AChE and serum BChE has been used initially as a readout for toxicity and a basis from which to analyse protection by CHO-derived rMaBChE. Thus, paraoxon which is not neutralized in the lung will enter the blood and can be measured by the inhibition of AChE and BChE activity in lysed whole blood using using a modified assay (Ellman et al, 1961) with 5,5'-dithiobis(2-nitrobenzoic acid), the substrate acetyl-thiocholine (ATC) and 20uM etho-propazine to inhibit BChE activity.

Initially, the dose of aerosolized paraoxon required to produce ~50% inhibition of red blood cell (RBC) AChE and BChE in the circulation was first determined in mice prior to the macaque studies. The LD_{50} of paraoxon in rodents has been established using oral, percutaneous (pc) and subcutaneous (sc) routes (mice: 760 ug/kg orally; 270 - 800 ug/kg sc and for rats: 1800 ug/kg orally and 200 - 430 sc (reviewed in Levine, 2006; Villa et al., 2007). Milatovic et al. (1996) showed that a single acute injection of 0.09, 0.12, or 0.19 mg/kg paraoxon in rats, representing 40% LD_{50}, 52% LD_{50} and 83% LD_{50} respectively, did not produce signs of cholinergic hyperactively. In the present study, the effective dose of aerosolized paraoxon resulting in 50% inhibition in mice was found to be 150-180 ug/kg which is less toxic than the parenteral route. In addition, aerosolized BChE given 24 hr prior to the paraoxon significantly reduced the AChE inhibition (our unpub. data). Rodents contain a high endogenous levels of CaE, another stoichiometric OP scavenger (Dirnhuber et al. 1979) and are known to be ~10-fold less sensitive to soman than non-human primates (Maxwell et al., 2006). Accordingly, a dose of 15 ug/kg of aerosolized paraoxon has been shown to result in 50-60% RBC AChE inhibition and preliminary data indicate that PEG-rMaBChE , delivered as a pre-exposure aerosol one hour prior to exposure, can totally reduce this inhibition in a dose –dependent manner.

4.4.3 Intravenous PEG-rMaBChE protection against subcutaneous paraoxon exposure

These studies are being formed to compare routes of delivery with efficacy of protection and indicate that while paraoxon delivered sc is also more toxic than as an aerosol, complete protection can be achieved by PEG-rMaBChE pretreatment.

5. References

Alavanja, M. C. (2009). Pesticides Use and Exposure Extensive Worldwide. *Rev Environ Health*, 24(4):303-9.

Altamirano, C. V., Lockridge, O. (1999). Association of tetramers of human butyrylcholinesterase is mediated by conserved aromatic residues of the carboxy terminus. *Chem Biol Interact*. 119-120:53-60.

Altmann, F. (2007). The role of protein glycosylation in allergy. *Int Arch Allergy Immunol*. 142(2):99-115.

Ashani, Y., & Pistinner, S. (2004). Estimation of the upper limit of human butyrylcholinesterase dose required for protection against organophosphates toxicity: a mathematically based toxicokinetic model. *Toxicol Sci*. 77(2):358-67.

Blong, R. M., Bedows, E., & Lockridge, O. (1997). Tetramerization domain of human butyrylcholinesterase is at the C-terminus. *Biochem J*. 327 (Pt 3):747-57.

Boeck, A. T., Schopfer, L. M., & Lockridge, O. (2002). DNA sequence of butyrylcholinesterase from the rat: expression of the protein and characterization of the properties of rat butyrylcholinesterase. *Biochem Pharmacol*. 63(12):2101-10.

Boelaert, J. R., Schurgers, M. L., Matthys, E. G., Belpaire, F. M., Daneels, R. F., De Cre, M. J., & Bogaert, M. G. (1989). Comparative pharmacokinetics of recombinant erythropoietin administered by the intravenous, subcutaneous, and intraperitoneal routes in continuous ambulatory peritoneal dialysis (CAPD) patients. *Perit Dial Int*. 9(2):95-8.

Buckley, N. A., Eddleston, M., Li, Y., Bevan, M., & Robertson, J. (2011). Oximes for acute organophosphate pesticide poisoning. *Cochrane Database Syst Rev*. 16;(2):CD005085. Review.

Cao, J. L., Varnell, A. L., & Cooper, D. C. (2011). Gulf War Syndrome: A role for organophosphate induced plasticity of locus coeruleus neurons. *Nature Precedings*: hdl:10101/npre.2011.6057.1: Posted 23 Jun 2011.

Cascio, C., Comite Ghiara, M., Lanza, G., & Ponchione, A. (1988). Use of serum cholinesterases in severe organophosphorus poisioning. Our experience. *Minerva Anestesiol*. 54(7-8):337-8.

Castilho, A., Strasser, R., Stadlmann, J., Grass, J., Jez, J., Gattinger, P., Kunert, R., Quendler, H., Pabst, M., Leonard, R., Altmann, F., & Steinkellner, H. (2010). In planta protein sialylation through overexpression of the respective mammalian pathway. *J Biol Chem.* 285(21):15923-30.

Chilukuri, N., Sun, W., Naik, R. S., Parikh, K., Tang, L., Doctor, B. P., & Saxena, A. (2008a). Effect of polyethylene glycol modification on the circulatory stability and immunogenicity of recombinant human butyrylcholinesterase. *Chem Biol Interact.* 175(1-3):255-60.

Chilukuri, N., Sun, W., Parikh, K., Naik, R. S., Tang, L., Doctor, B. P., & Saxena, A. (2008b). A repeated injection of polyethyleneglycol-conjugated recombinant human butyrylcholinesterase elicits immune response in mice. *Toxicol Appl Pharmacol.* 231(3):423-9.

Chung, C. H., Mirakhur, B., Chan, E., Le, Q. T., Berlin, J., Morse, M., Murphy, B. A., Satinover, S.M., Hosen, J., Mauro, D., Slebos, R. J., Zhou, Q., Gold, D., Hatley, T., Hicklin, D. J., & Platts-Mills, T. A. (2008). Cetuximab-induced anaphylaxis and IgE specific for galactose-alpha-1,3-galactose. *N Engl J Med.* 358(11):1109-17.

Cohen, O., Kronman, C., Chitlaru, T., Ordentlich, A., Velan, B., & Shafferman, A. (2001). Effect of chemical modification of recombinant human acetylcholinesterase by polyethylene glycol on its circulatory longevity. *Biochem J.* 357(Pt 3):795-802.

Cohen, O., Kronman, C., Velan, B., & Shafferman, A. (2004). Amino acid domains control the circulatory residence time of primate acetylcholinesterases in rhesus macaques (Macaca mulatta). *Biochem J.* 378(Pt 1):117-28.

Colosio, C., Tiramani, M., Brambilla, G., Colombi, A., & Moretto, A. (2009). Neurobehavioural effects of pesticides with special focus on organophosphorus compounds: which is the real size of the problem? *Neurotoxicology.* 30(6): 1155-61.

Dauterman, W. C. (1971). Biological and nonbiological modifications of organophosphorus compounds. *Bull World Health Organ.* 44(1-3):133-50. Review.

Deshpande, L. S., Carter, D. S., Blair, R. E., & DeLorenzo, R.J. (2010). Development of a prolonged calcium plateau in hippocampal neurons in rats surviving status epilepticus induced by the organophosphate diisopropylfluorophosphate. *Toxicol Sci.* 116(2):623-31.

Diaz, S. L., Padler-Karavani, V., Ghaderi, D., Hurtado-Ziola, N., Yu, H., Chen, X., Brinkman-Van der Linden, E. C., Varki, A., & Varki, N. M. (2009). Sensitive and specific detection of the non-human sialic Acid N-glycolylneuraminic acid in human tissues and biotherapeutic products. *PLoS One.* 4(1):e4241.

Dirnhuber, P., French, M. C., Green, D.M., Leadbeater, L., & Stratton, J. A. (1979). The protection of primates against soman poisoning by pretreatment with pyridostigmine. J Pharm Pharmacol. 31(5):295-9.

Doctor, B. P., Maxwell, D. M., Ashani, Y., Saxena, A., & Gordon, R. K. (2001). New approaches to Medical Protection against Chemical Warfare Nerve Agents. Somani, SM and Romano, JA, Eds. CRC Press, New York, p191-214.

Ellman, G. L., Courtney, K. D., Andres, V. Jr., & Feather-stone, R. M. (1961). A new and rapid colorimetric determination of acetylcholinesterase activity. *Biochem Pharmacol.* 7:88-95.

Geyer, B. C., Kannan, L., Garnaud, P.E., Broomfield, C. A., Cadieux, C. L., Cherni, I., Hodgins, S.M., Kasten, S.A., Kelley, K., Kilbourne, J., Oliver, Z. P., Otto, T. C., Puffenberger, I., Reeves, T. E., Robbins, N. 2nd., Woods, R. R., Soreq, H., Lenz, D. E., Cerasoli, D. M., & Mor, T. S. (2010). Plant-derived human butyrylcholinesterase, but not an organophosphorous-compound hydrolyzing variant thereof, protects rodents against nerve agents. *Proc Natl Acad Sci U S A.* 107(47):20251-6.

Gleba, Y., Klimyuk, V., & Marillonnet, S. (2005). Magnifecti on--a new platform for expressing recombinant vaccines in plants. *Vaccine.* 23(17-18):2042-8. Review.

Goodin, M. M., Zaitlin, D., Naidu, R. A., & Lommel, S. A. (2008). Nicotiana benthamiana: its history and future as a model for plant-pathogen interactions. *Mol Plant Microbe Interact.* 21(8):1015-26. Review.

Grunwald, J., Marcus, D., Papier, Y., Raveh, L., Pittel, Z., & Ashani, Y. (1997). Large-scale purification and long-term stability of human butyrylcholinesterase: a potential bioscavenger drug. *J Biochem Biophys Methods.* 34(2):123-35.

Güven, M., Sungur, M,, Eser, B., Sari, I., & Altuntaş, F. (2004). The effects of fresh frozen plasma on cholinesterase levels and outcomes in patients with organophosphate poisoning. *J Toxicol Clin Toxicol.* 42(5):617-23.

Heatherington, A. C., Schuller, J., & Mercer, A. J. (2001). Pharmacokinetics of novel erythropoiesis stimulating protein (NESP) in cancer patients: preliminary report. Br J Cancer. 84 Suppl 1:11-6.

Huang, Y.J., Huang, Y., Baldassarre, H., Wang, B., Lazaris, A., Leduc, M., Bilodeau, A. S., Bellemare, A., Côté, M., Herskovits, P., Touati, M., Turcotte, C., Valeanu, L., Lemée, N., Wilgus, H., Bégin, I., Bhatia, B., Rao, K., Neveu, N., Brochu, E., Pierson, J., Hockley, D. K., Cerasoli, D. M., Lenz, D. E., Karatzas, C. N., & Langermann, S. (2007). Recombinant human butyrylcholinesterase from milk of transgenic animals to protect against organophosphate poisoning. *Proc Natl Acad Sci U S A.* 104(34):13603-8.

Jenkins, T., Balinsky, D., & Patient, D. W. (1967). Cholinesterase in plasma: first reported absence in the Bantu; half-life determination. *Science.* 156(783):1748-50.

Jeyaratnam, J. (1990). Acute pesticide poisoning: a major global health problem.*World Health Stat Q.,* 43(3):139-44.

Kang, J. S., Deluca, P. P., & Lee, K. C. (2009). Emerging PEGylated drugs. *Expert Opin Emerg Drugs.* 14(2):363-80. Review.

Kolarich, D., Weber, A., Pabst, M., Stadlmann, J., Teschner, W., Ehrlich, H., Schwarz, H. P., & Altmann, F. (2008). Glycoproteomic characterization of butyrylcholinesterase from human plasma. *Proteomics.* 8(2):254-63.

Kronman, C., Chitlaru, T., Elhanany, E., Velan, B., & Shafferman, A. (2000). Hierarchy of post-translational modifications involved in the circulatory longevity of glycoproteins. Demonstration of concerted contributions of glycan sialylation and subunit assembly to the pharmacokinetic behavior of bovine acetylcholinesterase. *J Biol Chem.* 275(38):29488-502.

Kronman, C., Cohen, O., Raveh, L., Mazor, O., Ordentlich, A., & Shafferman, A. (2007). Polyethylene-glycol conjugated recombinant human acetylcholinesterase serves as an efficacious bioscavenger against soman intoxication. *Toxicology.* 233(1-3): 40-6.

Lenz, D. E., Broomfield, C. A., Maxwell, D. M., & Cerasoli, D. M. (2001). Nerve Agent Bioscavengers: Protection against High- and Low- Dose Organophosphorus Exposure. Somani, SM and Romano, JA, Eds. CRC Press, New York, p215-243.

Lenz, D. E., Maxwell, D. M., Koplovitz, I., Clark, C. R., Capacio, B. R., Cerasoli, D. M., Federko, J. M., Luo, C., Saxena, A., Doctor, B. P., & Olson, C. (2005). Protection against soman or VX poisoning by human butyrylcholinesterase in guinea pigs and cynomolgus monkeys. *Chem Biol Interact.* 157-158:205-10.

Lenz, D. E., Yeung, D., Smith, J. R., Sweeney, R. E., Lumley, L. A., & Cerasoli, D. M. (2007). Stoichiometric and catalytic scavengers as protection against nerve agent toxicity: a mini review. *Toxicology.* 233(1-3):31-9. Review.

Levine, E. S. (2006). Nerve Agent Simulants: Can They Be Used as Substitutes for Nerve Agents in biomedical Research? Prepared for the U.S. Army Medical Research Institute of Chemical Defense under Contract No. GS-23F-8006H.

Li, H., Schopfer, L. M., Masson, P., & Lockridge, O. (2008). Lamellipodin proline rich peptides associated with native plasma butyrylcholinesterase tetramers. *Biochem J.* 411(2):425-32.

Litchfield, M. H. (2005). Estimates of acute pesticide poisoning in agricultural workers in less developed countries. *Toxicol Rev.*, 24(4):271-8. Review.

Lockridge, O. (1990). Genetic variants of human serum cholinesterase influence metabolism of the muscle relaxant succinylcholine.*Pharmacol Ther* 47: 35-60.

Lotti, M., & Moretto, A. (2005). Organophosphate-induced delayed polyneuropathy. *Toxicol Rev.* 24(1):37-49. Review.

Luo, C., Tong, M., Maxwell, D. M., & Saxena, A. (2008). Comparison of oxime reactivation and aging of nerve agent-inhibited monkey and human acetylcholinesterases. *Chem Biol Interact.* 175(1-3):261-6.

Mager, D. E., Neuteboom, B., & Jusko, W. J. (2005). Pharmacokinetics and pharmacodynamics of PEGylated IFN-beta 1a following subcutaneous administration in monkeys. *Pharm Res.* 22(1):58-61.

Mager, D. E., Woo, S., & Jusko, W. J. (2009). Scaling pharmacodynamics from in vitro and preclinical animal studies to humans. *Drug Metab Pharmacokinet.* 24(1):16-24. Review.

Matzke, S. M., Oubre, J. L. Caranto, G. R., Gentry, M. K., & Galbicka, G. (1999). Behavioral and immunological effects of exogenous butyrylcholinesterase in rhesus monkeys. *Pharmacol Biochem Behav.* 62(3):523-30.

Maxwell, D. M., Brecht, K. M., Koplovitz, I., & Sweeney, R. E. (2006). Acetylcholinesterase inhibition: does it explain the toxicity of organophosphorus compounds? *Arch Toxicol.* 80(11):756-60.

McLennan, D. N., Porter, C. J., Edwards, G.A., Heatherington, A. C., Martin, S. W., & Charman, S. A. (2006). The absorption of darbepoetin alfa occurs predominantly via the lymphatics following subcutaneous administration to sheep. *Pharm Res.* 23(9):2060-6.

Milatovic, D., & Dettbarn, W. D. (1996). Modification of acetylcholinesterase during adaptation to chronic, subacute paraoxon application in rat. *Toxicol Appl Pharmacol.* 136(1):20-8.

Millard, C. B., Kryger, G., Ordentlich, A., Greenblatt, H. M., Harel, M., Raves, M. L., Segall, Y., Barak, D., Shafferman, A., Silman, I., & Sussman, J. L. (1999). Crystal structures of aged phosphonylated acetylcholinesterase: nerve agent reaction products at the atomic level. *Biochemistry,* 38:7032-9.

Mumford, H., Price, M. E., Cerasoli, D.M., Teschner, W., Ehrlich, H., Schwarz, H.P., & Lenz, D. E. (2010). Efficacy and physiological effects of human butyrylcholinesterase as a post-exposure therapy against percutaneous poisoning by VX in the guinea-pig. *Chem Biol Interact.* 187(1-3):304-8.

Nachon, F., Nicolet, Y., Viguié, N., Masson, P., Fontecilla-Camps, J. C., & Lockridge, O. (2002). Engineering of a monomeric and low-glycosylated form of human butyrylcholinesterase: expression, purification, characterization and crystallization. *Eur J Biochem.* 269(2):630-7.

Ohayo-Mitoko, G. J., Kromhout, H., Simwa, J. M., Boleij, J. S., & Heederik, D. (2000). Self reported symptoms and inhibition of acetylcholinesterase activity among Kenyan agricultural workers. *Occup Environ Med.* 57(3):195-200.

Paccalet, T., Bardor, M., Rihouey, C., Delmas, F., Chevalier, C., D'Aoust, M. A., Faye, L., Vézina, L., Gomord, V., & Lerouge, P. (2007). Engineering of a sialic acid synthesis pathway in transgenic plants by expression of bacterial Neu5Ac-synthesizing enzymes. *Plant Biotechnol J.* 5(1):16-25.

Park, E. I., Mi, Y., Unverzagt, C., Gabius, H. J., & Baenziger, J. U. (2005). The asialoglycoprotein receptor clears glycoconjugates terminating with sialic acid alpha 2,6GalNAc. *Proc Natl Acad Sci U S A.* 102(47):17125-9.

Pepinsky, R. B., LePage, D. J., Gill, A., Chakraborty, A., Vaidyanathan, S., Green, M., Baker, D. P., Whalley, E., Hochman, P. S., & Martin, P. (2001). Improved pharmacokinetic properties of a polyethylene glycol-modified form of interferon-beta-1a with preserved in vitro bioactivity. *J Pharmacol Exp Ther.* 297(3):1059-66.

Porter, C. J., Edwards, G. A.,& Charman, S.A. (2001). Lymphatic transport of proteins after s.c. injection: implications of animal model selection. *Adv Drug Deliv Rev.* 50(1-2):157-71. Review.

Radwanski, E., Perentesis, G., Jacobs, S., Oden, E., Affrime, M., Symchowicz, S., & Zampaglione, N. (1987). Pharmacokinetics of interferon alpha-2b in healthy volunteers. *J Clin Pharmacol.* 27(5):432-5.

Ramakrishnan, R., Cheung, W. K., Farrell, F., Joffee, L., & Jusko, W. J. (2003). Pharmacokinetic and pharmacodynamic modeling of recombinant human erythropoietin after intravenous and subcutaneous dose administration in cynomolgus monkeys. . J Pharmacol Exp Ther. 306(1):324-31.

Raveh, L., Ashani, Y., Levy, D., De La Hoz, D., Wolfe, A. D., & Doctor, B.P. (1989). Acetylcholinesterase prophylaxis against organophosphate poisoning. Quantitative correlation between protection and blood-enzyme level in mice. *Biochem Pharmacol.* 38(3):529-34.

Raveh, L., Grauer, E., Grunwald, J., Cohen, E., & Ashani, Y. (1997). The stoichiometry of protection against soman and VX toxicity in monkeys pretreated with human butyrylcholinesterase. *Toxicol Appl Pharmacol.* 145(1):43-53.

Rosenberg, Y. J., Luo, C., Ashani, Y., Doctor, B. P., Fischer, R., Wolfee, G., & Saxena, A. (2002). Pharmacokinetics and immunologic consequences of exposing macaques to purified homologous butyrylcholinesterase. *Life Sci.* 72(2):125-34.

Rosenberg, Y. J., Saxena, A., Sun, W., Jiang, X., Chilukuri, N., Luo, C., Doctor, B. P., & Lee, K. D. (2010). Demonstration of in vivo stability and lack of immunogenicity of a polyethyleneglycol-conjugated recombinant CHO-derived butyrylcholinesterase bioscavenger using a homologous macaque model. *Chem Biol Interact.* 187(1-3):279-86.

Rosenberry, T. L., Mallender, W. D., Thomas, P. J., & Szegletes, T. (1999). A steric blockade model for inhibition of acetylcholinesterase by peripheral site ligands and substrate. *Chem Biol Interact.* 119-120:85-97.

Sainsbury, F., Lavoie, P. O., D'Aoust, M. A., Vézina, L.P., & Lomonossoff, G. P. (2008). Expression of multiple proteins using full-length and deleted versions of cowpea mosaic virus RNA-2. *Plant Biotechnol J.* 6(1):82-92.

Saxena, A., Ashani, Y., Raveh, L., Stevenson, D., Patel, T., & Doctor, B. P. (1998). Role of oligosaccharides in the pharmacokinetics of tissue-derived and genetically engineered cholinesterases. *Mol Pharmacol.* 53(1):112-22.

Saxena, A., Sun, W., Fedorko, J. M., Koplovitz, I., & Doctor, B. P. (2011). Prophylaxis with human serum butyrylcholinesterase protects guinea pigs exposed to multiple lethal doses of soman or VX. *Biochem Pharmacol.* 81(1):164-9.

Stoger, E., Sack, M., Nicholson, L., Fischer, R., & Christou, P. (2005). Recent progress in plantibody technology. *Curr Pharm Des.* 11:2439-57.

Strasser, R., Altmann, F., Mach, L., Glössl, J., & Steinkellner, H. (2004). Generation of Arabidopsis thaliana plants with complex N-glycans lacking beta1,2-linked xylose and core alpha1,3-linked fucose. *FEBS Lett.* 561(1-3):132-6.

Strasser, R., Stadlmann, J., Svoboda, B., Altmann, F., Glössl, J., & Mach, L. (2005). Molecular basis of N-acetylglucosaminyltransferase I deficiency in Arabidopsis thaliana plants lacking complex N-glycans. *Biochem J.* 387(Pt 2):385-91.

Sun, W., Doctor, B. P., & Saxena, A. (2005). Safety and pharmacokinetics of human serum butyrylcholinesterase in guinea pigs. Chem Biol Interact. 157-158:428-9.

Sun, W., Doctor, B. P., Lenz, D. E., & Saxena, A. (2008). Long-term effects of human butyrylcholinesterase pretreatment followed by acute soman challenge in cynomolgus monkeys. *Chem Biol Interact.* 175(1-3):428-30.

Sun, W., Luo, C., Naik, R. S., Doctor, B. P., & Saxena, A. (2009). Pharmacokinetics and immunologic consequences of repeated administrations of purified heterologous and homologous butyrylcholinesterase in mice. *Life Sci.* 85(17-18): 657-61.

Supersaxo, A., Hein, W. R., & Steffen, H. (1990). Effect of molecular weight on the lymphatic absorption of water-soluble compounds following subcutaneous administration. *Pharm Res.* 7(2):167-9.

Vale, J. (1998). Toxicokinetic and toxicodynamic aspects of organophosphorus (OP) insecticide poisoning. Toxicol Lett. 102-103:649-52.

Varki, A. (2001). N-glycolylneuraminic acid deficiency in humans. *Biochimie.* 83(7):615-22. Review.

Villa, A. F., Houze, P., Monier, C., Risède, P., Sarhan, H., Borron, S. W., Mégarbane, B., Garnier, R., & Baud, F. J. (2007). Toxic doses of paraoxon alter the respiratory pattern without causing respiratory failure in rats. *Toxicology.* 232(1-2): 37-49.

Wenderoth, I., & von Schaewen, A. (2000). Isolation and characterization of plant N-acetyl glucosaminyltransferase I (GntI) cDNA sequences. Functional analyses in the Arabidopsis cgl mutant and in antisense plants. *Plant Physiol.* 123(3):1097-108.

Worek, F., Aurbek, N., Herkert, N. M., John, H., Eddleston, M., Eyer, P., & Thiermann, H. (2010). Evaluation of medical countermeasures against organophosphorus compounds: the value of experimental data and computer simulations. *Chem Biol Interact.* 187(1-3):259-64.

Yang, M.X., Shenoy, B., Disttler, M., Patel, R., McGrath, M., Pechenov, S., & Margolin, A. L. (2003). Crystalline monoclonal antibodies for subcutaneous delivery. *Proc Natl Acad Sci U S A.* 100(12):6934-9.

7

Production of Insecticidal Baculoviruses in Insect Cell Cultures: Potential and Limitations

Juan D. Claus[1], Verónica V. Gioria[1],
Gabriela A. Micheloud[1] and Gabriel Visnovsky[2]
*[1]Laboratory of Virology; Facultad de Bioquímica y Ciencias Biológicas;
Universidad Nacional del Litoral,
[2]Chemical and Process Engineering, University of Canterbury
[1]República Argentina;
[2]New Zealand*

1. Introduction

The potential of baculoviruses to be employed as insecticides is known since more than 75 years ago (Benz, 1986). To date, over 30 different baculoviruses are used to control several insect plagues in agriculture, horticulture and forestry (Moscardi, 1999). The use of baculovirus as insecticides is based on a set of useful properties, such as pathogenicity, specificity, narrow host range, environmental persistence, ability to act synergistically with other natural enemies of the pest and ability to induce artificial epizootics. Despite these advantages, very few baculoviruses have become widely used as insecticides, standing out as some successful examples the use of the *Anticarsia gemmatalis multiple nucleopolyhedrovirus* (AgMNPV) to control the velvetbean caterpillar in soybean crops in Latin America, *Cidia pommonella granulovirus* (CpGV) to fight the codling moth attacks in fruit orchards, and *Spodoptera exigua multiple nucleopolyhedrovirus* (SeMNPV) to control the armyworm in vegetable crops under cover in Europe (Moscardi, 1999). The causes of the limited acceptance of baculoviruses as insecticides are diverse, including slow speed of action, problems to register and market these biological insecticides and difficulties to produce them at an appropriate scale.

The technologies currently used to produce insecticidal baculoviruses are based on the infection of susceptible insect larvae (Black et al., 1997). However, the implementation of processes to produce baculovirus in insect larvae is hampered by several limitations: high labour requirements, lack of expertise in standardization and validation of such processes, difficulties in scaling production to levels consistent with the profitability of the process and difficulties to properly control both the process production and product quality. While several improvements in production systems in insect larvae have been described in the last years which could help overcome some of the problems described above (van Beek & Davis, 2007), it has been also proposed that the adoption of an alternative technology based on the viral propagation in insect cell cultures could enable the development of well standardized, controlled and scalable production processes for insecticidal baculoviruses (Szewczyk et al., 2006).

The purpose of this chapter is to review the current state of the art about insect cell culture technology and its application to the production of viral insecticides belonging to the family *Baculoviridae*. The several restrictions still existing to develop feasible processes as well as the prospects for overcoming these limitations will be also reviewed.

2. The baculoviruses

2.1 Structure and classification

The baculoviruses (family *Baculoviridae*) are a group of arthropod-specific viral pathogens that have a circular and supercoiled double-stranded DNA genome (Rohrmann, 2011). The size of the genome ranges from 80 to 180 kbp. The genome is contained in a rod-shaped nucleocapsid with helical symmetry. The baculoviruses generate two different progenies, called budded virus (BVs) and occluded virus (OVs) that share the same nucleocapsid. Both viral progenies play different roles in the natural cycle of these viruses. BVs, consisting of a single nucleocapsid surrounded by an envelope derived from the cellular plasmatic membrane, are responsible for the transmission of the infection from cell to cell in an infected animal. In OVs, on the other hand, one or more nucleocapsids are contained by an envelope synthesized *de novo* in the nucleus of infected cells. These virions are then included in a crystalline protein matrix, consisting mainly of a single polypeptide which is product of the hyperexpression of a very late viral gene, resulting in the so-called occlusion bodies (OBs). The OBs, whose polypeptide structure gives protection to the OVs contained therein, are responsible for the transmission of the infection between susceptible animals in nature and, in fact, constitute the viral progeny useful as insecticide. The structures of BVs, OVs and OBs are shown in Figure 1.

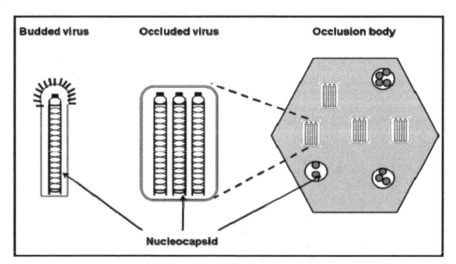

Fig. 1. schematic structures of budded virus, occluded virus and occlusion bodies of baculoviruses.

The members of the *Baculoviridae* family are classified into four different genera (Jehle et al., 2006). The viruses classified into the genus *a-baculovirus* are able to infect lepidopteran insects to produce nuclear polyhedrosis. The viruses belonging to the genus *β-baculovirus* can also infect lepidopteran insects, but they produce granulosis. On the other hand, the viruses classified into the genus *γ-baculovirus* can infect hymenopteran insects to produce nuclear polyhedrosis, while the viruses of the genus *δ-baculovirus* are associated to the production of nuclear polyhedrosis in dipteran insects. As most baculoviruses used as insecticides so far belong to the genus *a-baculovirus*, the subject henceforth will be focused on lepidopteran nucleopolyhedroviruses.

2.2 Natural cycle and pathogenicity

The main route of infection of lepidopteran larvae with nucleopolyhedrovirus is the ingestion of food contaminated with viral OBs (Granados & Williams, 1986). Once ingested, OBs are transported to the larvae´s midgut, where they are dissolved to release the occluded virions, due to the combined action of the alkaline environment and the presence of alkaline proteases. The released OVs pass through the midgut peritrophic matrix and find the brush border membrane of the columnar midgut epithelial cells, which fuse with the viral envelope to enter the viral nucleocapsids within the cytoplasm. The ability of OVs to infect midgut epithelial cells is dependent on the expression of a set of genes whose products are denominated "*per os* infectivity factors" (PIFs) (Rohrmann, 2011). Most nucleocapsids are then transported to the nucleus through a process that is dependent on actin polymerization. Once the nucleocapsid has been entered into the cell nucleus, the viral DNA is naked and starts the transcriptional cascade that lead ultimately to the assembly of progeny nucleocapsids. A distinctive feature of the primary replication of nucleopolyhedroviruses in the midgut is that the nucleocapsids that were assembled in the nucleus are almost totally exported to the basal cytoplasmatic membrane, from where they finally egress as BVs. The budding of BVs occurs at regions of the plasmatic membrane that have been modified by insertion of the glycoproteins characteristic of the BV progeny, GP64 and/or F. Then, BVs would cross the basal lamina to begin the dissemination process that leads to secondary systemic infections. Alternatively, it has been proposed that baculoviruses can reach the main insect cavity through previous infection of tracheal cells (Pasarelli, 2011). Secondary infections, that affect almost all insect tissues, start when BVs penetrate into cells through a receptor-mediated endocytosis process. After the fusion of the viral envelope with the membrane of acidified endosomes, viral nucleocapsids are released into the cytoplasm and then transported to the nucleus, where the viral genome is naked. Differently from what occur in midgut epithelial cells, the transcriptional cascade in secondary infections drives the replication process to the production of nucleocapsids that, besides of feeding the budding of BVs, are assembled into the nucleus to form OVs. The assembly of OVs implies the retention of nucleocapsids inside the nucleus and the acquisition of an envelope synthesized *de novo* at the expense of the inner nuclear membrane. The OVs are finally occluded inside a crystalline matrix consisting mainly of the viral protein polyhedrin , whose gene is expressed at very high levels during the very late stage of the transcriptional cascade. The products of the occlusion process are the OBs, which are retained inside the nucleus until the death and lysis of the infected cell. At the end of the pathogenic process, the infected insect is full of OBs that, after its death and liquefaction of its tissues, are released into the environment to restart the cycle again.

The symptoms of the disease associated to baculovirus infection are not usually apparent during the first days post-infection (Granados & Williams, 1986). The change of color and the altered behavior of infected insects are often the earliest signs of infection with α-baculovirus. The lack of appetite, which after several days culminates in the total interruption of feeding, is another sign of infection. The growth of infected larvae is delayed with respect to uninfected controls, and the death occurs after several days post-infection. The length of the interval of time between infection and death of the insect varies between 3 days and 3 weeks, and depends on many factors, including larval age, nutritional status, dose of virus and virulence of the viral strain, as well as on environmental factors. Anyway, nor the cessation of ingestion or death occur immediately to infection, facts that constitute strong constraints to the acceptance of baculoviruses as insecticides.

2.3 The transcriptional regulation of the baculovirus replication cycle and the production of viral progenies

The replication cycle of baculoviruses is mainly regulated at the transcriptional level. The baculovirus transcriptional program occurs in three stages, called early, late and very late, respectively, which are coordinated in a cascade (Table 1) (Blissard & Rohrmann, 1990; Rohrmann, 2011).

Time post-infection *	RNA polymerase	Transcriptional stage	DNA replication	BVs production	OVs production	OBs assembly
0 – 8 hs.	Cellular RNA polymerase II	Early	-	-	-	-
8 – 18 hs.	Viral RNA polymerase	Late	+++	+++	-	-
From 18 hs.	Viral RNA polymerase	Very late	-	+	+++	+++

Table 1. Cascade of transcriptional events during the replication of baculoviruses and temporal distribution of production of viral progenies. *AcMNPV replication.

Each transcriptional stage is characterized by the expression of a specific group of genes. Early transcription begins immediately after the parental viral DNA is naked into the nucleus. It is carried out by the cellular RNA polymerase II, and includes a set of genes whose products are trans-activators and enzymes that will then have a role in viral DNA replication. As an exception, in the early stage is also transcribed the *gp64* gene whose expression´s product is the principal glycoprotein of the group I nucleopolyhedrovirus´ envelope. The end of the early stage is determined by the onset of viral DNA synthesis, and involves a change in the pattern of transcription by which some of the genes initially transcribed are no longer expressed, and a new set of genes begins to be transcribed by a viral RNA polymerase, starting the late stage of transcription. The late stage involves the expression of genes whose products are proteins and glycoproteins that are part of the structure of the budded virions, which are assembled and released from the infected cell during this time. Finally, there occurs a further change in the pattern of transcription, whose most notable feature is the hyperexpression of genes whose products are proteins and glycoproteins involved in the assembly of occluded virions and occlusion bodies, such as polyhedrin and P10. The RNA polymerase involved in the very late transcription is the same viral enzyme in charge of the late transcription. Although some BVs are produced during the very late stage, the hallmark of this period is the assembly of occlusion bodies, which extends up to the cell death.

3. Technological applications of baculoviruses

Baculoviruses are arthropod-specific pathogens, with a host range that is generally very narrow, and lack the ability to replicate and produce pathogenic effects in other animals and plants, properties that have promoted their use as safe insecticides with reduced environmental impact (Huber, 1986). In addition, the expression of very late genes in the baculovirus genome is under the control of regulatory elements with a very high promoter activity, a property that has allowed the development of one of the most widely used expression systems for the production of recombinant proteins, the baculovirus vector expression system (Luckow & Summers, 1988). Also, baculoviruses are able to penetrate into mammalian cells, although they can not replicate into them. This property permits the use of these viruses as vectors for gene delivery (Kost & Condreay, 2002). Baculoviruses also exhibit a potent immunostimulating activity in mammals, opening the possibility of their use as adjuvants in the formulation of novel vaccines (Abe & Matsuura, 2010). Some of these applications have yet to demonstrate its market potential, but others are a reality and products based on these viruses are used today in agriculture, veterinary medicine and human medicine, among others.

3.1 Baculoviruses as insecticides

Baculoviruses presents a number of advantages over traditional synthetic chemical insecticides (Moscardi, 1999). Their high specificity makes them safe for other insects, and thus helps to preserve and even enhance the natural mechanisms of plagues control. In addition, although baculoviruses can infect mammalian cells, including human cells, they can not replicate in them and therefore they lack of pathogenicity for the human being and other animals, making safe their use. The multiplication in their natural hosts, and their capacity to persist in the environment make them suitable for the inoculative control of plagues in forestry. In addition, the same properties coupled with the aforementioned preservation of natural enemies, permit the reduction of the number of the applications needed to keep the insect plague under control in annual crops, thus contributing to reduce the costs of protection. Finally, its use in replacement of synthetic insecticides helps to reduce the overall levels of chemical pollution.

The use of baculoviruses as insecticides also has limitations. Their high specificity is also a disadvantage to their widespread use, since they are only useful when the damage to the crop to be protected is produced predominantly by a single insect, and they are not effective in controlling pest complexes. The insecticidal effect of baculoviruses is not evident immediately after application, and the delay usually is accompanied by an increase of the level of crop damage. This defect can be counteracted by an earlier application of the virus, but it requires a close quantitative following of the insect population. Also, since baculovirus production processes are based on viral replication in living hosts, their yields can not match the high yields obtained at relatively low costs in the synthesis of chemical insecticides. Table 2 presents a list of selected baculoviruses belonging to the genus *a-baculovirus* registered for their use as insecticides. As can be observed, most of them are rather specific for one lepidopteran specie, except AcMNPV that has a wider host range.

Virus	Insect target	Crops
AgMNPV	*Anticarsia gemmatalis*	Soybean
AcMNPV	*Autographa californica, Trichoplusia ni, Pseudoplusia includens*, etc.	Cotton, cabbage, tomato, broccoli
HzSNPV	*Helicoverpa zea, Heliothis virescens*	Cotton, corn, tomato, vegetables
HaNPV	*Helicoverpa armigera*	Cotton, tomato
LdMNPV	*Lymantria dispar*	Forest
MbMNPV	*Mamestra brassicae*	Cabbage
OpMNPV	*Orgya pseudotsugata*	Forest
SeMNPV	*Spodoptera exigua*	Vegetables
SlNPV	*Spodoptera littoralis*	Cotton
SpltMNPV	*Spodoptera litura*	Vegetables, cotton

Table 2. Examples of *a-baculovirus* registered for insecticidal use.

3.2 Genetically modified baculoviruses as improved insecticides

As explained above, the adoption of baculoviruses as insecticides is limited by some of its pathogenic properties. One of the strategies developed to overcome this limitation is the modification of the viral genome in order to improve the insecticidal capabilities of the modified virus. To this end, two alternative routes have been followed: the insertion of foreign genes whose expression gives the virus an increased virulence, and the deletion of viral genes responsible for the delay in the evolution of the viral pathogenic process (reviewed by Szewczyk et al., 2006). The genes corresponding to several specific insect toxins, hormones or enzymes have been cloned and expressed in different baculoviruses, resulting in most cases in increased virulence, decreased time to insect death, and decreased plant damage. Besides, the deletion of the viral gene codifying for the ecdysteroid glycosil transferase (*egt*) - involved in the metabolism of the hormone ecdysone - also resulted in reduction of food consumption and faster killing of infected larvae.

Although the genetic modification has demonstrated to be a promissory strategy to improve the insecticidal ability of these viruses, the public perception about the risks that would involve the field release of recombinant viruses has limited the interest in developing novel insectides based on genetically modified baculoviruses. In fact, although have elapsed 20 years since the first publications that described the development of genetically modified baculovirus with enhanced insecticidal activity, no one product based on these recombinant viruses has yet come to market, and companies that were involved initially in these developments have canceled the processes for obtaining approvals for its use.

4. Insect cell culture technology

4.1 A brief history

The first studies conducted *in vitro* on tissues of invertebrate animals were made by Goldschmidt in 1915 (Day & Grace, 1959). Thereafter, and for about 40 years, attempts to multiply insect cells and tissues *in vitro* have had limited success. After completion of the Second World War, and already having the air filtration technology that permitted the safe handling of animal cell cultures in sterile environments, the work of Wyatt et al. (1956) on the chemical composition of the insect hemolymph allowed the development of the first culture media specifically designed for the cultivation of lepidopteran cells (Grace, 1958). The establishment of the first insect cell lines obtained from tissues of lepidopteran insects was an achievement reached by Grace (1962). Since then until now, at least half thousand cell lines, from different insects and distinct tissues have been established. A milestone in this process was the establishment of the cell line IPLB-Sf21 (Vaughn et al., 1977). This cell line, used to plaque baculovirus for the first time, exhibited relevant technological properties, such as the ability to grow indistinctly in static cultures and in agitated suspension cultures. Also, IPLB-Sf21 was the insect cell line where the clone Sf9 was produced from. This clone was closely linked to the development of the baculovirus – insect cell expression system for recombinant proteins (Summers & Smith, 1983). At the same time, new culture media were developed, such as MM, TC-100, TNM-FH and IPL-41, and the insect haemolymph that was initially used to supplement them, was replaced by fetal calf serum. At the end of the 80's, two important developments opened the possibility of expanding the cultivation of insect cells to an industrial scale: first, the development of microemulsions of lipids and sterols allowed the formulation of serum-free media, and second, the demonstration of the protective effect of surfactant poli-alcohols on the integrity of insect cells in suspension cultures aerated by sparging permitted the scaling-up to large stirred tank reactors and airlift reactors (Maiorella et al, 1988). In recent years, the main contributions to the technology of cultivation of insect cells have come from the development of genetically modified cell lines, capable, for example, to produce proteins with humanized molecular structures (Shi & Jarvis, 2007).

4.2 Insect cell lines

In a process of baculovirus production in cell cultures is crucial to make a proper selection of the cell line to be used as substrate for virus multiplication. The selected cell line must be susceptible and permissive to the virus, which must replicate in abundance to produce high yields of both budded virus and occlusion bodies. Preferably, nutritional requirements and metabolism should be well characterized, and the cell line should show relevant technological properties such as adaptability to suspension cultures, capability to grow in a low-cost serum-free medium and ability to grow in industrial bioreactors. Furthermore, it should be genetically stable, and should not be a source of viral variability.

Currently there are hundreds of cell lines established from tissues and organs of lepidopteran insects (Lynn, 2007), but very few meet the requirements described above. Table 3 shows a list of the lepidopteran insect cell lines more used for producing wild-type and/or recombinant baculovirus. The IPLB-Sf21 cell line and its clone Sf9 have been used intensively and they are well characterized. They can grow in suspension cultures at high cell concentration in bioreactors and there are several serum-free media available for them. Both wild-type and recombinant AcMNPV replicate very well in Sf cell lines

(O'Reilly et al., 1994). In addition, these cell lines had shown to be susceptible and permissive to the replication of other baculoviruses (Claus et al, 1993). The cell line BTI-TN-5B1-4, known commercially as High Five®, is also being used widely to produce recombinant proteins due to its susceptibility to AcMNPV and elevated specific productivity (Chung & Shuler, 1993).

Cell line	Insect	Tissue of origin	Susceptibility to baculoviruses
IPLB-Sf21 / Sf9	*Spodoptera frugiperda*	Ovarioles	AcMNPV, AgMNPV, SfMNPV, SlNPV, TnSNPV
BTI-TN-5B1-4	*Trichoplusia ni*	Embryos	AcMNPV, AgMNPV, TnSNPV
BCIRL-HZ-AM1	*Heliothis zea*	Ovarioles	AcMNPV, HaNPV, HzSNPV
Bm5	*Bombyx mori*	Ovarioles	AcMNPV, BmNPV
IPLB-LdEIta	*Lymantria dispar*	Embryos	AcMNPV, LdMNPV
saUFL-AG-286	*Anticarsia gemmatalis*	Embryos	AcMNPV, AgMNPV

Table 3. Lepidopteran insect cell lines used frequently to produce wild-type and recombinant baculoviruses.

Besides the widely used cell lines that were mentioned in the preceding paragraph, other cell lines have been used more narrowly in processes involving the multiplication of other baculoviruses. In general, these lines were established from tissues of the natural host of the baculovirus to replicate, because viral yields tend to be higher in infected cultures of these homologous cell lines than in the heterologous ones. For instance, the cell lines saUFL-AG-286 (*Anticarsia gemmatalis*) (Sieburth & Maruniak, 1988), BM5 (*Bombyx mori*) (Grace, 1967), BCIRL-HZ-AM1 (*Heliothis zea*) (McIntosh & Ignoffo, 1981) and IPLB-LdEIta (*Lymantria dispar*) (Lynn et al, 1988) have been used to produce specifically the viruses AgMNPV, BmNPV, HaSNPV and LdMNPV, respectively. However, these cell lines are not as well characterized as the most widely used lines, and their technological properties (adaptation to suspension, ability to grow in serum-free culture media) are less remarkable or yet unknown.

4.3 Nutrition and metabolism of lepidopteran insect cells
Most existing data on nutrition and metabolism of lepidopteran insect cells in culture has been obtained from studies of a few cell lines, mostly Sf9 and BTI-TN-5B1-4. The comparative analysis of that information permits to conclude that each cell line is considerably flexible for satisfying their nutritional requirements. However, there are marked differences in metabolic behavior between different cell lines. Carbohydrates and amino acids are the most important nutrients, and the knowledge about their quantitative demands and metabolism will be briefly reviewed below. The information available about

the quantitative requirements of other important nutrients, such as lipids, sterols, vitamins and mineral salts, is much scarce.

Carbohydrates are essential components of all culture media for insect cells, due to their role as main sources of carbon and energy. Insect cells are capable to grow in culture media containing glucose as the unique carbohydrate, but insect cells can also consume other monosaccharides and disaccharides (Mitsuhashi, 1989). Sf9 cells consume glucose without production of lactate. This behavior has been attributed to the existence of an active tricarboxylic acid cycle, where 70 – 80% of the consumed glucose would be totally oxidized (Neermann & Wagner, 1996). However, these results are still controversial, because other studies have found that the percentage of glucose consumed that is derived to the tricarboxylic acid cycle is much lower (Benslimane et al., 2005). The production of other metabolites such as alanine, glycerol and ethanol, as well as fatty acid synthesis, could explain the fate of a significant fraction of carbon incorporated to Sf9 cells through glucose consumption (Drews et al., 2000; Bernal et al., 2009). BTI-TN-5B1-4 and BM5 cell lines have a different behavior: both cell lines produce lactate, displaying a behavior similar to that of mammalian cells (Rhiel et al., 1997; Stavroulakis et al., 1991). The flux of glucose was studied in BTI-TN-5B1-4 cells, resulting that the proportion oxidized in the tricarboxylic acids cycle was lower than in Sf9 cells (Benslimane et al., 2005). The influence of baculovirus infection on the metabolism of insect cells, and specifically on glucose metabolism, is a topic that has been scarcely addressed, and the information is contradictory (Bernal et al., 2009; Gioria et al., 2006).

According to Mitsuhashi (1989), 15 amino acids are essential to insect cells. Glutamine is the amino acid consumed faster and in a greater extension in cultures of most lepidopteran insect cell lines characterized to date. However, it has been demonstrated that glutamine is dispensable for the growth of Sf9 cells, providing that the cells have ammonium as nitrogen source (Öhman et al., 1996). Other amino acids that can be consumed in significant quantities are asparagine, aspartate, glutamate and serine, and precisely asparagine is consumed faster than glutamine in cultures of the BTI-TN-5B1-4 cell line (Rhiel et al., 1997). The demand of other amino acids is much lower. Besides serving as precursors for the synthesis of proteins and nucleic acids, amino acids that are consumed faster are used as sources of energy, such as glutamine and asparagine. The metabolism of glutamine has been studied in Sf9 cells, where the utilization pathways are depending on the availability of glucose (Drews et al., 2000). It has been proposed that, in glucose excess, the cytoplasmatic enzyme glutamate synthase transfers the amidic nitrogen of glutamine to the amine-position in glutamate, from where it is transaminated to alanine, the main product of glutamine metabolism. But when glucose is exhausted, glutamine is metabolized in mitochondria, where the amide-nitrogen and the amine-nitrogen of glutamine are sequentially released as ammonium ion, which accumulates as the main product of glutamine metabolism under glucose limitation in Sf9 cells. The metabolism of glutamine and other amino acids used as energy sources in other cell lines, such as BTI-TN-5B1-4 and saUFL-AG-286, is probably regulated differently, because they produce ammonia even in the presence of excess glucose (Gioria et al., 2006; Rhiel et al., 1997). The information about the influence of baculovirus infection on glutamine requirement and metabolism is scarce, but most results appear to indicate that insect cells tend to reduce the demand after infection (Gioria et al., 2006; Bernal et al., 2009).

4.4 Culture media for lepidopteran insect cells

Almost all media used for cultivation of insect cells have a chemical composition partially defined. They consist of a basal medium composed of chemically defined mixtures of carbohydrates, amino acids, vitamins, salts and, in some cases, organic acids. In addition, the media must be supplemented with compounds of undefined chemical composition that contribute to cell´s proliferation, such as fetal calf serum, microbial extracts and/or protein hydrolysates.

4.4.1 Basal culture media

TC-100, IPL-41, Grace and TNM-FH are the most commonly used basal media for *in vitro* culture of lepidopteran insect cells (Schlaeger, 1996).TC-100 and TNM-FH have almost the same amino acids composition as the Grace medium, from which they originated. IPL-41 has the same qualitative composition of amino acids, but with higher concentration for the ones consumed faster: glutamine, asparagine, glutamic acid, aspartic acid and cystine. In any case, it has been demonstrated for several different insect cell lines that the concentrations of most amino acids in basal culture media are oversized with respect to cellular requirements (Bédard et al., 1993; Ferrance et al., 1993; Lua & Reid, 2003) . TC-100 and TNM-FH contain also protein hydrolysates, and the last is additionally supplemented with yeast extract. With regard to carbohydrates, all media contain glucose, although at different concentrations, and IPL-41 contains also an additional monosaccharide, fructose. Grace, TNM-FH and IPL-41 media also contain disaccharides, all of them contain sucrose and the last also maltose. IPL-4 is richer in vitamins and also contains organic acids, not present in the original formulations of the media Grace, TNM-FH and TC-100. The composition of mineral salts of IPL-41 differs from the other three, and is also enriched with oligoelements.   In spite of the differences in the chemical composition described above, the four basal media can support the growth of Sf9 and BTI-TN-5B1-4 cell cultures, provided they are supplemented with fetal calf serum. This fact highlights the nutritional plasticity of these cell lines. On the other hand, the richer medium IPL-41 has demonstrated to be better suited than the other three to formulate serum-free media (Ikonomou et al., 2001; Maiorella et al., 1988).

4.4.2 Fetal calf serum

Blood serum obtained from bovine fetuses is the most used undefined supplement for animal cell cultures, including both mammalian and insect cells. It concentrates, in a single component, several essential functions for cultured cells (Barnes & Sato, 1980). Serum provides, in a water-soluble vehicle, lipids and cholesterol, and it is also a rich source of growth factors, vitamins´ and mineral oligoelements. Its proteins´ transporters allow the supply of poorly soluble ions, such as Fe^{+++}. In addition, serum has detoxifying activity, and their proteins can contribute to the preservation of the structural integrity of cells when they are subjected to mechanical stress.

In spite of the advantages of fetal calf serum, they are accompanied by several disadvantages. Its composition is undefined and variable from batch to batch. It is also a possible vehicle for the introduction of chemical and biological contaminants, such as plaguicides, virus and prions, among others. The high concentration of serum proteins may involve interference with the extraction and purification of products. Besides, serum proteins are a source of foam in culture processes where aeration is made by sparging. The cost of using fetal bovine serum as a supplement of a basic medium is so high that it can reach 90% of the final cost of complete culture medium, which can be unacceptable for the development of a production process for a viral bioinsecticide. Finally, the regulatory agencies are becoming increasingly restrictive in relation to the use of raw materials of animal origin.

4.4.2.1 Serum-free media for insect cell cultures

The problems about the use of serum as a supplement in animal cell cultures, described above, have driven the development of new culture media capable to support the growth of insect cell cultures in a serum-free environment. As the serum is a very complex substance, and its functions are diverse, it is necessary to use a mixture of various components to replace it. The supply of hydrophobic nutrients is replaced by adding microemulsions that contain a source of lipids and cholesterol (Maiorella et al, 1988; Ikonomou et al., 2001). The most used sources of lipids are the methyl esters of fatty acids isolated from the liver of marine fishes, but recently it has been described the use of cooking soybean oil, a cheaper and more abundant source of lipids (Micheloud et al., 2009). In addition, microemulsions also contain the surfactant Pluronic F68, whose presence protects the cells from the detrimental influence of bubbles in sparged bioreactors.

On the other hand, the contribution of growth factors that made the serum is replaced by mixtures of enzymatic hydrolysates of proteins and yeast extract (Schlaeger, 1996). Hydrolysates from milk proteins such as lactalbumin and casein, and meat proteins, are commonly used as cheap replacements of serum, but the peptides responsible for the growth factor activity have not been identified. The effects of the addition of yeast extract to culture media for insect cells are similar to that of fetal calf serum, modifying the specific growth rate and increasing the maximum cell density (Eriksson & Häggström, 2005).

There are currently several commercial serum-free media available that were specifically designed to cultivate either Sf9 or BTI-TN-5B1-4 cell lines, but eventually can also support the growth of other insect cell lines. The growth parameters of cultures in these media are remarkable, as well as the yields of baculovirus and recombinant proteins obtained in infected cultures. However, the cost of the commercial serum-free media is, at least, as high as the cost of complete medium supplemented with fetal calf serum, precluding its utilization for the economically feasible production of insecticidal baculoviruses.

New serum-free media were specifically designed in recent years for culturing a few cell lines of interest, due to their potential application to the development of production processes for insecticidal baculoviruses in insect cell cultures. A prototype low cost medium (LCM) was developed to both cultivate the cell line BCIRL-HZ-AM1 and to produce the baculovirus HaSNPV (Lua & Reid, 2003). Comparable maximum cell densities and growth rates were obtained in both the LCM and a commercial serum-free medium, but lower specific virus yields were reached in LCM. The composition of the LCM medium was not disclosed. On the other hand, the low-cost medium UNL-10 was developed to grow the saUFL-AG-286 cell line, useful to produce the baculovirus AgMNPV (Micheloud et al, 2009). The yields of occlusion bodies in suspension cultures, using optimized parameters of infection, were as high as 3×10^{11} OBs L^{-1}, with specific yields higher than 600 OBs/cell. The composition of the UNL-10 medium, that was optimized to improve the yield of OBs of AgMNPV, has glucose as the only source of carbohydrates, a lower concentration of most amino acids, an improved mixture of vitamins and a lipid emulsion made with cooking oil. The growth factor activity is exerted by an optimized mixture of an enzymatic hydrolysate of casein, tryptose broth and yeast extract.

4.5 Physicochemical conditions in insect cell cultures

The pH of the medium in insect cell cultures is determined by its chemical composition, depending mainly on the buffer activity of salts, although the mixture of amino acids can also play a role in pH regulation. The optimal pH for all lepidopteran insect cell cultures is acid, between 6.2 and 6.4 (Schlaeger, 1996). While the pH can be modified through the evolution of an insect cell culture, the changes tend to be limited and usually do not compromise the cellular physiology.

The optimum osmolarity can differ for distinct lepidopteran insect cell lines, which can react also differently according to the agents utilized to modify it. The osmolarity of most culture media for insect cells varies between 250 and 350 mOsm kg^{-1}, but the initial value is habitually modified through the evolution of the culture, usually without consequences on the cellular physiology (Kurtti & Munderloh, 1984).

Insect cells are cultivated in vitro at temperatures ranging between 25 and 30 °C. The optimum temperature for most lepidopteran cell lines is 28 °C (O'Reilly et al., 1994). Cells cultured in serum-free media are less tolerant of temperature changes that cells grown in media supplemented with serum (Mitsuhashi & Goodwin, 1989).

The dissolved oxygen concentration is a critical parameter in insect cell cultures due to the reduced solubility of oxygen in aqueous culture media (Palomares & Ramírez, 1996). Dissolved oxygen levels between 40 and 70% are usually appropriate to keep acceptable growth parameters in insect cell cultures. In addition, most of the available information indicates that it is also critical to keep a proper level of dissolved oxygen in baculovirus-infected cultures, because oxygen deprivation is a cause of low yield of virus or recombinant protein.

5. Bioreactors in insect cell cultures

Although the most used insect cell lines can grow in suspension or as static cultures indistinctly, the scaling-up of static cultures is not a feasible alternative for the production of insecticidal baculoviruses. Thus, in this section it will only be reviewed the information about insect cell suspension cultures, from agitated erlenmeyers and spinner-flasks for low-scale cultures up to stirred tank and airlift reactors at larger scale.

The suspension cell cultures at low scale in agitated erlenmeyers or spinner-flasks usually do not offer significant technological difficulties, being able to reach cell densities as high as 1×10^7 viable cells per milliliter, with doubling times from 18 to 30 hours during the exponential growth phase (Bédard et al., 1993; Benslimane et al., 2007; Gioria et al., 2006; Lua & Reid, 2003; Rhiel et al., 1997). This is true for most insect cell lines as long as the ratio between the areas of the gas phase –usually air- to the liquid phase –culture medium- is large enough to ensure that the superficial supply of oxygen is adequate. In practical terms this means that the volume of culture should never exceed 25% of the total volume of the container. In addition, the stirring speed should be adjusted to 60 - 80 rpm in spinner-flasks and 100 - 120 rpm in flasks with orbital shaking.

The large scale culture of animal cells in suspension requires an adequate mixing through agitation, either mechanical or pneumatic, in order to keep cells in suspension, as well as to ensure physicochemical homogeneity and adequate mass transference. But these requirements of scaling collide with certain morphological characteristics of the insect cells, as their large size and lack of cell wall, that make them fragile and sensitive to the effects of agitation and gas sparging (Trinh et al., 1994). A successful scaling-up of suspension cultures of insect cells will be always the result of a compromise between the satisfaction of an adequate mixing and the preservation of the structural and functional cellular integrity. Stirred tank reactors, where agitation is performed mechanically through impellers, have demonstrated to be useful to cultivate insect cells at large scale (Maranga et al., 2004). The agitation rate should be carefully controlled in stirred reactors, especially when cultures are aerated by sparging, due to deleterious effects on cell functionality and viability that occurs when cultures are stirred at speeds over 300 rpm with Rushton turbines (Cruz et al., 1998). This is because the shear in the zone near to the impeller - where the energy of agitation is introduced into the reactor- is very high. Three different ways to aerate insect cell cultures in stirred tank reactors have been used: surface aeration (Kamen et al., 1991), bubble-free aeration (Chico & Jáger, 2000) and sparging (Cruz et al., 1998). Without a doubt, the aeration method that offers a better chance of escalation in stirred reactors is through direct gas bubbling. Two process parameters should be carefully considered in stirred tank reactors aerated through sparging in order to keep a proper cellular viability and functionality: the aeration rate and the bubble size. High aeration rates, as well as low bubble size are main causes of impaired growth and functionality in insect cell cultures (Trinh et al., 1994).

Airlift reactors have been much less used than stirred tank reactors, in spite they offer advantageous characteristics to cultivate large and fragile cells, as insect cells. One of the advantages of airlift reactors, when compared to classical stirred tank reactors is that they have the potential to provide high mass transfer rates with low and homogeneously distributed shear. Thus, adequate oxygen transfer rates from the gas to the liquid phase and from this one to the suspended cells can be obtained within a homogeneous environment and with a reduced exposure to sources of mechanical stress (Merchuk, 1991). On the other hand, these reactors improve their performance as it increases in size, provided that the optimal relations between the reactor geometrical parameters are preserved. In addition, due to its simplicity of design and construction, airlift reactors require less capital investment, and its operation and maintenance costs are also lower than in stirred tank reactors. While some articles have been published on the cultivation of insect cell lines in airlift reactor and its application to the production of baculovirus and recombinant proteins, there are few systematic studies for optimization of the geometrical parameters and operation in processes involving these reactors, nor models for their scaling-up (King et al., 1992; Maiorella et al., 1988; Visnovsky et al.; 2003).

In addition to the classical airlift and stirred tank reactors, other reactor designs were applied successfully to the cultivation of lepidopteran insect cells, such as the rotating-wall vessel (Cowger et al., 1999). However, given the level of production scale that should be achieved for the development of economically feasible processes, only stirred tank and airlift reactors appear to have the potential to be used for the large scale production of insecticidal baculoviruses in insect cell cultures. This is also because of the experience existing in the scale-up of cultures of microorganisms and animal cells in these reactors.

6. Baculovirus infection in lepidopteran insect cell cultures

Figure 3 outlines an infection process of infection with baculovirus in an insect cell culture. The preceding sections have addressed the characteristics of the cells, culture media and reactors. This section will deal with those aspects of the process that relate specifically to infection: the viral inoculums, the parameters of infection, the operation strategy and the product.

6.1 BVs: the viral inoculums in insect cell cultures

The baculovirus inoculum is composed by BVs that are added to the insect cell culture at the time of infection. The quality of the seed virus is a critical factor to determine the quality of the final product, the occlusion bodies, as well as the process productivity. First, the selected strain of virus should be virulent for the insect to be controlled. It should be also capable to replicate in the cell culture, yielding a high productivity of OBs. In addition, OBs produced from that inoculums should have a high biological activity. To meet these requisites, the viral inoculums should be free of genomic variants capable to reduce either the product yield or the biological activity of OBs. Two main types of genomic variants of baculovirus capable to reduce the OBs yield and biological activity have been described: the "few polyhedra" (FP) phenotype and the defective interfering particles (DIPs).

The mutations responsible for the FP phenotype, often associated with the inactivation of the gene 25k, are expressed through the following features: reduced yield of OBs, reduced content of OVs per occlusion body, reduced biological activity of OBs and increased yield of BVs (Beames & Summers, 1989; Harrison & Summers, 1995). The emergence of the FP phenotype is responsible for a sharp drop of the final yields of OBs, as well as its biological activity (Lua et al., 2002). Once emerged, the population of FP mutants tends to enrich through successive passages in insect cell cultures, due to its increased capability to produce BVs.

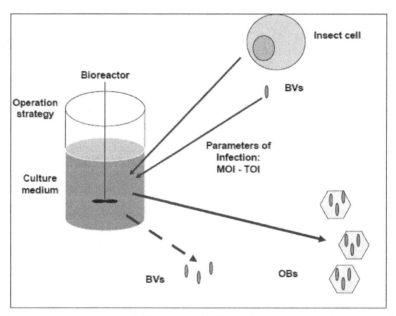

Fig. 3. Schematic representation of OBs insecticidal baculovirus production process in insect cell cultures. MOI: multiplicity of infection; TOI: time of infection.

The DIPs are generated as consequence of genomic deletions that originate shorter viral genomes (Kool et al., 1990). The DIP genomes can not replicate autonomously, but they can do it with the help of complete genomes. The replication of the DIPs competes and interferes with the replication of complete genomes, and leads to progressive enrichment of the population of defective virus, at the expense of the population of wild-type virus (van Lier et al., 1990). The generation of DIPs in cell cultures infected with baculovirus is favored by conditions that increase the probability of homologous recombination, such as in infections at high multiplicity of infection. The proportion of DIPs in a viral population, as well as the proportion of FP mutants, increases with the number of passages in cell cultures, a phenomenon known as "passage effect"(Krell, 1996). This effect impairs the amplification of baculovirus stocks necessary to infect large scale cultures of insect cells.

The viral stock is the most expensive raw material in processes destined to the production of insecticidal OBs in insect cell cultures. Obtaining the stock for infecting the production reactor usually requires the amplification of the seed virus through successive rounds of infection in insect cell cultures at progressively larger scales (Rhodes, 1996). The optimization of the conditions of infection for the production of BVs would have a double beneficial effect on the whole production process: on the one side, it would allow to reduce the consumption of an expensive raw material and, on the other side, it would permit to reduce the steps of the seed virus amplification process, and therefore to reduce the probabilities to generate and propagate deleterious genomic variants. Despite its importance, the information available about the optimization of infection conditions for the production of the BV progeny of baculovirus is very scarce (Carinhas et al., 2009).

The quality of baculovirus stocks may also be affected by reduced infectivity in relation to the total amount of viral particles (Dee & Shuler, 1997). Although several causes have been proposed to explain this phenomenon, the inactivation of BVs could be an important detrimental factor, especially in stocks prepared in serum-free media (Jorio et al., 2006). An optimized management of the preservation of serum-free baculovirus stocks could have an important impact on the feasibility of the scaling-up process.

6.2 Parameters of infection

The fate of a baculovirus infection in an insect cell culture is strongly dependent on the selection of the multiplicity of the infection (MOI) and the time of infection (TOI) (Carinhas et al., 2009; Licari & Bailey, 1991; Micheloud et al., 2009). MOI is defined as the ratio between the number of infectious viral particles and the number of cells of the infected culture. The selection of the MOI determines the proportion of cells that become initially infected, as well as the distribution of infectious particles per cell. In cultures infected at high MOI (greater than 5 infectious units per cell), all the cells are infected synchronously. This prompts the viral replicative process to follow the same temporal pattern in all cells, with the emergence of a unique peak of viral progeny. On the other hand, when cultures are infected at lower values of MOI, only a proportion of the cellular population is infected initially. Thus, at least two cellular sub-populations will coexist in a culture infected at low MOI: infected and uninfected cells. Viral replication takes place immediately in initially infected cells, producing a viral progeny after one generation time. Uninfected cells, on the other hand, proliferate and will be infected later with the viral progeny of the initially infected cells. An important difference is that in cultures infected synchronously, viral replication occurs in cells that are in similar physiological state and subjected to similar environmental conditions, while in asynchronously infected ones viral replication takes place through successive rounds of infection, in cells under different physiological states and subjected to different environmental conditions. The importance of this difference is that the replicative capacity of insect cells varies, depending on both their physiological condition and the quality of the culture medium at the time that they are actually infected.

One factor that has contributed strongly to limiting the development of feasible production processes of insecticidal baculoviruses in insect cell cultures is the so called "cell density effect" (Wood et al., 1982). This phenomenon is characterized by the reduction of the intrinsic ability of insect cells to replicate baculovirus as the infection is delayed in time, and the cell density increases. The phenomenon has been observed also for recombinant proteins produced by genetically modified baculovirus. Several causes have been proposed to explain the "cell density effect", including nutrient limitation (Bédard et al., 1994), accumulation of toxic by-products (Taticek & Shuler, 1997), autocrine factors and cell cycle distribution (Braunagel et al., 1998; Calles et al., 2006), and inhibition of the central energy metabolism (Bernal et al., 2009), among others, but the causes remain to be identified. The cell density effect, that could be observed in several insect cell lines infected with different baculoviruses, both wild-type and recombinant, can be overcome, at least partially in Sf9 cell cultures by adopting alternative strategies for cell culture and viral infection, such as fed-batch or perfusion, as will be explained hereinafter.

6.3 Operation strategy

The typical strategy for the production of viruses in cultured animal cells is the infection of batch cultures. This strategy implies that a cell culture is cultivated in a proper medium up to reach the desired cellular concentration, when it is added the viral inoculums and the infection is allowed to progress without ulterior modification of the system until the harvest of the product is made. The batch production of insecticidal baculoviruses can be easily implemented, and it is possible to obtain high yields of occlusion bodies (Lua & Reid, 2003; Micheloud et al., 2009; Rodas et al., 2005). However, the cell density effect is a strong limitation to reach economically significant yields of virus, so that the MOI and the time of infection should be optimized to avoid an increase in cell density of infected cultures. This limitation could be partially overcome through the medium exchange before infection (Bédard et al., 2004). Although this strategy is feasible for working at low scale in shaken-flasks or spinner-flasks, it is difficult to implement at larger scale.

Fed-batch is an alternative operation strategy to overcome the cell density effect in cultures of the *Spodoptera frugiperda* cell lines. Concentrated solutions containing the more demanded nutrients (glucose, glutamine and other amino acids, yeast extract and lipids´ emulsion) were added to high density cell cultures for obtaining high yields of recombinant proteins and occlusion bodies (Bédard et al., 1994; Elias et al., 2000). The main advantage of this strategy resides in its technological simplicity, which makes it proper to be implemented in both low and large scale. However, the use of this kind of operation is limited to insect cell lines that do not accumulate toxic by-products, such as IPLB-Sf21 and Sf9. For other, cell lines, like BTI-TN-5B1-4 and saUFL-AG-286, fed-batch would not be a suitable strategy, since ammonia accumulation could be the cause of inhibition of viral replication.

The perfusion strategy, that implies the continuous removal of spent medium and its replacement by fresh medium with retention of the cell mass inside the cultivation device, has been also used. It has been employed to obtain high yields of recombinant proteins in insect cell cultures of high density, infected with genetically modified baculoviruses (Chico & Jäger, 2000). The advantage of this strategy with regard to fed-batch resides in the possibility to use it with cell lines that accumulate toxic by products, which are continuously removed with the spent medium. However, its implementation requires sophisticated devices to remove spent cell-free medium, making it unlikely use in the development of economically feasible processes for the production of insecticidal baculoviruses in large scale.

The production in culture systems that operate continuously appears to be an attractive option to reduce the operation costs in fermentation processes. Given its lytic nature, a continuous process for baculovirus production should involve the use of at least two cultivation devices, one for cellular propagation and a second one for virus infection and replication. The first reactor is continuously fed with fresh medium, and the viral product is continuously harvested from the second reactor, which is in turn fed with cell culture from the first reactor, also in a continuous way and at the same rate. Although it has been demonstrated that is possible to operate this kind of continuous process to produce baculovirus OBs by a limited time, the viral yield is affected rapidly by the passage effect, invalidating this strategy as an effective approach (van Lier et al, 1990).

6.4 Product yield and quality

The feasibility of a production process for an insecticidal baculovirus will finally reside on the yield and the quality of viral occlusion bodies. Rhodes (1996), through a detailed economic analysis of an *in vitro* process for production of an insecticidal baculovirus, has established that the limit yield to reach the economical feasibility should be at least 2×10^{11} OBs.L^{-1}. More recently, Nguyen et al. (2011) have argued that the minimum volumetric yield for an economically feasible process of production of HaSNPV in insect cell cultures should be 40 times higher. The OBs yields effectively reached for several baculoviruses in serum-free cultures of different cell lines, according to the available information, ranged from 1×10^9 to 3×10^{11} OBsL^{-1}, with cell specific yields ranging from 3 to 700 OBs per cell (Chakraborty et al., 1999; Gioria et al., 2006; Lua & Reid, 2003; McKenna et al., 1997; Micheloud et al., 2009; Rodas et al., 2005).

The insecticidal potency of baculovirus occlusion bodies that are produced in insect cell cultures is a controversial topic. Some papers have shown that occlusion bodies produced in infected insect cell cultures are less potent than polyhedra produced by infection of insect larvae (Chakraborty et al., 1999; McKenna et al., 1997). However, this reduced activity may not be an intrinsic characteristic of the occlusion bodies produced in cell cultures, but related to the extraction method used. Extraction with a solution of sodium dodecyl sulfate (SDS), a widely used procedure for releasing the occlusion bodies accumulated in the nuclei of infected cells, alters the structure of the polyhedron envelope and would reduce the content of occluded virus per polyhedron, and thus its insecticidal ability (Lua et al., 2003). Consequently, the extraction with SDS should be avoided for the purpose of preserving the quality of occlusion bodies. On the other hand, the occlusion bodies produced by infection in larvae contain an alkaline protease that is not codified in the viral genome, and that would be incorporated from the tissues of the infected insect. This protease, which is not present in the occlusion bodies produced in cell cultures, could be an additional factor of virulence, accelerating the dissolution of occlusion bodies in the insect midgut and thus contributing to increase its biological activity (Rohrmann, 2011). Finally, it has been also reported that the composition of the culture medium could affect the activity of occlusion bodies, but the causes are unknown (Pedrini et al, 2006).

7. Scaling-up: limitations and possibilities

Although the production of insecticidal baculoviruses in insect cell cultures has been proposed as an alternative to overcome the limitations of the processes *in vivo*, so far no process *in vitro* could be even implemented on an industrial scale, and occlusion bodies are still produced in infected insect larvae. Some factors that 25 years ago have hindered the development of large-scale production processes for insecticidal baculoviruses in insect cell cultures, such as the sensitivity of insect cells to the stresses linked to the mechanical agitation in stirred tank reactors and to the bubble rupture in sparged bioreactors, have been resolved and several cell lines can be cultivated today in industrial bioreactors of large volume to produce occlusion bodies or recombinant proteins. However, other factors that still limit the development of feasible processes have not yet been satisfactorily resolved, and will be reviewed below.

Obtaining a cell line with relevant technological properties and with the ability to replicate the virus at a high yield of OBs, is a requirement to develop a feasible process for the production of an insecticidal baculovirus in insect cell cultures. Besides Sf9 and BTI-TN-5B1-4, there are few cell lines that fulfill these requisites. The cell line BCIRL-HZ-AM1, used to produce HaSNPV, is capable to grow in suspension cultures in a low-cost serum-free medium in stirred tank reactors. BCIRL-HZ-AM1 cells can produce high specific yields of HaSNPV OBs in infected cultures (Lua & Reid, 2003), but its ability to produce high yields of BVs, a property that is important for the scaling-up, is more limited (Pedrini et al., 2011). The cell line saUFL-AG-286, of election to produce AgMNPV, can generate high specific yields of OBs in serum-free suspension cultures, but the production of OBs is strongly inhibited at cell densities higher $8x10^5$ cells mL^{-1}, thus limiting the possibility to reach very high volumetric yields of OBs (Micheloud et al., 2009). As these cell lines are heterogeneous, the isolation of cell clones with improved ability to produce baculovirus OBs appears to be a reliable possibility to enhance the productivity of viral insecticides (Nguyen et al., 2011; Pasumarthy & Murhammer, 1994). For the production of other insecticidal baculoviruses will be necessary to establish new cell lines, obtained preferably from tissues of their respective target insects.

Another requisite that must be resolved before confronting the scaling-up of an insecticide baculovirus production process is the development of a low cost serum-free culture medium for the selected cell line. It has been indicated that the cost of the culture medium for an economically feasible process should not be higher than U$S 2.5 per liter (Rhodes et al., 1996), or it even should be lower than U$S 1 (Gong et al., 1997). Commercial serum-free media for Sf9 and BTI-TN-5B1-4 cells are sold at prices that are 30 times greater, and therefore are not useful for producing insecticidal baculoviruses at industrial scale. Besides, the cost of media specifically developed for producing insecticidal baculoviruses are yet above the acceptable limit for an economically feasible process. The rational approach to further reduce the cost of culture media for insect cell cultures is the simplification of the chemical composition, based on the deep knowledge of the nutritional demands and metabolism of insect cells, both uninfected and infected. However, most cell lines have not been sufficiently characterized as to progress towards a simplification of the composition of the culture medium. A more empirical approach to reduce the cost of the culture medium is the replacement of costly ingredients, such as amino acids and lipids, by optimized mixtures of raw materials of lower cost such as protein hydrolysates and cooking oil.

The usual strategy to produce baculovirus occlusion bodies in insect cell cultures has been the infection of batch cultures. However, the possibility to obtain high volumetric yields of viral OBs in batch cultures is impaired by the "cell density effect". Whenever possible, the adoption of alternative strategies of infection could be a way to overcome the cell density effect and thus improve the viral productivity. The fed-batch culture, which has proven to be a feasible alternative to increase the yield of recombinant proteins and BVs in Sf9 cell cultures at high density, could also be an alternative strategy to increase the yield of occlusion bodies. A deeper understanding of the causes that lead to the manifestations of the cell density effect could help to design more rational feeding schedules than those used to date, and thus increase the viral productivity. However, the usefulness of the fed-batch strategy is restricted to cell lines that do not accumulate toxic by-products.

A large-scale process to produce insecticidal baculovirus OBs in insect cell cultures requires the completion of successive steps of viral amplification in growing scale (Rhodes, 1996). OBs are the final product of the whole process, but BVs are the product for each of the intermediate steps of scaling. Despite the importance to improve the yield of BVs, few studies have systematically explored the optimization of the production of this viral progeny (Carinhas *et al*, 2009). The optimization of BVs production could help to reduce the number of scaling steps necessary to get the number of virions needed to feed the OBs production reactor, and therefore reduce scaling cost. Furthermore, the reduction in the number of stages of scaling would contribute to limiting the probability of emergence of unproductive viral variants, such as FP mutants and DIPs. The approach patented by Lua and Reid (2005), using occluded virions extracted from occlusion bodies as seed, could alleviate the need for viral inoculum at the beginning of the scaling-up process, but does not prevent the need to improve the yields of BVs in the later stages. Additionally, the improvement of the ratio infectious particles/total particles, through better BVs preservation, could mean significant savings in the demand for seed virus, and therefore a step towards the feasible scaling-up of the viral insecticide production process in insect cell cultures.

8. Concluding remarks and perspectives

Baculoviruses are a group of arthropod-specific pathogens which have a significant potential to be used as safe and environmentally friendly insecticides in agriculture, horticulture and forestry. The replication of baculoviruses produces two viral progenies, budded and occluded viruses. The last are included into proteinaceous structures called occlusion bodies, which display insecticidal activity when ingested by susceptible insects. The current technology to produce insecticidal occlusion bodies is based on the viral infection of susceptible insects, but an alternative technology based on the viral replication in insect cell cultures could aid to overcome some of the limitations of the former. The insect cell line, the culture medium, the bioreactor, the virus, the infection parameters and the culture strategy are elements of the insect cell culture technology that must be optimized in order to develop *in vitro* production processes for insecticidal baculoviruses. While it is now possible to grow insect cells in large-scale industrial reactors using serum-free media to produce high yields of occlusion bodies for several baculoviruses, the current technology is still insufficient to achieve economic feasibility. To do that, in the next future the efforts should be mainly orientated:

- to gain deep insight over the insect cell biology in order to identify the factors responsible of the cell density effect;
- to improve the composition and to reduce the cost of serum-free culture media for insect cell cultures;
- to increase, through the manipulation of the infection parameters in batch cultures or, whenever possible, through optimized fed-batch strategies, the volumetric yield of occlusion bodies;

-   to optimize the production of budded virus in order to reduce the length and the cost of the viral inoculum scaling-up, and to minimize the risk of generating and selecting unproductive viral variants.

Only obtaining satisfactory solutions for these remaining problems will make possible to establish economically viable processes for the production of insecticidal baculoviruses in insect cell cultures on an industrial scale.

9. References

Abe, T., Matsuura, Y. Host innate immune responses induced by baculovirus in mammals. *Current Gene Therapy.* Vol.10, (2010), ISSN 15665232, pp. 226-231.

Barnes, D. & Sato, G. Serum-free cell culture: a unifying approach. *Cell,* Vol.22, (1980), ISSN 0092-8674 , pp. 649-55.

Beames, B. & Summers, M.D, .Location and nucleotide sequence of the 25K protein missing from baculovirus few polyhedra (FP) mutants. *Virology.* Vol.168, (1989), ISSSN 0042-6822, pp. 344-53.

Bédard, C., Kamen, A.A., Tom, R. & Massie, B. Maximization of recombinant protein yield in the insect cell/baculovirus system by one-time addition of nutrients to high density batch cultures. *Cytotechnology.* Vol.15, (1994), ISSN 0920-9069, pp. 129-138.

Bédard, C., Tom, R., Kamen, A. Growth, nutrient consumption and end-product accumulation in Sf-9 and BTI-EAA insect cell cultures: insights into growth limitation and metabolism. *Biotechnology Progress.* Vol. 9, (1993), ISSN 8756-7938, pp. 615-624.

Benslimane, C., Elias, C.B., Hawari, J., Kamen A. Insights into the central metabolism of Spodoptera frugiperda (Sf-9) and Trichoplusia ni BTI-Tn-5B1-4 (Tn-5) insect cells by radiolabeling studies. *Biotechnology Progress.* Vol.21, (2005), ISSN 8756-7938, pp. 78-86.

Benz, G.A. (1986) Introduction: Historical Perspectives. In: *The Biology of Baculoviruses. Volume I: Biological Properties and Molecular Biology*, R. R. Granados & B.A. Federici, (Eds.), 1-36, CRC Press, ISBN 0849359872, Boca Raton, Florida, USA.

Bernal, V., Carinhas, N., Yokomizo, A.Y., Carrondo, M.T.J. & Alves, P.M. Cell density effect in the baculovirus-insect cells system: a quantitative analysis of energetic metabolism. *Biotechnology and Bioengineering.* Vol.104, ISSN 0006-3592, pp. 162-180.

Black, B.C., Brennan, L.S., Dierks, P.M. & Gard, I.E. (1997). Commercialization of Baculoviral Insecticides. In: *The Baculoviruses*, L.K. Miller (Ed.), 341-397, Plenum Press, ISBN 0306456419, New York.

Blissard, G.W. & Rohrmann, G.F. Baculovirus diversity and molecular biology. *Annual Review of Entomology.* Vol.35, (1990), ISSN 00664170, pp. 127-155.

Braunagel, S.C., Parr, R., Belyavskyi, M. & Summers, M.D. *Autographa californica nucleopolyhedrovirus* infection results in Sf9 cell cycle arrest at G2/M phase. *Virology.* Vol. 244, (1998), ISSN 0042-6822, pp. 195-211.

Calles, K., Erikson, U. & Häggström, L. Effect of conditioned medium factors on productivity and cell physiology in *Trichoplusia ni* insect cell cultures. *Biotechnology Progress.* Vol.22, (2006), ISSN 8756-7938, pp. 653-659.

Carinhas N., Bernal, V., Yokomizo, A.Y., Carrondo, M.T.J., Oliveira, R. & Alves, P.M. Baculovirus production for gene therapy: the role of cell density, multiplicity of infection and medium exchange. *Applied Microbiology and Biotechnology*. Vol.81, (2009), ISSN 1432-0614, pp. 1041-1049.

Chakraborty, S., Monsour, C., Teakle, R. & Reid, S. Yield, biological activity, and field performance of a wild-type *Helicoverpa* nucleopolyhedrovirus produced in *H. zea* cell cultures. *Journal of Invertebrate Pathology*. Vol.73, (1999), ISSN 0022-2011, pp. 199-205.

Chico, E. & Jäger, V. Perfusion culture of baculovirus-infected BTI-Tn-5B1-4 insect cells: a method to restore cell-specific beta-trace glycoprotein productivity at high cell density. *Biotechnology and Bioengineering*. Vol.70, ISSN 0006-3592, pp. 574-586.

Chung, I.S. & Shuler, M.L. Effect of Trichoplusia ni BTI-Tn-5B1-4 cell density on human secreted alkaline phosphatase production. *Biotechnology Letters*. Vol.15, (1993), ISSN 0141-5492, pp. 1007-1012.

Claus, J.D., Remondetto, G.E., Guerrero, S.A., Demonte, A.M., Murguía, M. & Marcipar, A.J. *Anticarsia gemmatalis* nuclear polyhedrosis virus replication in serum-free and serum-reduced insect cell cultures. *Journal of Biotechnology*. Vol.31, (1993), ISSN 0168-1656, pp. 1-15.

Cowger, N.L., O'Connor, K.C., Hammond, D.J., Lacks, D.J. & Navar, G.L. Characterization of bimodal cell death of insect cells in a rotating-wall vessel and shaker-flasks. *Biotechnology and Bioengineering*. Vol.64, (1999), ISSN 0006-3592, pp. 14-26.

Cruz, P.E., Cuhna, A., Peixoto, C.C., Clemente, J., Moreira, J.L. & Carrondo, M.J.T. Optimization of the production of virus-like particles in insect cells. *Biotechnology and Bioengineering*. Vol.60, ISSN 0006-3592, pp. 408–418.

Day, M.F & Grace, T.D.C. Culture of insect tissues. *Annual Review of Entomology*. Vol.4, (1959), ISSN 00664170, pp. 17-38.

Dee, K.U. & Shuler, M.L. Optimization of an assay for baculovirus titer and design of regimens for the synchronous infection of insect cells. *Biotechnology Progress*. Vol.13, (1997), ISSN 8756-7938, pp. 14-24.

Drews, M., Doverskog, M., Öhman, L., Chapman, B.E., Jacobson, U., Kuchel, P.W. & Häggström, L. Pathways of glutamine metabolism in *Spodoptera frugiperda* (Sf9) insect cells: evidence for the presence of the nitrogen assimilation system, and a metabolic switch by $^{1}H/^{15}N$ NMR. *Journal of Biotechnology*. Vol.78, (2000), ISSN 0168-1656, pp. 23-37.

Elias, C.B., Zeiser, A., Bédard, C. & Kamen, A. Enhanced growth of Sf-9 cells to a máximum density of 5.2 x 107 cells per mL and production of β-galactosidase ast high cell density by fed batch culture. *Biotechnology and Bioengineering*. Vol.68, (2000), ISSN 0006-3592, pp. 381-388.

Eriksson, U. & Häggström, L. Yeast extract from Express Five serum-free medium contains factors at about 35 kDa, essential for growth of Trichoplusia ni insect cells. *Biotechnology Letters*. Vol.27, (2005), ISSN 0141-5492, pp. 1623-1627.

Ferrance, J.P., Goel, A. & Ataai, M.M. Utilization of glucose and amino acids in insect cell cultures: Quantifying the metabolic flows within the primary pathways and medium development. *Biotechnology & Bioengineering*. Vol.42, (1993), ISSN 0006-3592, pp. 697-707.

Gioria, V.V., Jäger, V. & Claus, J.D. Growth, metabolism and baculovirus production in suspension cultures of an *Anticarsia gemmatalis* cell line. *Cytotechnology*.Vol. 52, ISSN 0920-9069, pp. 113-124.

Grace, T.C. The prolonged growth and survival of ovarian tissue of the promethea moth (*Callosamia promethea*) *in vitro*. *Journal of General Physiology*. Vol.41, (1958), ISSN 0022-1295, pp. 1027-1034.

Grace, T.D. Establishment of four strains of cells from insect tissues grown in vitro. *Nature*. Vol.195, (1962), ISSN 0028-0836, pp. 788-789.

Grace, T.D. Establishment of a line of cells from the silkworm *Bombyx mori*. *Nature*, Vol.216, (1967), ISSN 0028-0836, p. 613.

Granados, R.R. & Williams, K.A. (1986) In Vivo Replication of Baculoviruses. In: *The Biology of Baculoviruses. Volume I: Biological Properties and Molecular Biology*, R. R. Granados & B.A. Federici, (Eds.), 89-108, CRC Press, ISBN 0849359872, Boca Raton, Florida, USA.

Harrison, R.L., Summers, M.D. Mutations in the Autographa californica multinucleocapsid nuclear polyhedrosis virus 25kDa protein gene result in reduced virion occlusion, altered intranuclear envelopment and enhanced virus production. *Journal of General Virology*.Vol. 76, (1995), ISSN 0022-1317, pp. 1451-1459.

Huber, J. (1986) Use of Baculovirus in Pest Management Programs. In: *The Biology of Baculoviruses. Volume II: Practical Application for Insect Control*, R. R. Granados & B.A. Federici, (Eds.), 181-202, CRC Press, ISBN 0849359872, Boca Raton, Florida, USA.

Ikonomou, L., Bastin, G., Schneider, Y.J. & Agathos, S.N. Design of an efficient medium for insect cell growth and recombinant protein production. *In Vitro Cellular & Developmental Biology Animal*.Vol. 37, (2001), ISSN 1071-2690 , pp. 549-559.

Jehle, J.A., Blissard, G.W., Bonning, B.C., Cory, J.S., Herniou, E.A., Rohrmann, G.F., Theilmann, D.A., Thiem, S.M. & Vlak, .J.M. On the classification and nomenclature of baculoviruses: a proposal for revision. *Archives of Virology*, Vol.151, (2006), ISSN 0304-8608, pp. 1257-1266.

Jorio, H., Tran, R. & Kamen, A. Stability of serum-free and purified baculovirus stocks under various storage conditions. *Biotechnology Progress*. Vol.22, (1996), ISSN 8756-7938, pp. 319-325.

Kamen, A.A., Tom, R.L., Caron, A.W., Chavarie, C., Massie, B & Archambault, J. Culture of insect cells in a helical ribbon impeller bioreactor. *Biotechnology and Bioengineering*. Vol.38, (1991), ISSN 0006-3592, pp. 619-628.

King, G.A.., Daugulis, A.J., Faulkner, P., Goosen, M.F.A. Recombinant beta-galactosidase production in serum-free medium in a 14-liter airlift bioreactor. *Biotechnology Progress*. Vol.8, (1992), ISSN 8756-7938, pp. 567-571.

Kool, M., Voncken, J.W., van Lier, F.L.J., Tramper, J., Vlak, J.M. Detection and analysis of Autographa californica nuclear polyhedrosis virus mutants with defective interfering properties. *Virology*. Vol.183, (1990), ISSN 0042-6822, pp. 739-746.

Kost, T.A. & Condreay, J.P. Recombinant baculoviruses as mammalian cell gene-delivery vectors. *Trends in Biotechnology*, Vol.20, (2002), ISSN 01677799, pp. 173-180.

Krell, P. J. Passage effect of virus infection in insect cells. *Cytotechnology*. Vol.20, (1996), ISSN 0920-9069, pp. 125-137.

Kurtti, T.J. & Munderloh, U.G. (1984) Mosquito cell culture. In: *Advances in Cell Culture, Vol.3*, K Maramorosch (Ed.), 259-302, Academic Press, ISBN 0120079038, New York, USA.

Licari, P. & Bailey, J.E. Factors influencing recombinant protein yields in an insect cell-baculovirus expression system: multiplicity of infection and intracellular protein degradation. *Biotechnology and Bioengineering*. Vol.37, (1991), ISSN 0006-3592, pp. 238-246.

Lua, L.H., Nielsen, L.K. & Reid, S. Sensitivity of Helicoverpa armigera nucleopoly-hedrovirus polyhedra to sodium dodecyl sulfate. *Biological Control*. Vol.26, (2003), ISSN 1049-9644, pp. 57-67.

Lua, L.H., Pedrini, M.R., Reid. S., Robertson, A. & Tribe, D.E. Phenotypic and genotypic analysis of Helicoverpa armigera nucleopolyhedrovirus serially passaged in cell culture. *Journal of General Virology*. Vol.83, (2002), ISSN 0022-1317, pp. 945-955.

Lua, L.H. & Reid, S. Growth, viral production and metabolism of a Helicoverpa zea cell line in serum-free culture. *Cytotechnology*, Vol.42, (2003), ISSN 0920-9069, pp: 109-120.

Lua, L.H. & Reid, S. Method of producing baculovirus. PCT N° WO2005/045014. Retrieved from http://ip.com/pdf/patent/US7521219.pdf.

Luckow, V.L. & Summers, M.D. Trends in the development of baculovirus expression vectors. *Bio/Technology*, Vol.6, (1988), ISSN 0733-222X, pp. 47-55.

Lynn, D.E. (2007). Available Lepidopteran Insect Cell Lines. In: *Baculovirus and Cell Expresion Protocols: Second Edition*. Murhammer, D.W. (Ed.), 117-137, Humana Press, ISBN 1588295370, Totowa, USA.

Lynn, D.E., Dougherty, E.M., McClintock, J.T. & Loeb, M.(1988). Development of cell lines from various tissues of Lepidoptera. In: *Invertebrate and Fish Tissue Culture*. Y. Kuroda, E. Kurstak, & K. Maramorosch (Eds.), 239-242, Japan Scientific Societies Press/Springer-Verlag, ISBN 9780387192086, Tokyo/Berlin.

Maiorella, B., Inlow, D., Shauger, A. & Harano, D. Large-scale insect cell culture for recombinant protein production. *Bio/Technology*. Vol.6, (1988), ISSN 0733-222X, pp. 1406-1410.

Maranga, L., Cunha, A., Clemente, C., Cruz, P. & Carrondo, M.J.T. Scale-up of virus-like particles production: effects of sparging, agitation and bioreactor scale on cell growth, infection kinetics and productivity. *Journal of Biotechnology*. Vol.107, (2004), ISSN 0168-165 , pp. 55-64.

McIntosh, A.H. & Ignoffo, C.M. Replication and infectivity of the single-embedded nuclear polyhedrosis virus, baculovirus Heliothis, in homologous cell line. *Journal of Invertebrate Pathology*. Vol. 37, (1981), ISSN 0022-2011, pp. 258-264.

McKenna, K.A., Shuler, M.L. & Granados, R.R. Increased virus production in suspension culture by a Trichoplusia ni cell line in serum-free media. *Biotechnology Progress*. Vol. 13, (1997), ISSN 8756-7938, pp. 605-609.

Merchuk, J.C. Shear effects on suspended cells. *Advances in Biochemical Engineering and Biotechnology*.Vol. 44,.(1991), ISSN 0724-6145, pp. 65-95.

Micheloud, G.A., Gioria, V.V., Pérez, G. & Claus, J.D. Production of occlusion bodies of *Anticarsia gemmatalis multiple nucleopolyhedrovirus* in serum-free suspension cultures of the saUFL-AG-286 cell line: influence of infection conditions and statistical optimization. *Journal of Virological Methods*. Vol.162, (2009), ISSN 0166-0934, pp 258-266.

Mitsuhashi, J. (1989). Nutritional Requirements of Insect Cells *In Vitro*. In: *Invertebrate Cell System Applications*. Mitsuhashi, J. (Ed.), 3-21, CRC Press, ISBN 0849343739, Boca Raton, Florida, USA.

Mitsuhashi, J. & Goodwin, R.H. (1989) Serum-free culture of insect cells in vitro. In: *Invertebrate Cell System Applications*. Mitsuhashi, J. (Ed.), 31-44, CRC Press, ISBN 0849343739, Boca Raton, Florida, USA.

Moscardi, F. Assessment of the application of baculoviruses for control of Lepidoptera. *Annual Review of Entomology*, Vol. 44, (1999), ISSN 0066-4170, pp. 257-289.

Neermann J, Wagner R. Comparative analysis of glucose and glutamine metabolism in transformed mammalian cell lines, insect and primary liver cells. *Journal of Cellular Physiology*. Vol.166, (1996), ISSN 0021-9541, pp. 152-169.

Öhman, L., Alarcón, M., Ljunggren, J., Ramqvist, A.K. & Häggström, L. (1996) Glutamine is not an essential amino acid for Sf-9 insect cells. *Biotechnology Letters*. Vol.18, ISSN 0141-5492, pp. 765-770.

O'Reilly, D.R., Miller, L.K., Luckow, V.A. (1994). *Baculovirus Expression Vectors. A Laboratory Manual*. Oxford University Press, ISBN 0795091310, New York, USA.

Pasarelli, A.L. Barriers to success: how baculoviruses establish efficient systemic infections. *Virology*, Vol. 411, (2011), ISSN 0042-6822, pp. 383-392.

Pasumarthy, M.K. & Murhammer, D.W. Clonal variation in the *Spodoptera frugiperda* IPLB-SF21-AE insect cell population. *Biotechnology Progress*. Vol. 10, (1994), ISSN 8756-7938, pp. 314-319.

Pedrini, M.R., Christian, P., Nielsen, L.K., Reid, S. & Chan, L.C. Importance of virus-medium interactions on the biological activity of wild-type Heliothine nucleopolyhedroviruses propagated via suspension insect cell cultures. *Journal of Virological Methods*. Vol.136, (2006), ISSN 0166-0934, pp. 267-72.

Pedrini, M.R., Reid, S., Nielsen, L.K. & Chan, L.C. Kinetic characterization of the group II *Helicoverpa armigera nucleopolyhedrovirus* propagated in suspension cell cultures: Implications for development of a biopesticides production process. *Biotechnology Progress*. Vol.27, (2011), ISSN 8756-7938, pp. 614-624.

Nguyen, Q., Qi, Y.M, Wu, Y., Chan, L.C.L., Nielsen, L.K. & Reid, S. In vitro production of Helicoverpa baculovirus biopesticides — Automated selection of insect cell clones for manufacturing and systems biology studies. *Journal of Virological Methods*. Vol.175, (2011), ISSN 0166-0934, pp. 197-205.

Rhiel, M., Mitchell-Logean, C.M. & Murhammer, D.W. Comparison of *Trichoplusia ni* BTI-Tn-5B1-4 (High Five™) and *Spodoptera frugiperda* Sf-9 insect cell line metabolism in suspension cultures. *Biotechnology and Bioengineering*. Vol.55, (1997), ISSN 0006-3592, pp. 909-920.

Rhodes, D.J. Economics of baculovirus – insect cell production systems. *Cytotechnology*. Vol.20, (1996), ISSN 0920-9069, pp. 291-297.

Rodas, V., Marques, F., Honda, M., Soares, D., Jorge, S.; Antoniazzi, M., Medugno, C., Castro, M., Ribeiro, B., Souza, M., Tonso, A. & Pereira, C. Cell culture derived AgMNPV bioinsecticide: biological constraints and bioprocess issues. *Cytotechnology.* Vol. 48, (2005), ISSN 0920-9069, pp. 27-39.

Palomares, L.A. & Ramírez, O.T. The effect of dissolved oxygen tension and the utility of oxygen uptake rate in insect cell culture. *Cytotechnology.* Vol.22, (1996), ISSN 0920-9069, pp. 225-237.

Rohrmann, G.F. (2011). *Baculovirus Molecular Biology: Second Edition.* National Library of Medicine (US), National Center for Biotechnology Information, retrieved from http://www.ncbi.nlm.nih.gov/books/NBK49500/

Schlaeger, E-J. Medium design for insect cell culture. *Cytotechnology.* Vol.20, (1996), ISSN 0920-9069, pp. 57-70.

Shi, X. & Jarvis, D.L. Protein N-glycosylation in the baculovirus-insect cell system. *Current Drug Targets.*Vol.8, (2007), ISSN 1389-4501, pp. 1116-1125.

Sieburth, P., Maruniak, J. Growth characteristics of a continuous cell line from the velvetbean caterpillar, *Anticarsia gemmatalis* Hübner (Lepidoptera: Noctuidae*). In Vitro Cellular and Developmental Biology.*Vol.24, (1988), ISSN 1071-2690, pp. 195-198.

Stavroulakis, D.A., Kalogerakis, N. & Behie, L.A. Kinetic data for the BM-5 insect cell line in repeated-batch suspension cultures. *Biotechnology and Bioengineering.* Vol.38, (1991), ISSN 0006-3592, pp. 116-126.

Szewczyk, B., Hoyos-Carvajal, L., Paluszek, M., Skrzecz, I. & Lobo de Souza, M.. Baculoviruses: re-emerging biopesticides. *Biotechnology Advances*, Vol. 24, (2006), ISSN 07349750, pp. 143-160.

Taticek, R. & Shuler, M. Effect of elevated oxygen and glutamine levels on foreign protein production at high cell densities using the insect cell-baculovirus expression system. *Biotechnology and Bioengineering.* Vol.54, (1997), ISSN 0006-3592, pp. 142-152.

Trinh, K., García-Briones, M., Hink, F. & Chalmers, J. Quantification of damage to suspended insect cells as a result of buble rupture. *Biotechnology and Bioengineering.* Vol.43, (1994), ISSN 0006-3592, pp. 37-45.

van Beek, N. & Davis, D.C. Baculovirus insecticide production in insect larvae. *Methods in Molecular Biology*, Vol.388, (2007), ISSN 1064-3745 , pp. 367-378.

van Lier, F.L., van den End, E.J., de Gooijer, C.D., Vlak, J.M. & Tramper, J. Continuous production of baculovirus in a cascade of insect-cell reactors. *Applied Microbiology and Biotechnology.* Vol.33, (1990), ISSN 1432-0614, pp. 43-47.

Vaughn, J.L., Goodwin, R.H., Tompkins, G.J., McCawley, P. The establishment of two cell lines from the insect *Spodoptera frugiperda* (Lepidoptera; Noctuidae). *In Vitro.* Vol.13, (1977), ISSN 0073-5655, pp. 213-177.

Visnovsky, G., Claus, J. & Merchuk, J.C. Airlift reactors as a tool for insect cells and baculovirus mass production. *Latin-America Applied Research.* Vol 33, (2003), ISSN 0327-0793, pp. 117-121.

Wood, H.A., Johnston, L.B., Burand, J.P. Inhibition of Autographa californica nuclear polyhedrosis virus in high-density Trichoplusia ni cell cultures. *Virology.* Vol.119, (1982), ISSN 0042-6822, pp. 245-254.

Wyatt, G.R., Lougheed, T.C., Wyatt, S.S. The chemistry of insect hemolymph; organic components of the hemolymph of the silkworm, Bombyx mori, and two other species. *Journal of General Physiology*, Vol.39, (1956), ISSN 0022-1295, pp. 853-868.

8

Alternatives to Chemical Control of Insect Pests

Eric J. Rebek[1], Steven D. Frank[2],
Tom A. Royer[1] and Carlos E. Bográn[3]
[1]Oklahoma State University
[2]North Carolina State University
[3]Texas A&M University
United States of America

1. Introduction

In 2011, practitioners and advocates of Integrated Pest Management (IPM) find themselves addressing agricultural, societal, and political pressures worldwide resulting from human population growth. This growth brings simultaneous burdens of sustaining a steady food supply; these include preventing losses from pests, dealing with increased human global travel, which in turn intensifies opportunities for the establishment of non-endemic pests into new ecosystems, and addressing global climate change that potentially will shift pest distributions into new areas. Concurrently, societal concerns about pesticide presence in our food and environment have resulted in political and economic pressures to reduce chemical pesticide use, or at a minimum, emphasize the development and use of products that are less toxic and more environmentally safe. These concerns drive the discovery and development of alternatives to chemical control of plant pathogens, weeds, and insect pests. The term Integrated Pest Management has, more often than not, been identified with entomologists. Stern et al. (1959) first used the term "integrated control" to describe the potential for integration of chemical and biological control tactics. Yet from a historical view, the concept of integrating chemical control with other tactics was proposed much earlier (Hoskins et al., 1939). Furthermore, integrating multiple non-chemical tactics to control a pest has been a cornerstone of the discipline of plant pathology throughout much of its early history (Jacobsen, 1997). In fact, because plant pathologists did not have an array of corrective pesticides available to them, the development and integration of control methods that emphasized non-pesticide controls (e.g., genetic host resistance, crop rotations, tillage, and plant sanitation) for plant diseases was a necessity, not simply an option for plant disease management. In contrast, entomologists and weed scientists were more insulated from that necessity due to the availability of relatively inexpensive pesticides to correct a problem.

Several events stimulated the necessity for developing IPM programs in entomology, including those that emphasized development of non-chemical methods of insect control (e.g., cultural, biological, and physical control described herein). The chlorinated hydrocarbon, DDT, had been used for control of various insects since the 1950's. Soon after its use began, some pests began to develop resistance to DDT, including house flies, mosquitos, bed bugs, and body lice (Metcalf, 1989). The publication of Rachel Carson's book, "Silent Spring", in 1962 also generated public concern. Carson highlighted the negative impacts that widespread use of insecticides could have on the environment and ultimately, human health. What followed was a passionate global reaction that generated intense economic and political pressure to regulate pesticide use and monitor their relative impacts on biological systems. In the United States, the Environmental Protection Agency was created and charged with regulating the registration of all pesticides through the Federal Insecticide, Fungicide and Rodenticide Act (as amended in 1972). Concerns over pesticide use also stimulated the political thrust necessary for support of IPM programs. In the United States and worldwide, IPM flourished in the following three decades and was adopted as policy by various governments (Kogan, 1998).

Today, IPM has attained many successes but fallen short on some issues. Due to the awareness and biological understanding of how insecticide resistance develops, and because insecticides are so expensive to develop, in 1984 the manufacturers of insecticides created the Insecticide Resistance Action Committee (IRAC) to encourage the responsible use of their products in a manner that minimizes the risk of insecticide in target pest populations (IRAC, 2010). New calls have been made for changing the direction of IPM in response to waning political support for funding IPM programs. Frisbie & Smith (1989) coined the term "biologically intensive" IPM, which involves reliance on ecological methods of control based on knowledge of a pest's biology. Benbrook et al. (1996) promoted the idea of moving IPM along a continuum from simple to complex, or 'biointensive'. The National Research Council officially introduced the term, "Ecologically Based Pest Management", calling for a new paradigm for IPM in the 21st Century (National Research Council, 1996); eight years earlier, however, Horn (1988) outlined how principles of insect ecology could be incorporated into insect pest management strategies. More recently, Koul & Cuperus (2007) published "Ecologically Based Integrated Pest Management", essentially capturing the breadth and depth of the evolution that IPM has undergone over the past 60 years. While the scope of the "New Solutions" aspect of the NRC's charge has been challenged (Kogan, 1998; Royer et al., 1999), the term "ecologically based" has become infused into the IPM lexicon.

2. Cultural control methods to reduce insecticide applications

Cultural controls are management tools and activities that make the crop habitat less favorable for pests to survive and cause damage (Horne & Page, 2008). Cultural management practices may make the crop or habitat inhospitable to pests directly, for example, by planting cultivars resistant to pest feeding or rotating crops to deny overwintering pests their preferred food source. Cultural management practices can also make the habitat less hospitable to pests in an indirect manner by encouraging natural enemies (predators and parasitoids) to enhance biological control (see Section 3).

Cultural control is a key pest management tool available to growers because the crop variety, habitat, and selected inputs set the stage for future pest fitness and abundance. Thus, implementing preventive cultural control tactics that slow pest population growth can delay or negate the need for insecticide applications and significant plant damage. In this section we outline the major types of cultural control tactics available to growers and other pest management personnel. Our objective is to demonstrate the breadth of tactics that are used, although we do not have the space to consider them in depth. We draw examples from a diversity of well-studied plant systems from field crops to ornamental landscapes to provide examples of how they affect plant-herbivore-natural enemy interactions to reduce pest abundance and damage.

2.1 Cultural control via plant resistance

Plant resistance to herbivores is a cultural control strategy having the most direct influence on herbivore behavior, fitness, and damage. Plant resistance is achieved through three general mechanisms: antibiosis, antixenosis, and tolerance. Antibiosis is the adverse effect of plant physical or chemical traits on arthropod biology (Painter, 1951). This may include reduced size, survival, fecundity, or longevity and increased development time or mortality. Antixenosis is the effect of plant traits on herbivore behavior that reduces herbivore interactions with the plant (Painter, 1951). These effects can include reduced feeding, preference, residence time, or oviposition on plants having particular traits such as trichomes or defensive compounds. Tolerance is a plant trait that reduces the impact of herbivory on plant growth, allowing tolerant plants to sustain herbivore damage but maintain yields similar to undamaged plants (Painter, 1951).

Physical plant traits such as leaf pubescence, trichomes, and epicuticular wax, and chemical traits such as alkaloids and terpenoids have antibiotic and antixenotic effects on herbivores (Kennedy & Barbour, 1992; Painter, 1951). In the well-studied tomato production system, effects of leaf trichomes as a plant resistance trait are well documented (Kennedy, 2003; Simmons & Gurr, 2005). Trichomes and associated chemicals confer resistance to some tomato varieties against mites, aphids, whiteflies, beetles, and caterpillars (Gentile & Stoner, 1968; Heinz & Zalom, 1995; Kennedy, 2003; Kennedy & Sorenson, 1985; Simmons & Gurr, 2005). Trichomes are stiff hairs that sometimes contain chemical glands. Glandular trichomes have chemical exudates that confer resistance through antibiosis and kill or reduce longevity of pests feeding on them and entrap pests that forage on the leaves (Simmons & Gurr, 2005). Trichomes also have antixenotic effects on herbivore pests. Increasing trichome density can reduce oviposition by many species of beetles, caterpillars, true bugs, and mites. Of particular relevance is the effect of trichome density on whitefly and mites pests (Simmons & Gurr, 2005). The antibiotic and antixenotic effects of leaf pubescence on whitefly behavior and fitness have been studied in depth in a number of systems such as tomato, tobacco, cucurbits, and ornamental plants (Hoddle et al., 1998; Inbar & Gerling, 2008).

The soybean aphid offers a current example of how identifying pest resistance in crop plants can benefit IPM. Soybean aphid arrived in the U.S. from Asia in 2000 (Ragsdale et al., 2011). Since that time plant resistance conferred through antibiosis and antixenosis mechanisms has played an important role in mediating the economic impact of this pest on soybean yield (Ragsdale et al., 2011). Aphid fitness is negatively affected in resistant lines because it takes twice as long for aphids to probe into the phloem and initiate feeding (Diaz-Montano et al., 2007). Further, feeding bouts are reduced by more than 90% from less than 7 minutes per bout in resistant lines compared to greater than 60 minutes in susceptible lines (Diaz-Montano et al., 2007). Likewise, production of nymphs was reduced by 50-90% in resistant versus susceptible varieties, confirming antibiosis in resistant lines (Diaz-Montano et al., 2006; Hill et al., 2004). Antixenosis was also demonstrated in resistant varieties wherein adult aphids preferred to colonize susceptible over some resistant lines (Diaz-Montano et al., 2006; Hill et al., 2004).

In contrast to conventional breeding programs, plants can now be genetically modified to include lethal traits from other organisms, such as the bacterium, *Bacillus thuringiensis* (Bt). Bt genes are now used in many crop species to confer antibiosis in otherwise susceptible crops. Although we do not focus on this mode of plant resistance here, transgenic traits have had a tremendous effect on modern crop production and yield. However, like any management tactic, Bt crops do not function in a vacuum and effects on natural enemies and other non-targets, secondary pest outbreaks, and evolution of pest resistance have been intensely studied (Gould, 1998; O'Callaghan et al., 2005; Shelton et al., 2002).

2.1.1 Interaction of plant resistance traits and biological control

Effects of plant resistance and biological control can be contradictory, complementary, or synergistic (Cai et al., 2009; Farid et al., 1998; Johnson & Gould, 1992). Plant resistance can work in conjunction with natural enemies to maintain low pest abundance and damage. In general, natural enemies have slower population growth rates than pests. Thus, by reducing pest population growth rates, plant resistance may help natural enemies better regulate pest populations. For example, research in wheat systems has shown that aphid-resistant wheat varieties do not have negative effects on parasitoid life history parameters such as size and development time (Farid et al., 1998). Parasitism rates may be equal or greater on resistant varieties, which when combined with reduced aphid population growth due to host plant resistance, can improve pest management dramatically (Cai et al., 2009).

Just as trichome exudates reduce herbivore survival they can also have negative effects on natural enemies. Survival and development of natural enemies may be reduced by poisoning or entrapping them, and natural enemy foraging efficiency, predation, or parasitism rate may be inhibited (Kaufman & Kennedy, 1989a, b; Obrycki & Tauber, 1984; Simmons & Gurr, 2005). For example, increasing trichome density and related changes in chemical composition of tomato leaves reduced the walking speed, parasitism rate, and survival of the egg parasitoid, *Trichogramma pretiosum* (Kashyap et al., 1991a, b). Tiny whitefly parasitoids in the genera, *Eretmocerus* and *Encarsia*, are highly affected by plant pubescence and trichome density (Hoddle et al., 1998; van Lenteren et al., 1995). Biological control may be disrupted because these parasitoids avoid highly pubescent plants. Once on the plants, pubescence reduces walking speed, foraging efficiency, oviposition, and parasitism rate (De Barro et al., 2000; Headrick et al., 1996; Hoddle et al., 1998; Inbar & Gerling, 2008).

Trichomes and other plant resistance traits also affect predator behavior and efficacy. Predatory mites used in biological control of spider mite, *Tetranychus urticae*, are readily trapped by trichomes and forage less efficiently due to reduced mobility (Nihoul, 1993a; van Haren et al., 1987). The consequence of mortality and reduced foraging efficiency is reduced biological control on cultivars with high trichome density, although the effect is also dependent on environmental factors such as temperature (Nihoul 1993a, b). Likewise, foraging efficiency of the spotted lady beetle, *Coleomegilla maculata*, and the bigeyed bug, *Geocoris punctipes*, was reduced by high trichome density, resulting in less predation of *Heliothis zea* eggs (Barbour et al., 1993, 1997). Increasing pubescence on poinsettia leaves by just 15% reduced oviposition and whitefly predation by *Delphastus catalinae* and other predators (Heinz & Parrella, 1994).

2.1.2 Herbivore resistance to plant resistance traits

Herbivores are in a constant evolutionary arms race with plants to overcome resistance traits (Ehrlich & Raven, 1964). It is not surprising then that pests have developed physiological resistance to genetically modified and conventional plant resistance traits (Gould, 1998). For example, certain soybean aphid biotypes are resistant to Rag1 or Rag2 genes that confer resistance to soybean plants (Hill et al., 2009, 2010). Evidence from theoretical and empirical research suggests that multiple resistance traits or genes and a combination of different modes of action such as antibiosis and antixenosis should confer more stable resistance to crops. In addition, mixing resistant and susceptible varieties in the same field can reduce evolution of resistance by insect pest populations (Gould, 1986, 1998).

2.2 Cultural control via fertility management

Plant fertility and water stress play a major role in plant susceptibility to herbivore feeding, tolerance to herbivore damage, and herbivore population growth. Nitrogen can be a limiting nutrient for herbivorous insects due to the nitrogen-poor quality of their host plants (Mattson, 1980). Therefore, increasing nitrogen concentration within plants by applying fertilizer has a tendency to increase plant quality for herbivores (Mattson, 1980). Increasing foliar nitrogen can reduce pest development time and increase survival and fecundity, leading to more rapid population growth (Mattson, 1980). Research in potato crops has found that increasing nitrogen fertilization increases leaf consumption, reduces development time, and increases abundance of Colorado potato beetles (Boiteau et al., 2008). Likewise, in greenhouse ornamental production, increasing fertilization increases the fecundity, body size, and development rate of citrus mealybug (Hogendorp et al., 2006), and through similar mechanisms increases population growth rates of whiteflies, thrips, aphids, and spider mites (Bentz et al., 1995; Chau et al., 2005; Chau & Heinz, 2006; Chow et al., 2009).

In ornamental landscapes, fertilizer is often applied to improve the growth of trees and other plants and increase their resistance to abiotic and biotic stress, including herbivore feeding. However, nitrogen fertilization of trees has been shown to reduce plant resistance to many arthropod pests (Herms, 2002; Kytö et al., 1996). This reduced resistance occurs through a combination of fertilizer effects on plant nutrition for herbivores and defense against herbivores (Herms & Mattson, 1992). Herms & Mattson (1992) hypothesized that as nitrogen fertilization stimulates rapid plant growth, carbon available for production of defensive compounds is limited. Thus, over-fertilization of trees, shrubs, and other plants provides a dual benefit to many herbivores via increased nitrogen availability and decreased defensive compounds (Raupp et al., 2010).

2.3 Cultural control via pesticide selection and management

Pesticide applications are often an essential aspect of plant culture. Managing the type and frequency of applications is a cultural control tactic with well-documented implications. Insecticides can disrupt natural enemy communities and biological control via several mechanisms. Direct toxicity of pyrethroids and organophosphates to natural enemies has been documented frequently (Desneux et al., 2004b; see Galvan et al., 2005). Direct toxicity of insecticides to natural enemies results in smaller natural enemy populations on crop and landscape plants (Frank & Sadof, *in press*; Raupp et al., 2001). Insecticides also cause sublethal effects in parasitoids and predators. For example, the pyrethroid, lambda-cyhalothrin, disrupts the host location and oviposition behavior of *Aphidius ervi*, resulting in lower parasitism rates of aphids (Desneux et al., 2004a).

Non-target impacts on natural enemy communities are not limited to contact insecticides. Systemic neonicotinoids such as imidacloprid and thiamethoxam have lethal and sublethal effects on natural enemy development, fitness, and efficacy (Cloyd & Bethke, 2009; Desneux et al., 2007). These compounds can reduce survival of developing parasitoids and intoxicate predators such as lady beetles and lacewing larvae exposed to the chemicals topically or by feeding on exposed prey (Moser & Obrycki, 2009; Papachristos & Milonas, 2008; Smith & Krischik, 1999; Szczepaniec et al., 2011). Parasitoids are also affected negatively via feeding on plant nectar or hosts exposed to the chemicals (Krischik et al., 2007; Rebek & Sadof, 2003). The consequence of disrupting natural enemy populations can be outbreaks of primary or secondary pests due to the loss of underlying biological control services (Raupp et al., 2010). Considerable work has documented this effect in field crops, orchards, vineyards, and landscape ornamentals (Penman & Chapman, 1988; Raupp et al., 2010). The effect is particularly prevalent among spider mites and scale insects that are not killed as easily as their natural enemies by insecticide applications. Pyrethroids and other broad-spectrum insecticides have direct and indirect effects on spider mites that can promote mite outbreaks. First, pyrethroids promote spider mite dispersal from treated to untreated areas of reduced competition (Iftner & Hall, 1983; Penman & Chapman, 1983). Spider mites have many predators including lady beetles, predatory bugs, lacewing larvae, and predatory mites. Pyrethroids can promote outbreaks of spider mites indirectly by killing the natural enemies that otherwise help suppress spider mite populations (Penman & Chapman, 1988).

Predatory mites in the family Phytoseiidae feed on spider mite eggs, juveniles, and adults and are effective at reducing spider mite abundance and damage on plants (McMurtry & Croft, 1997). In addition, phytoseiid mites often respond with a numerical increase to burgeoning spider mite populations via aggregation and increased reproduction. However, the abundance and efficacy of phytoseiid mites depends in large part on plant culture practices and plant characteristics. Phytoseiid mites are extremely susceptible to insecticides such as pyrethroids, organophosphates, and carbamates (Hardman et al., 1988). In many cases, phytoseiids have been found to be more vulnerable to these insecticides than spider mites (e.g., Sanford, 1967; Wong and Chapman, 1979). Therefore, by killing a disproportionate number of predatory mites compared to target pests, broad-spectrum insecticides frequently lead to spider mite outbreaks (Hardman et al., 1988). Similar dynamics have been demonstrated for scale insects, which are generally not killed by cover sprays of contact insecticides due to their protective cover. Moreover, by drastically reducing natural enemy abundance and efficacy, these insecticide applications create enemy-free space for scales, which can result in outbreak populations (McClure, 1977; Raupp et al., 2001).

Insecticide applications can directly benefit pest reproduction and survival through a process known as hormoligosis. Increased spider mite fecundity has been demonstrated after exposure to sublethal doses of pyrethroids (Iftner & Hall, 1984; Jones & Parrella, 1984). However, this is most evident in spider mites that frequently outbreak after applications of the neonicotinoid, imidacloprid (Gupta & Krischik, 2007; Raupp et al., 2004; Sclar et al., 1998; Szczepaniec et al., 2011). Outbreaks are triggered in part by negative effects on predators, but also by greater fecundity of spider mites that feed on imidacloprid-treated foliage (Szczepaniec et al., 2011). Although not commonly observed, this phenomenon points out another reason for proper insecticide management as a cultural control strategy.

2.4 Cultural control via crop rotation and planting practices

Exploiting the biological limitations of the pest to minimize insecticide applications is the essence of cultural control tactics such as crop rotation. This strategy has been used successfully to control corn rootworm for over 100 years (Forbes, 1883). Crop rotation has been highly effective as a tool to reduce Western corn rootworm, *Diabrotica virgifera virgifera*, and Northern corn rootworm *Diabrotica barberi*, damage in corn (Levine & Oloumi-Sadeghi, 1991; Spencer et al., 2009). Corn rootworm eggs overwinter in corn fields and larvae are present to feed on corn roots the following year (Spencer et al., 2009). Therefore, rotating to a different crop such as soybeans denies food to hatching rootworm larvae (Spencer et al., 2009). Likewise, corn planted after soybeans or other crops has less rootworm damage because the field is free of overwintering eggs and larvae. However, Western and Northern corn rootworm populations eventually developed resistance to this strategy (Gray et al., 2009; Levine et al., 2002; Spencer & Levine, 2008). Northern corn rootworms circumvent crop rotation by prolonging egg diapause for two winters instead of one (Chiang, 1965; Levine et al., 1992). Therefore, larvae hatch when fields are replanted in corn two years after the eggs were laid. Western corn rootworm has become resistant to crop rotation by a behavioral rather than physiological mechanism. Western corn rootworm adults move from corn fields to soybean and other crop fields, feeding on soybean leaves and ovipositing in soybean fields (Levine et al., 2002). Selection pressure imposed by rotation of two primary crops, corn and soybeans, strongly rewarded female beetles that strayed from corn for oviposition.

Other planting practices such as delayed planting dates can also benefit pest control. Hessian fly is an introduced pest of winter wheat that has been in the U.S. since the 1700's. Prior to the development of resistant wheat varieties, growers exploited the fly's life cycle to reduce damage to winter wheat crops. Hessian fly adults become active in the fall when they oviposit in wheat and other grasses. By planting after a "fly free date" when fly activity subsides, winter wheat is protected from oviposition from the fall hessian fly generation (Buntin et al., 1991). This is a perfect example of how simple changes in plant culture can reduce the need for insecticide applications, increase yield, and provide economic benefit to growers (Buntin et al., 1992).

3. Biological control of insect pests

Many definitions of biological control have been published in the literature since the term was first used by H.S. Smith more than 90 years ago (Caltagirone & Huffaker, 1980; Cook, 1987; Coppel & Mertins, 1977; DeBach & Rosen, 1991; Garcia et al., 1988; see Huffaker & Messenger, 1976; Perkins & Garcia, 1999; Rabb, 1972; Smith, 1919). In its strictest sense, biological control is the use of beneficial organisms to reduce the relative abundance of, and damage caused by, noxious ones. This definition attributes economic rather than biological characters to organisms that fall into two categories, beneficial and noxious, based on their positive or negative impact on human-valued resources. It is also important to distinguish *biological* from *natural* control, which does not require human intervention, and from similar methods of pest control that do not involve whole (living) organisms (Huffaker et al., 1984). In fact, biological control involves interspecific, population-level processes by way of predation, parasitism, competition, or a combination of these mechanisms (van Driesche & Bellows, 1996). In practice, the effectiveness and appropriateness of biological control methods rely on real-time evolutionary forces that shape the beneficial organism's genotype, phenotype, and performance. This is not the case for similar, biologically based methods such as the application of insecticides formulated with pathogens, antagonists, or their byproducts. Furthermore, in its strictest definition, biological control does not include the deployment of pest-tolerant organisms, regardless of the source or origin of the resistance-conferring characters (e.g., Bt crops) (see Perkins & Garcia, 1999).

The history and origins of biological control have been extensively covered in previous volumes (Caltagirone & Doutt, 1989; DeBach & Rosen, 1991; van Driesche & Bellows, 1996) and is not the subject of this review. However, it is significant to note that early theory and application of biological control principles pre-date the modern insecticide era (Smith, 1919). Therefore, it is modern insecticides that became an alternative to biological control and not the other way around. In this context, biological control should not be viewed as a novel tactic but as the foundation of a successful pest management strategy involving, at minimum, the conservation of ecosystem resources to facilitate the process of pest-natural enemy colonization, host/prey finding, and ultimately, damage reduction. Although what constitutes biological control (or not) continues to be a subject of discussion and will likely evolve with new technologies, the recognition of three principal biological control methods remains unchanged. These three approaches are importation (a.k.a., classical biological control), augmentation, and conservation biological control (Smith, 1919).

3.1 Importation biological control

Importation biological control is the oldest of the three approaches (hence its alternative name, 'classical'). The first successful case of importation biological control occurred over a century ago in the control of cottony cushion scale in California citrus following importation of the vedalia beetle (Horn, 1988). The classical approach involves re-establishing the interspecific interactions (and their impact on population regulation) between pests and their natural enemies (i.e., predators, parasitoids, or insect-killing pathogens) as they occur in the pest's endemic range (Howarth, 1983). The need to re-establish these interactions arises because pests are commonly introduced into areas outside their native range where they lack natural enemies, or those that are present do not significantly impact the pest's abundance and local distribution. Since its inception, importation biological control has been used with varying degrees of success against noxious pests like cassava mealybug in Africa, Rhodesgrass mealybug in Texas, walnut aphid in California, and southern green stink bug in Australia, New Zealand, and Hawaii (Hokkanen, 1997).

The technical expertise, time commitment, and considerable expense necessary to carry out importation biological control require the involvement of specially trained university and government scientists. Importation is highly regulated in many countries, largely due to growing concern over the introduction of exotic, invasive species into new environments. In the U.S., the Animal and Plant Health Inspection Service (APHIS) oversees and coordinates importation biological control programs. The agency's charge is to preserve the safety and effectiveness of biological control primarily through post-release monitoring of biological control agents (USDA APHIS, 2011). Although there are a few documented cases of introduced biological control agents causing economic or ecological harm, societal perceptions that importation biological control is too risky are often influenced by subjectivity and misinformation (Delfosse, 2005). To minimize risk, researchers must provide evidence that introduced natural enemies are unlikely to harm crops, humans, and ecosystems. This requires substantial analysis of host feeding preference and other biological traits of prospective biological control agents (see Briese, 2005).

3.2 Augmentation biological control

The aim of augmentation biological control is to improve the numerical ratio between pest and natural enemy to increase pest mortality. It involves the release of natural enemies, typically mass reared in an insectary, either to inoculate or inundate the target area of impact (Obrycki et al., 1997; Parrella et al., 1992; Ridgway, 1998). Inoculative releases involve relatively low numbers of natural enemies, typically when pest populations are low or at the beginning of a growth cycle or season. Inundation involves relatively high numbers of natural enemies released repeatedly throughout the growth cycle or season. Thus, inundative release of natural enemies is similar to insecticide use in that releases are made when pests achieve high enough density to cause economic harm to the crop. In both types of release, the objective is to inflict high mortality by synchronizing the life cycles of the pest and natural enemy. Hence, an effective monitoring program of pest populations is essential to the success of augmentation biological control.

Augmentation biological control has been used successfully against key pests of field and greenhouse crops. A well-known example of augmentation biological control is the use of the parasitoid, *Encarsia formosa*, for control of greenhouse whitefly (Hoddle et al., 1998). Indeed, augmentation plays an important role in greenhouse production, especially in Europe, and many natural enemies are commercially available for control of perennial greenhouse pests such as spider mites, aphids, scales, and whiteflies (Grant, 1997; Pottorff & Panter, 2009). The success of augmentative releases in greenhouses, and elsewhere, depends on the compatibility of cultural practices such as insecticide use with natural enemies (see Section 2.3). Greenhouses are often ideal sites for augmentation biological control because of the relative stability of the enclosed environment. In contrast, a critical review of augmentation biological control in field crops revealed that augmentation was typically less effective and more expensive than conventional control with pesticides (Collier & van Steenwyk, 2004). The authors found that the low success rate of augmentation biological control in field crops is influenced by ecological limitations such as unfavorable environmental conditions, natural enemy dispersal, and refuge for herbivores from released natural enemies.

3.3 Conservation biological control

Conservation biological control involves any practice that increases colonization, establishment, reproduction, and survival of native or previously established natural enemies (Landis et al., 2000). Conservation biological control can be achieved in two ways: modifying pesticide use and manipulating the growing environment in favor of natural enemies. Conservation practices have proven effective in a wide variety of growing situations ranging from small garden plots to large fields, agricultural to urban environments, and commercial to private settings (Frank & Shrewsbury, 2004; Landis et al., 2000; Rebek et al., 2005, 2006; Sadof et al., 2004; Tooker & Hanks, 2000).

3.3.1 Conserving natural enemies via modified pesticide use

Modifications to pesticide regimens include reducing or eliminating insecticide use, using pest-specific insecticides when needed, making applications when beneficial arthropods are not active, and making treatment decisions based on monitoring and the presence of vulnerable life stages. While total independence from chemical control is not feasible for most situations, reductions in insecticide use are possible through IPM programs based on rigorous pest monitoring, established treatment thresholds, and/or insect population models (see Horn, 1988; Pimental, 1997). Thus, insecticides are used only when needed to prevent crop damage that results in economic loss. When insecticide use is warranted, adverse effects on natural enemies can be minimized by using selective, pest-specific products that are only effective against the target pest and its close relatives. Selective chemistries include microbial insecticides, insect growth regulators, botanicals, and novel insecticides with specific modes of action against target insects. Alternatively, insecticide applications can be timed so they not coincide with natural enemy activity; dormant or inactive predators and parasitoids are not exposed to broad-spectrum insecticides applied when they are dormant or inactive (van Driesche & Bellows, 1996). This strategy requires a thorough understanding of the crop, agroecosystem, and the biology and life cycle of important natural enemies in the system.

3.3.2 Conserving natural enemies via habitat manipulation
Natural enemies are attracted to habitats rich in food, shelter, and nesting sites (Landis et al., 2000; Rabb et al., 1976). Many perennial plants can provide these resources when incorporated into the system. Ellis et al. (2005) and Rebek et al. (2005) independently observed significantly enhanced parasitism of two key ornamental pests, bagworm and euonymus scale, in experimental plots containing nectar and pollen sources (i.e., resource plants). Resource plants also served as refuge for vertebrate predators of bagworms as evidenced by increased predation rates (Ellis et al., 2005). Resource plants can harbor alternative prey/host species, which sustain adult and immature natural enemies when primary prey/hosts are scarce. For example, many studies have focused on the influence of banker plants, which contain alternative prey species, on natural enemy effectiveness (see Frank, 2010).

Resource plants provide more than food to enhance natural enemy abundance in impoverished landscapes. Suitable changes in microclimate are afforded by many plants, tempering environmental extremes by providing improved conditions for natural enemy survival (Rabb et al., 1976). Candidate plants include small trees, shrubs, bushy perennials, and tall ornamental grasses with dense canopies or complex architecture. Similarly, organic mulches and ground cover plants can support large numbers of ground-dwelling predators like spiders and ground beetles (Bell et al., 2002; Mathews et al., 2004; Rieux et al., 1999; Snodgrass & Stadelbacher, 1989), which may enhance biological control of key pests (Brust, 1994). Finally, resource plants can enhance reproduction of natural enemies and provide refuge from their own enemies (Landis et al., 2000; Rabb et al., 1976).

The effectiveness of habitat manipulation to improve biological control requires careful planning and selection of plant attributes that are appropriate for the natural enemy complex present in the system (Landis et al., 2000). For example, flower morphology can significantly impact nectar accessibility by foraging parasitoids (Patt et al., 1997; Wäckers, 2004). Also important is coincidence of floral bloom with natural enemy activity. Selected resource plants should overlap in blooming periods to ensure a continuous supply of nectar and pollen to natural enemies (Bowie et al., 1995; Rebek et al., 2005). Other considerations that exceed the scope of this chapter include the influence of landscape-level attributes on biological control at different spatial scales (Kruess & Tscharntke, 1994; Marino & Landis, 1996; Roland & Taylor, 1997).

3.4 Factors affecting success of biological control
While there have been some tremendous successes, the worldwide rate of effective biological control is estimated to be between 16-25% (Hall et al., 1980; Horn, 1988; van

Lenteren, 1980). In practice, the successful application of biological control usually requires a combination of at least two of the three approaches, importation, augmentation, and conservation of natural enemies (DeBach & Rosen, 1991; van Driesche & Bellows, 1996). What drives the success or failure of biological control programs in plant crops has been the subject of many analyses, either using historical records or theoretical approaches (Andow et al., 1997; Murdoch et al., 1985; Murdoch & Briggs, 1996; van Lenteren, 1980). In general terms, biological control programs are more likely to succeed under certain production systems and environmental conditions (Clausen, 1978; van Driesche & Heinz, 2004). Biological control has been more successful in crops that: 1) are perennial versus annual; 2) grow in areas with few pests versus many pests; 3) the harvested portion is not damaged by the target pest; 4) the target pest is not a disease vector; and 5) the aesthetic damage is acceptable (e.g., some food and fiber crops versus ornamentals).

Failures in biological control programs, especially those recorded in the literature, also involve cases where the biology and ecology of the natural enemy or the pests are not well understood or altogether unknown. Historically, failures in importation biological control have occurred after errors in identification of a pest or natural enemy at the level of species, biotype, or even local strain; a mismatch in micro-environmental requirements for natural enemy growth and development; incorrectly timing natural enemy release when the production system is not conducive to establishment; or when socioeconomic or regulatory barrier have prevented adoption or implementation (Clausen, 1978; Greathead, 1976; Hall & Ehler, 1979; Knutson, 1981). Similarly, failures in augmentation and conservation biological control, although not commonly recorded in the literature, may be due to a lack of understanding of the basic biology and ecology of the species involved, the basic requirements of the production system, and any socioeconomic barriers including real or perceived costs and benefits (Murdoch et al., 1985; Perkins & Garcia, 1999; Collier & van Steenwyk, 2004). The success of biological control programs involves integrated efforts at many levels ranging from biology to economics, from research to implementation and experience, and from the farm to the community and region.

4. Physical control strategies to reduce pest incidence

Plant health can benefit greatly from preventing or limiting injury from arthropod pests from the start. Indeed, the cornerstone of an effective IPM program is prevention, which can be achieved, in part, through physical control. Physical control strategies include methods for excluding pests or limiting their access to crops, disrupting pest behavior, or causing direct mortality (Vincent et al., 2009). Physical control methods can be categorized as active and passive (Vincent et al., 2009). Active methods involve the removal of individual pests by hand, pruning out infested plant tissues, and rogueing out heavily infested plants. Passive methods usually include the use of a device or tool for excluding or removing pests from a crop. Typically, these devices serve as barriers between plants and insect pests, thus protecting plants from injury and damage. Other passive tools include repellents and traps. While traps are often used for monitoring pest abundance and distribution, many are designed as "attract and kill" technologies, which attract insect pests through color, light, shape, texture, and scent, or a combination of these factors.

The greatest disadvantage to physical control is that these methods can be laborious and time consuming, especially for crops grown in large areas. Also, a moderate degree of specialization or training is often required due to the highly technical nature of some physical control methods. Physical control methods may also be difficult or practically impossible in some crops like large trees grown in extensive monocultures (e.g., timber production). For many crops, however, physical control of certain pests can be incorporated into established routines for managing crops. Despite the drawbacks and considering the costs, regulations, and limitations of insecticide use, physical control methods are likely candidates for inclusion in many pest management programs, especially for high-value crops (see Vincent et al. 2003). Here, we discuss briefly some examples of physical control classified by their primary function: exclusion, behavior modification, and destruction of pests.

4.1 Physical control via exclusion
Pest exclusion is a key factor in preventing pests from accessing crops, thereby reducing the economic impact of insects. Both passive and active exclusion methods have been implemented in various agricultural systems including fields, greenhouses, and postharvest facilities. Physical control via exclusion devices is perhaps most important in protected environments such as greenhouses and grain bins, where optimal temperatures and humidity, a readily available food source, and a general lack of natural enemies contribute to the proliferation of pest populations. Screens are common passive exclusion devices used in greenhouse production. Screens can prevent pest migration into greenhouses through vents and other openings, especially when insect populations build up in weeds and crops in the surrounding environment (Gill et al., 2006; Pottorff & Panter, 2009). However, screen mesh size is an important concern as fine materials with small openings inhibit entry of tiny arthropods such as thrips and mites but also restrict air flow for cooling (Pottorff & Panter, 2009). Other active methods of physical control are necessary components of greenhouse IPM. Specifically, crops should be inspected for pests prior to moving new plant materials into production areas; discovered pests are removed by hand, pruned out, or discarded and destroyed with heavily infested plants.

In the field, floating row covers can exclude important vegetable pests such as cabbage maggot, flea beetles, and cabbageworm (Rekika et al., 2008; Theriault et al., 2009). Adhesives and burlap have been used to trap caterpillar pests such as gypsy moth and cankerworms as they migrate vertically along tree trunks (Potter, 1986). Other barriers include fences, ditches, moats, or trenches. For example, V-shaped trenches have been used around potato fields to prevent movement of Colorado potato beetle into the crop from adjacent, overwintering habitat (Boiteau & Vernon, 2001; Misener et al., 1993; see Vincent et al., 2003). Efficacy of this technique relies on trench design and knowledge of the pest, specifically, the population size and the ratio of crawling to flying individuals (Weber et al., 1994; Vincent et al., 2003).

4.2 Physical control via behavior modification
IPM programs often consist of physical control methods that alter the behavior of insect pests. Behaviors such as reproduction, aggregation, oviposition, feeding, alarm, and defense can be modified in two ways: "push-pull" strategies and mating disruption (Cook et al., 2007; Zalom, 1997). The former are designed to repel (push) or attract (pull) insect pests away from a crop by exploiting their reproductive, feeding, or aggregation behavior. Although many repellents and attractants are chemically based, here we treat their use in IPM as a form of non-chemical (non-insecticidal) control.

Pheromones, or chemical lures, are used in IPM programs to monitor pest populations and modify their behavior. Specifically, pheromone traps are used to detect pest activity in a crop and estimate their relative abundance in order to properly time an insecticide application or natural enemy release. Pheromones and other olfactory stimuli are receiving increased attention as repellents and attractants in push-pull strategies for modifying pest behavior (see Cook et al., 2007). Repellents include synthetic chemicals (e.g., DEET), non-host volatiles that mask host plant odors (e.g., essential oils), anti-aggregation and alarm pheromones, anti-feedants (e.g., neem oil), and oviposition deterrents (e.g., oviposition-deterring pheromones) (Cook et al., 2007). Herbivore-induced plant volatiles are host plant semiochemicals that induce plant defense from herbivores and attract natural enemies (James, 2003). Non-chemical repellents include reflective mulches, which have been shown to reduce damage and population density of tarnished plant bug in strawberry fields (Rhainds et al., 2001). Attractants include sex and aggregation pheromones, host plant volatiles, and feeding stimulants (e.g., baits), and oviposition stimulants (Cook et al., 2007). Other attractants are based on visual cues. For example, apple maggots are effectively controlled in apple orchards with 8-cm, red, spherical traps covered in adhesive. The attractiveness of these traps is enhanced by adding butyl hexanoate and ammonium acetate, synthetic olfactory stimulants (Prokopy et al., 1994).

Another common tactic is to use sex pheromones for mating disruption. Many insect pests rely on a species-specific, sex pheromone produced by females for mate location and recognition. Mating disruption is achieved by flooding the crop environment with the chemical signal, thus confusing males and reducing mate-finding success. This approach has been used with varying degrees of success for management of orchard and vineyard pests including codling moth, oriental fruit moth, grape berry moth, and peachtree borer (see Zalom, 1997).

4.3 Physical control via pest destruction

Insects can be killed directly through mechanical, thermal, or other means. Vincent et al. (2009) list several strategies that inflict mortality on pests including freezing, heating, flaming, crushing, and irradiating. One of the most common mechanical methods requires no specialized equipment – many gardeners derive great satisfaction from hand picking pests from a plant and crushing them. Hand removal can be used effectively for a myriad of relatively sessile landscape pests including bagworms, tent caterpillars, and sawfly larvae. Galls, egg masses, and web-making insects can also be pruned out of infested landscape plants (Potter, 1986). However, this tactic may be impractical for large trees or shrubs and dense populations of the pest. Other mechanical control options require specialized machinery. Pneumatic control involves removing pests from crops by use of a vacuum or blower and subsequently destroying them. Field crop pests such as Colorado potato beetle and lygus bug have been controlled in this manner, although care must be taken to avoid negatively impacting natural enemies (Vincent et al., 2003, 2009). Another example of mechanized destruction is the entoleter, an impact machine that is used in mills to remove and kill all life stages of insect pests (Vincent et al., 2003).

Modifying the microclimate can be effective in killing many insect pests, which cannot survive outside of optimal temperature and humidity ranges. Heat has been shown to be a very effective control method for bed bugs, which are difficult to control and are becoming more prevalent in domestic dwellings worldwide (Pereira et al., 2009). A wide variety of stored product pests can be controlled by pumping hot or cold air into the food storage facility, or modifying the storage environment with elevated temperatures and carbon dioxide (Vincent et al., 2003, 2009). Hot-water immersion, flaming, steaming, and solar heating are other thermal control options (Vincent et al., 2003). Electromagnetic energy has been studied for its effectiveness at killing insects (Vincent et al., 2009). Ionizing radiation has been used in quarantine facilities to treat fruit and other commodities suspected of carrying serious agricultural pests (Vincent et al., 2003). Targets of other electromagnetic methods, especially microwave treatments, include stored product pests. However, electromagnetic treatments may be limited by government regulations, cost, and the need for specialized equipment and training (Vincent et al., 2009).

5. Conclusions

Crop culture sets the stage for interactions between plants, pests, and natural enemies, and has a strong influence on the outcome of these interactions. In many cases, implementing effective cultural controls can be the most economical pest management tactic available to growers because labor and expense are incurred regardless of whether an effective cultural tactic is used. Understanding and implementing cultural practices can reduce other production expenses such as insecticides and fertilizer. Cultural control can be compatible with biological control if the myriad interactions among plants, pests, and natural enemies are well defined. Improving the predictability of biological control will rely on elevating the discipline to its proper place in applied evolutionary ecology and further refinement of the art and practice of biological control (van Lenteren, 1980; Heinz et al., 1993; Heinz, 2005). Fortunately, the organic and sustainable agriculture movements that are gaining both societal and political momentum seem to embrace the art and science of biological pest control (Edwards, 1990; Raynolds, 2000). While various physical control techniques have been used successfully in production systems, this strategy is limited by the significant labor, time, cost, and specialization required for successful control (Vincent et al., 2009). Further refinements and developments in physical control technologies hold promise for enhanced efficacy, compatibility with cultural and biological control, and profits.

As we move into the future of pest management, new challenges await. Crops are now genetically modified to produce their own "insecticides" for protection. Newly registered insecticides tend to be more target specific and often, more expensive. Older chemistries are being removed both voluntarily and involuntarily from the market. There is increasing demand for organically grown food, or food perceived as "safe" for consumption. Yet we must still feed a growing human population. More than ever, IPM researchers need to develop programs that use effective alternatives to insecticides whenever possible. We also must intensify efforts to truly integrate insecticides selectively into our IPM programs, so that they are not the predominant tool in our IPM toolbox. As such, we need to further develop principles and methods of cultural, biological, and physical control as relevant pest management tools for sustainable agricultural production.

6. References

Andow, D. A., Ragsdale, D. W., & Nyvall, R. F. (Eds.) (1997). *Ecological Interactions and Biological Control*. Intercept Ltd., ISBN 0813387582, Andover, Hants, UK

Barbour, J. D., Farrar, R. R., Jr., & Kennedy, G. G. (1993). Interaction of *Manduca sexta* resistance in tomato with insect predators of *Helicoverpa zea*. *Entomologia*

Experimentalis et Applicata, Vol. 68, No. 2, (August 1993), pp. 143–155, ISSN 0013-8703

Barbour, J. D., Farrar, R. R., Jr., & Kennedy, G. G. (1997). Populations of predaceous natural enemies developing on insect-resistant and susceptible tomato in North Carolina. *Biological Control*, Vol. 9, No. 3, (July 1997), pp. 173–184, ISSN 1049-9644

Bell, J. R., Johnson, P. J., Hambler, C., Haughton, A. J., Smith, H., Feber, R. E., Tattersall, F. H., Hart, B. H., Manley, W., & Macdonald, D. W. (2002). Manipulating the abundance of *Lepthyphantes tenuis* (Araneae: Linyphiidae) by field margin management. *Agriculture, Ecosystems and Environment*, Vol. 93, No. 1-3, (December 2002), pp. 295-304, ISSN 0167-8809

Benbrook, C. M., Groth, E., III, Halloran, J. M., Hansen, M. K., & Marquardt, S. (1996). *Pest Management at the Crossroads*. Consumers Union of the United States, ISBN 0890439001, New York, New York, USA

Bentz, J. A., Reeves, J., Barbosa, P., & Francis, B. (1995). Nitrogen fertilizer effect on selection, acceptance, and suitability of *Euphorbia pulcherrima* (Euphorbiaceae) as a host plant to *Bemisia tabaci* (Homoptera: Aleyrodidae). *Environmental Entomology*, Vol. 24, No. 1, (February 1995), pp. 40-45, ISSN 0046-225X

Boiteau, G., Lynch, D. H., & Martin, R. C. (2008). Influence of fertilization on the Colorado potato beetle, *Leptinotarsa decemlineata*, in organic potato production. *Environmental Entomology*, Vol. 37, No. 2, (April 2008), pp. 575-585, ISSN 0046-225X

Boiteau, G., & Vernon, R. S. (2001). Physical Barriers for the Control of Insect Pests, In: *Physical Control Methods in Plant Protection*, C. Vincent, B. Panneton, & F. Fleurat-Lessard (Eds.), pp. 224-247, Springer, ISBN 3540645624, Berlin, Germany

Bowie, M. H., Wratten, S. D., & White, A. J. (1995). Agronomy and phenology of "companion plants" of potential for enhancement of insect biological control. *New Zealand Journal of Crop and Horticultural Science*, Vol. 23, No. 4, (December 1995), pp. 423-427, ISSN 0114-0671

Briese, D. T. (2005). Translating host-specificity test results into the real world: the need to harmonize the yin and yang of current testing procedures. *Biological Control*, Vol. 35, No. 3, (December 2005), pp. 208-214, ISSN 1049-9644

Brust, G. E. (1994). Natural enemies in straw-mulch reduce Colorado potato beetle populations and damage in potato. *Biological Control*, Vol. 4, No. 2, (June 1994), pp. 163-169, ISSN 1049-9644

Buntin, G. D., & Hudson, R. D. (1991). Spring control of the hessian fly (Diptera: Cecidomyiidae) in winter wheat using insecticides. *Journal of Economic Entomology*, Vol. 84, No. 6, (December 1991), pp. 1913-1919, ISSN 0022-0493

Buntin, G. D., Ott, S. L., & Johnson, J. W. (1992). Integration of plant resistance, insecticides, and planting date for management of the hessian fly (Diptera: Cecidomyiidae) in winter wheat. *Journal of Economic Entomology*, Vol. 85, No. 2, (April 1992), pp. 530-538, ISSN 0022-0493

Cai, Q. N., Ma, X. M., Zhao, X., Cao, Y. Z., & Yang, X. Q. (2009). Effects of host plant resistance on insect pests and its parasitoid: a case study of wheat-aphid-parasitoid system. *Biological Control*, Vol. 49, No. 2, (May 2009), pp. 134–138, ISSN 1049-9644

Caltagirone, L. E. & Huffaker, C. B. (1980). Benefits and Risks of Using Natural Enemies for Controlling Pests, In: *Environmental Protection and Biological Forms of Control of Pest*

Organisms, Ecological Bulletins No. 31, B. Lundsholm & M. Stackerud (Eds.), pp. 103-109, Swedish Natural Science Research Council, Stockholm, Sweden.

Caltagirone, L. E., & Doutt, R. L. (1989). The history of the vedalia beetle importation to California and its impact on the development of biological control. *Annual Review of Entomology*, Vol. 34, pp. 1–16, ISSN 0066-4170

Carson, R. (1962). *Silent Spring*. Houghton Mifflin Co., New York, New York, USA

Chau, A., & Heinz, K. M. (2006). Manipulating fertilization: a management tactic against *Frankliniella occidentalis* on potted chrysanthemum. *Entomologia Experimentalis et Applicata*, Vol. 120, No. 3, (September 2006), pp. 201-209, ISSN 0013-8703

Chau, A., Heinz, K. M., & Davies, F. T. (2005). Influences of fertilization on *Aphis gossypii* and insecticide usage. *Journal of Applied Entomology*, Vol. 129, No. 2, (March 2005), pp. 89-97, ISSN 0931-2048

Chiang, H. C. (1965). Survival of northern corn rootworm eggs through one and two winters. *Journal of Economic Entomology*, Vol. 58, No. 3, (April 1965), pp. 470–472, ISSN 0022-0493

Chow, A., Chau, A., & Heinz, K. M. (2009). Reducing fertilization for cut roses: effect on crop productivity and twospotted spider mite abundance, distribution, and management. *Journal of Economic Entomology*, Vol. 102, No. 5, (October 2009), pp. 1896-1907, ISSN 0022-0493

Clausen, C. P., (Ed.) (1978). *Introduced Parasites and Predators of Arthropod Pests and Weeds: a World Review*, Agriculture Handbook No. 480, Agricultural Research Service, United States Department of Agriculture, Washington, D.C., USA

Cloyd, R. A., & Bethke, J. A. 2009. Pesticide use in ornamental production: what are the benefits? *Pest Management Science*, Vol. 65, No. 4, (April 2009), pp. 345-350, ISSN 1526-498X

Collier, T., & van Steenwyk, R. (2004). A critical evaluation of augmentative biological control. *Biological Control*, Vol. 31, No. 2, (October 2004), pp. 245-256, ISSN 1049-9644

Cook, R. J. (1987). *Report of the Research Briefing Panel on Biological Control in Managed Ecosystems*. National Academy Press, Washington, D.C., USA

Cook, S. M., Kahn, Z. R., & Pickett, J. A. (2007). The use of push-pull strategies in integrated pest management. *Annual Review of Entomology*, Vol. 52, pp. 375-400. ISSN 0066-4170

Coppel, H. C., & Mertins, J. W. (1977). *Biological Insect Pest Suppression*, Advanced Series in Agricultural Sciences, Vol. 4, Springer-Verlag., Berlin, Germany

DeBach, P. D., & Rosen, D. (1991). *Biological Control by Natural Enemies* (2nd edition), Cambridge University Press, ISBN 0521391911, London, UK

De Barro, P. J., Hart, P. J., & Morton, R. (2000). The biology of two *Eretmocerus* spp. (Haldeman) and three *Encarsia* spp. Forster and their potential as biological control agents of *Bemisia tabaci* biotype B in Australia. *Entomologia Experimentalis et Applicata*, Vol. 94, No. 1, (January 2000), pp. 93–102, ISSN 0013-8703

Delfosse, E. S. (2005). Risk and ethics in biological control. *Biological Control*, Vol. 35, No. 3, (December 2005), pp. 319-329, ISSN 1049-9644

Desneux, N., Decourtye, A., & Delpuech, J. M. (2007). The sublethal effects of pesticides on beneficial arthropods. *Annual Review of Entomology*, Vol. 52, pp. 81–106, ISSN 0066-4170

Desneux, N., Pham-Delègue, M. H., & Kaiser, L. (2004a). Effect of a sublethal and lethal dose of lambdacyhalothrin on oviposition experience and host searching behaviour of a parasitic wasp, *Aphidius ervi*. *Pest Management Science*, Vol. 60, No. 4, (April 2004), pp. 381–389, ISSN 1526-498X

Desneux N., Rafalimanana, H., Kaiser, L. (2004b). Dose-response relationship in lethal and behavioural effects of different insecticides on the parasitic wasp *Aphidius ervi*. *Chemosphere*, Vol. 54, No. 5, (February 2004), pp. 619–627, ISSN 0045-6535

Diaz-Montano, J., Reese, J. C., Louis, J., Campbell, L. R., & Schapaugh, W. T. (2007). Feeding behavior by the soybean aphid (Hemiptera: Aphididae) on resistant and susceptible soybean genotypes. *Journal of Economic Entomology*, Vol. 100, No. 3, (June 2007), pp. 984–989, ISSN 0022-0493

Diaz-Montano, J., Reese, J. C., Schapaugh, W. T., & Campbell, L. R. (2006). Characterization of antibiosis and antixenosis to the soybean aphid (Hemiptera: Aphididae) in several soybean genotypes. *Journal of Economic Entomology*, Vol. 99, No. 5, (October 2006), pp. 1884-1889, ISSN 0022-0493

Edwards, C. A., Lal, R., Madden, P., Miller, R. H., & House, G. (Eds.) (1990). *Sustainable Agricultural Systems*, Soil and Water Conservation Society, ISBN 093573421X, Ankeny, Iowa, USA

Ehrlich, P. R., & Raven, P. H. (1964). Butterflies and plants: a study in coevolution. *Evolution*, Vol. 18, No. 4, (December 1964), pp. 586-608, ISSN 0014-3820

Ellis, J. A., Walter, A. D., Tooker, J. F., Ginzel, M. D., Reagel, P. F., Lacy, E. S., Bennett, A. B., Grossman, E. M., & Hanks, L. M. (2005). Conservation biological control in urban landscapes: manipulating parasitoids of bagworm (Lepidoptera: Psychidae) with flowering forbs. *Biological Control*, Vol. 34, No. 1, (July 2005), pp. 99-107, ISSN 1049-9644

Farid A., Johnson, J. B., Shafii, B., & Quisenberry, S. S. (1998). Tritrophic studies of Russian wheat aphid, a parasitoid, and resistant and susceptible wheat over three parasitoid generations. *Biological Control*, Vol. 12, No. 1, (May 1998), pp. 1–6, ISSN 1049-9644

Forbes , S. A. (1883). The corn root-worm. (*Diabrotica longicornis* Say) Order Coleoptera. Family Chrysomelidae. *Illinois State Entomologist Annual Report*, Vol. 12, pp. 10–31

Frank, S. D. (2010). Biological control of arthropod pests using banker plant systems: past progress and future direction. *Biological Control*, Vol. 52, No. 1, (January 2010), pp. 8-16, ISSN 1049-9644

Frank, S. D., & Sadof, C. S. *In press*. Reducing insecticide volume and non-target effects of ambrosia beetle management in nurseries. Submitted to *Journal of Economic Entomology*.

Frank, S. D., & Shrewsbury, P. M. (2004). Effect of conservation strips on the abundance and distribution of natural enemies and predation of *Agrotis ipsilon* (Lepidoptera: Noctuidae) on golf course fairways. *Environmental Entomology*, Vol. 33, No. 6, (December 2004), pp. 1662-1672, ISSN 0046-225X

Frisbie, R. E., & Smith, J. W., Jr. (1989). Biologically Intensive Integrated Pest Management: the Future. In: *Progress and Perspectives for the 21st Century*, J. J. Menn & A. L. Stienhauer (Eds.), pp. 156-184, Entomological Society of America, ISBN 0938522361, Lanham, Maryland, USA

Galvan, T. L., Koch, R. L., & Hutchison, W. D. (2005). Toxicity of commonly used insecticides in sweet corn and soybean to multicolored Asian lady beetle (Coleoptera: Coccinellidae). *Journal of Economic Entomology*, Vol. 98, No. 3, (June 2005), pp. 780-789, ISSN 0022-0493

Garcia, R., Caltagirone, L. E., & Gutierrez, A. P. (1988). Comments on a redefinition of biological control. *BioScience*, Vol. 38, No. 10, (November 1988), pp. 692-694, ISSN 0006-3568

Gentile, A. G., & Stoner, A. K. (1968). Resistance in *Lycopersicon* and *Solanum* species to potato aphid. *Journal of Economic Entomology*, Vol. 61, No. 5, (October 1968), pp. 1152-54, ISSN 0022-0493

Gill, S., Cloyd, R. A., Baker, J. R., Clement, D. L., & Dutky, E., (2006). *Pests and Diseases of Herbaceous Perennials: the Biological Approach*. Ball Publishing, ISBN 1883052508, Batavia, IL, USA

Gould, F. (1986). Simulation models for predicting durability of insect-resistant germ plasm - hessian fly (Diptera: Cecidomyiidae)-resistant winter wheat. *Environmental Entomology*, Vol. 15, No. 1, (February 1986), pp. 11-23, ISSN 0046-225X

Gould, F. (1998). Sustainability of transgenic insecticidal cultivars: integrating pest genetics and ecology. *Annual Review of Entomology*, Vol. 43, pp. 701-726, ISSN 0066-4170

Grant, J. A. (1997). IPM techniques for greenhouse crops. In: *Techniques for Reducing Pesticide Use: Economic and Environmental Benefits*, D. Pimental (Ed.), pp. 399-406, John Wiley and Sons, ISBN 0471968382, Chichester, West Sussex, UK

Gray, M. E., Sappington, T. W., Miller, N. J., Moeser, J., & Bohn, M. O. (2009). Adaptation and invasiveness of western corn rootworm: intensifying research on a worsening pest. *Annual Review of Entomology*, Vol. 54, pp. 303-321, ISSN 0066-4170

Greathead, D. J., (Ed.) (1976). *A Review of Biological Control in Western and Southern Europe*, Technical Communication Vol. 7, pp. 52-64, Commonweath Institute for Biological Control, Commonwealth Agricultural Bureau, ISBN 0851983693, Slough, UK

Gupta, G., & Krischik, V. A. (2007). Professional and consumer insecticides for the management of adult Japanese beetle on hybrid tea rose. *Journal of Economic Entomology*, Vol. 100, No. 3, (June 2007), pp. 830–837, ISSN 0022-0493

Hall, R. W., & Ehler, L. E. (1979). Rate of establishment of natural enemies in classical biological control. *Bulletin of the Entomological Society of America*, Vol. 25, No. 4, (15 December 1979), pp. 280-282, ISSN 0013-8754

Hall, R. W., Ehler, L. E., & Bisabri-Ershadi, B. (1980). Rate of success in classical biological control of arthropods. *Bulletin of the Entomological Society of America*, Vol. 26, No. 2, (15 June 1980), pp. 111-114, ISSN 0013-8754

Hardman, J. M., Rogers, R. E. L., & MacLellan, C. R. (1988). Advantages and disadvantages of using pyrethroids in Nova Scotia apple orchards. *Journal of Economic Entomology*, Vol. 81, No. 6, (December 1988), pp. 1737-1749, ISSN 0022-0493

Headrick, D. H., Bellows, T. S., Jr., & Perring, T. M. (1996). Behaviors of female *Eretmocerus* sp. nr. *californicus* (Hymenoptera: Aphelinidae) attacking *Bemisia argentifolii* (Homoptera: Aleyrodidae) on cotton, *Gossypium hirsutum* (Malavaceae), and melon, *Cucumis melo* (Cucurbitaceae). *Biological Control*, Vol. 6, No. 1, (February 1996), pp. 64–75, ISSN 1049-9644

Heinz, K. M., Nunney, L., & Parrella, M. P. (1993). Towards predictable biological control of *Liriomyza trifolii* (Diptera: Agromyzidae) infesting greenhouse cut

Chrysanthemums. *Environmental Entomology*, Vol. 22, No. 6, (December 1993), pp. 1217-1233, ISSN 0046-225X

Heinz, K. M., & Parrella, M. P. (1994). Poinsettia (*Euphorbia pulcherrima* Willd. ex Koltz.) cultivar-mediated differences in performance of five natural enemies of *Bemisia argentifolii* Bellows and Perring, n. sp. (Homoptera: Aleyrodidae). *Biological Control*, Vol. 4, No. 4, (December 1994), pp. 305-318, ISSN 1049-9644

Heinz, K. M., & Zalom, F. G. (1995). Variation in trichome-based resistance to *Bemisia argentifolii* (Homoptera: Aleyrodidae) oviposition on tomato. *Journal of Economic Entomology*, Vol. 88, No. 5, (October 1995), pp. 1494–1502, ISSN 0022-0493

Heinz, K. M. (2005). Evolutionary Pest Management: an Approach to the Twenty-First Century, In: *Entomology at the Land Grant University: Perspectives from the Texas A&M University Department Centenary*, K. M. Heinz, R. E. Frisbie, & C. E. Bográn (Eds.), pp. 305-315 , Texas A&M University Press, ISBN 1585444324 College Station, Texas, USA

Herms, D. A. (2002). Effects of fertilization on insect resistance of woody ornamental plants: reassessing an entrenched paradigm. *Environmental Entomology*, Vol. 31, No. 6, (December 2002), pp. 923-933, ISSN 0046-225X

Herms, D. A., & Mattson, W. J. (1992). The dilemma of plants - to grow or defend. *Quarterly Review of Biology*, Vol. 67, No. 3, (September 1992), pp. 283-335, ISSN 0033-5770

Hill, C. B., Crull, L., Herman, T., Voegtlin, D. J., & Hartman, G. L. (2010). A new soybean aphid (Hemiptera: Aphididae) biotype identified. *Journal of Economic Entomology*, Vol. 103, No. 2, (April 2010), pp. 509-515, ISSN 0022-0493

Hill, C. B., Kim, K. S., Crull, L., Diers, B. W., & Hartman G. L. (2009). Inheritance of resistance to the soybean aphid in soybean PI 200538. *Crop Science*, Vol. 49, No. 4, (July/August 2009), pp. 1193–1200, ISSN 0011-183X

Hill, C. B., Li, Y., & Hartman, G. L. (2004). Resistance of *Glycine* species and various cultivated legumes to the soybean aphid (Homoptera: Aphididae). *Journal of Economic Entomology*, Vol. 97, No. 3, (June 2004), pp. 1071-1077, ISSN 0022-0493

Hoddle, M. S., van Driesche, R. G., & Sanderson, J. P. (1998). Biology and use of the whitefly parasitoid *Encarsia formosa*. *Annual Review of Entomology*, Vol. 43, pp. 645-669, ISSN 0066-4170

Hogendorp, B. K., Cloyd, R. A., & Swiader, J. M. (2006). Effect of nitrogen fertility on reproduction and development of citrus mealybug, *Planococcus citri* Risso (Homoptera: Pseudococcidae), feeding on two colors of coleus, *Solenostemon scutellarioides* L. Codd. *Environmental Entomology*, Vol. 35, No. 2, (April 2006), pp. 201-211, ISSN 0046-225X

Hokkanen, H. M. T. (1997). Role of Biological Control and Transgenic Crops in Reducing Use of Chemical Pesticides for Crop Protection, In: *Techniques for Reducing Pesticide Use: Economic and Environmental Benefits*, D. Pimental (Ed.), pp. 103-127, John Wiley and Sons, ISBN 0471968382, Chichester, West Sussex, UK

Horn, D. J. (1988). *Ecological Approach to Pest Management*. The Guilford Press, ISBN 0898623022, New York, New York, USA

Horne, P., & Page, J. (2008). *Integrated Pest Management for Crops and Pastures*. Landlinks Press, ISBN 0643092579, Collingwood, Australia

Hoskins, W.M., Borden, A. D., & Michelbacher, A.E. (1939). Recommendations for a more discriminating use of insecticides, In: *Proceedings of the 6th Pacific Science Congress,* Vol. 5, pp. 119-123

Howarth, F. G. (1983). Classical biological control: panacea or Pandora's box? *Proceedings of the Hawaiian Entomological Society,* Vol. 24, pp. 239-244, ISSN 0073-134X

Huffaker, C. B., Berryman, A. A., & Laing, J. E. (1984). Natural Control of Insect Populations, In: *Ecological Entomology,* C. B. Huffaker & R.L. Rabb (Eds.), pp. 359-398, Wiley, ISBN 0471064939, New York, New York, USA

Huffaker, C. B., & Messenger, P. S. (Eds.) (1976). *Theory and Practice of Biological Control.* Academic Press, ISBN 0123603501, New York, New York, USA

Iftner, D. C., & Hall, F. R. (1983). Effects of fenvalerate and permethrin on *Tetranychus urticae* Koch (Acari: Tetranychidae) dispersal behavior. *Environmental Entomology,* Vol. 12, No. 6, (December 1983), pp. 1782-1786, ISSN 0046-225X

Iftner, D. C., & Hall, F. R. (1984). The effects of fenvalerate and permethrin residues on *Tetranychus urticae* Koch fecundity and rate of development. *Journal of Agricultural Entomology,* Vol. 1, pp. 191-200, ISSN 0735-939X

Inbar, M., & Gerling, D. (2008). Plant-mediated interactions between whiteflies, herbivores, and natural enemies. *Annual Review of Entomology,* Vol. 53, pp. 431-448, ISSN 0066-4170

Insecticide Resistance Action Committee (IRAC) (2010). 03.08.11, Available from http://www.irac-online.org/about/irac/

Jacobsen, B. J. (1997). Role of plant pathology in integrated pest management. *Annual Review of Phytopathology,* Vol. 35, pp. 373-391, ISSN 0066-4286

James, D. G. (2003). Synthetic herbivore-induced plant volatiles as field attractants for beneficial insects. *Environmental Entomology,* Vol. 32, No. 5, (October 2003), pp. 977-982, ISSN 0046-225X

Johnson, M. T., & Gould, F. (1992). Interaction of genetically engineered host plant-resistance and natural enemies of *Heliothis virescens* (Lepidoptera: Noctuidae) in tobacco. *Environmental Entomology,* Vol. 21, No. 3, (June 1992), pp. 587-597, ISSN 0046-225X

Jones, V. P., & Parella, M. P. (1984). The sublethal effects of selected insecticides on life table parameters of *Panonychus citri* (Acari: Tetranychidae). *Canadian Entomologist,* Vol. 116, pp. 1033-1040, ISSN 0008-347X

Kashyap, R. K., Kennedy, G. G., & Farrar, R. R., Jr. (1991a). Behavioral response of *Trichogramma pretiosum* Riley and *Telenomus sphingis* (Ashmead) to trichmone/methyl ketone-mediated resistance in tomato. *Journal of Chemical Ecology,* Vol. 17, No. 3, (March 1991), pp. 543–556, ISSN 0098-0331

Kashyap, R. K., Kennedy, G. G., & Farrar, R. R., Jr. (1991b). Mortality and inhibition of *Helicoverpa zea* egg parasitism rates by *Trichogramma* in relation to trichome/methyl ketone-mediated insect resistance of *Lycopersicon hirsutum f. glabratum* accession PI134417. *Journal of Chemical Ecology,* Vol. 17, No. 12, (December 1991), pp. 2381–2395, ISSN 0098-0331

Kaufman, W. C., & Kennedy, G. G. (1989a). Inhibition of *Campoletis sonorensis* parasitism of *Heliothis zea* and of parasitoid development by 2-tridecanone mediated insect resistance of wild tomato. *Journal of Chemical Ecology,* Vol. 15, No. 6, (June 1989), pp. 1919–1930, ISSN 0098-0331

Kaufman, W. C., & Kennedy, G. G. (1989b). Toxicity of allelochemicals from the wild insect-resistant tomato *Lycopersicon hirsutum f. glabratum* to *Campoletis sonorensis*, a parasitoid of *Helicoverpa zea*. *Journal of Chemical Ecology*, Vol. 15, No. 7, (July 1989), pp. 2051–60, ISSN 0098-0331

Kennedy, G. G. (2003). Tomato, pests, parasitoids, and predators: tritrophic interactions involving the genus *Lycopersicon*. *Annual Review of Entomology*, Vol. 48, pp. 51-72, ISSN 0066-4170

Kennedy, G. G., & Sorenson, C. F. (1985). Role of glandular trichomes in the resistance of *Lycopersicon hirsutum f. glabratum* to Colorado potato beetle (Coleoptera: Chrysomelidae). *Journal of Economic Entomology*, Vol. 78, No. 3, (June 1985), pp. 547-551, ISSN 0022-0493

Kennedy, G. G., & Barbour, J. D. (1992). Resistance Variation in Natural and Managed Systems, In: *Plant Resistance to Herbivores and Pathogens: Ecology, Evolution, and Genetics*, R. S. Fritz & E. L. Simms (Eds.), pp. 13-41, University of Chicago Press, ISBN 0226265536, Chicago, Illinois, USA

Knutson, L. (1981). Symbiosis on Biosystematics and Biological Control, In: *Biological Control in Crop Production*, G. C. Papavizas (Ed.), pp. 61-78, Allanheld, Osmun Publishers, London, UK

Kogan, M. (1998). Integrated pest management: historical perspectives and contemporary developments. *Annual Review of Entomology*, Vol. 43, pp. 243-270, ISSN 0066-4170

Koul, O., & Cuperus, G. W. (2007). *Ecologically Based Integrated Pest Management*. CAB International, ISBN 1845930649, Wallingford, UK

Krischik, V. A., Landmark, A. L., & Heimpel, G. E. (2007). Soil-applied imidacloprid is translocated to nectar and kills nectar-feeding *Anagyrus pseudococci* (Girault) (Hymenoptera: Encyrtidae). *Environmental Entomology*, Vol. 36, No. 5, (October 2007), pp. 1238–1245, ISSN 0046-225X

Kruess, A., & Tscharntke, T. (1994). Habitat fragmentation, species loss, and biological control. *Science*, Vol. 264, No. 5165, (10 June 1994), pp. 1581-1584, ISSN 0036-8075

Kytö, M., Niemela, P., & Larsson, S. (1996). Insects on trees: population and individual response to fertilization. *Oikos*, Vol. 75, No. 2, (March 1996), pp. 148-159, ISSN 0030-1299

Landis, D. A., Wratten, S. D., & Gurr, G. M. (2000). Habitat management to conserve natural enemies of arthropod pests in agriculture. *Annual Review of Entomology*, Vol. 45, pp. 175-201, ISSN 0066-4170

Levine, E. & Oloumi-Sadeghi, H. (1991). Management of diabroticite rootworms in corn. *Annual Review of Entomology*, Vol. 36, 229–255, ISSN 0066-4170

Levine, E., Oloumi-Sadeghi, H., & Fisher, J. R. (1992). Discovery of multiyear diapause in Illinois and South Dakota northern corn rootworm (Coleoptera: Chrysomelidae) eggs and incidence of the prolonged diapause trait in Illinois. *Journal of Economic Entomology*, Vol. 85, No. 1, (February 1992), pp. 262–267, ISSN 0022-0493

Levine, E., Spencer, J. L., Isard, S. A., Onstad, D. W., & Gray, M. E. (2002). Adaptation of the western corn rootworm to crop rotation: evolution of a new strain in response to a management practice. *American Entomologist*, Vol. 48, No. 2, pp. 94–107, ISSN 1046-2821

Marino, P. C., & Landis, D. A. (1996). Effect of landscape structure on parasitoid diversity and parasitism in agroecosystems. *Ecological Applications*, Vol. 6, No. 1, (February 1996), pp. 276-284, ISSN 1051-0761

Mathews, C. R., Bottrell, D. G., & Brown, M. W. (2004). Habitat manipulation of the apple orchard floor to increase ground-dwelling predators and predation of *Cydia pomonella* (L.) (Lepidoptera: Tortricidae). *Biological Control*, Vol. 30, No. 2, (June 2004), pp. 265-273, ISSN 1049-9644

Mattson, W. J. (1980). Herbivory in relation to plant nitrogen content. *Annual Review of Ecology and Systematics*, Vol. 11, pp. 119-161, ISSN 0066-4162

McClure, M. S. (1977). Resurgence of the scale, *Fiorinia externa* (Homoptera: Diaspididae), on hemlock following insecticide application. *Environmental Entomology*, Vol. 6, No. 3, (June 1977), pp. 480–484, ISSN 0046-225X

McMurtry, J. A., & Croft, B. A. (1997). Life-styles of phytoseiid mites and their role in biological control. *Annual Review of Entomology*, Vol. 42, pp. 291-321, ISSN 0066-4170

Metcalf, R. L. (1989). Insect resistance to insecticides. *Pesticide Science*, Vol. 26, pp. 333-358, ISSN 0031-613X

Misener, G. C., Boiteau, G., & McMillan, L. P. (1993). A plastic-lining trenching device for the control of Colorado potato beetle: beetle excluder. *American Potato Journal*, Vol. 70, No. 12, (December 1993), pp. 903–908, ISSN 0003-0589

Moser, S. E., & Obrycki, J. J. (2009). Non-target effects of neonicotinoid seed treatments; mortality of coccinellid larvae related to zoophytophagy. *Biological Control*, Vol. 51, No. 3, (December 2009), pp. 487–492, ISSN 1049-9644

Murdoch, W. W., Chesson, J., & Chesson, P. L. (1985). Biological control in theory and practice. *American Naturalist*, Vol. 125, No. 3, (March 1985), pp. 344-366, ISSN 0003-0147

Murdoch, W. W., & Briggs, C. J. (1996). Theory for biological control: recent developments. *Ecology*, Vol. 77, No. 7, (October 1996), pp. 2001-2013, ISSN 0012-9658

National Research Council (NRC). (1996). *Ecologically Based Pest Management: New Solutions for a New Century*, National Academy Press, ISBN 0309053307, Washington, D.C., USA

Nihoul, P. (1993a). Controlling glasshouse climate influences the interaction between tomato glandular trichome, spider mite and predatory mite. *Crop Protection*, Vol. 12, No. 6, (September 1993), pp. 443–447, ISSN 0261-2194

Nihoul, P. (1993b). Do light intensity, temperature and photoperiod affect the entrapment of mites on glandular hairs of cultivated tomatoes? *Experimental and Applied Acarology*, Vol. 17, No. 9, (September 1993), pp. 709–718, ISSN 0168-8162

Obrycki, J. J., Lewis, L. C., & Orr, D. B. (1997). Augmentative releases of entomophagous species in annual systems. *Biological Control*, Vol. 10, No. 1, (September 1997), pp. 30- 36, ISSN 1049-9644

Obrycki, J. J., & Tauber, M. J. (1984). Natural enemy activity on glandular pubescent potato plants in the greenhouse: an unreliable predictor of effects in the field. *Environmental Entomology*, Vol. 13, No. 3, (June 1984), pp. 679–83, ISSN 0046-225X

O'Callaghan, M., Glare, T. R., Burgess, E. P. J., & Malone, L. A. (2005). Effects of plants genetically modified for insect resistance on non-target organisms. *Annual Review of Entomology*, Vol. 50, pp. 271-292, ISSN 0066-4170

Papachristos, D. P., & Milonas, P. G. (2008). Adverse effects of soil applied insecticides on the predatory coccinellid *Hippodamia undecimnotata* (Coleoptera: Coccinellidae). *Biological Control*, Vol. 47, No. 1, (October 2008), pp. 77–81, ISSN 1049-9644

Painter, R. H. (1951). *Insect Resistance in Crop Plants*. The MacMillan Company, New York, New York, USA

Parrella, M. P., Heinz, K. M., & Nunney, L. (1992). Biological control through augmentative releases of natural enemies: a strategy whose time has come. *American Entomologist*, Vol. 38, No. 3, pp. 172–179, ISSN 1046-2821

Patt, J. M., Hamilton, G. C., & Lashomb, J. H. (1997). Foraging success of parasitoid wasps on flowers: interplay of insect morphology, floral architecture and searching behavior. *Entomologia Experimentalis et Applicata*, Vol. 83, No. 1, (April 1997), pp. 21-30, ISSN 0013-8703

Penman, D. R., & Chapman, R. B. (1983). Fenvalerate-induced distributional imbalances of 2-spotted spider-mite on bean plants. *Entomologia Experimentalis et Applicata*, Vol. 33, No. 1, (January 1983), pp. 71-78, ISSN 0013-8703

Penman, D. R., & Chapman, R. B. (1988). Pesticide-induced mite outbreaks: pyrethroids and spider mites. *Experimental and Applied Acarology*, Vol. 4, pp. 265-276, ISSN 0168-8162

Pereira, R. M., Walker, W., Pfeister, M., & Koehler, P. G. (2009). Lethal effects of heat and use of localized heat treatment for control of bed bug infestations. *Journal of Economic Entomology*, Vol. 102, No. 3, (June 2009), pp. 1182-1188, ISSN 0022-0493

Perkins, J. H., & Garcia, R. (1999). Social and Economic Factors Affecting Research and Implementation of Biological Control, In: *Handbook of Biological Control*, T. S. Bellows & T. W. Fisher (Eds.), pp. 993-1008, Academic Press, ISBN 0122573056, San Diego, California, USA

Pimental, D. (Ed.) (1997). *Techniques for Reducing Pesticide Use: Economic and Environmental Benefits*. John Wiley and Sons, ISBN 0471968382, Chichester, West Sussex, UK

Potter, D. A. (1986). Urban Landscape Pest Management, In: *Advances in Urban Pest Management*, G. W. Bennett & J. M. Owens (Eds.), pp. 219-251, Van Nostrand Reinhold Co., ISBN 0442209606, New York, New York, USA

Pottorff, L. P., & Panter, K. L. (2009). Integrated pest management and biological control in high tunnel production. *HortTechnology*, Vol. 19, No. 1, (January-March 2009), pp. 61-65, ISSN 1063-0198

Prokopy, R. J., Cooley, D. R., Autio, W. R., & Coli, W. M. (1994). Second-level integrated pest management in commercial apple orchards. *American Journal of Alternative Agriculture*, Vol. 9, No. 4, pp. 148-156, ISSN 0889-1893

Rabb, R. L. (1972). Principles and Concepts of Pest Management, In: *Implementing Practical Pest Management Strategies: Proceedings of a National Extension Pest Management Workshop*, pp. 6-29, Purdue University, West Lafayette, Indiana, USA

Rabb, R. L., Stinner, R. E., & van den Bosch, R. (1976). Conservation and Augmentation of Natural Enemies, In: *Theory and Practice of Biological Control*, C. B. Huffaker & P. S. Messenger (Eds.), pp. 233-254, Academic Press, ISBN 0123603501, New York, New York, USA

Ragsdale, D. W., Landis, D. A., Brodeur, J., Heimpel, G. E., & Desneux, N. (2011). Ecology and management of the soybean aphid in North America. *Annual Review of Entomology*, Vol. 56, pp. 375-399, ISSN 0066-4170

Raupp, M. J., Shrewsbury, P. M., & Herms, D. A. (2010) Ecology of herbivorous arthropods in urban landscapes. *Annual Review of Entomology*, Vol. 55, pp. 19-38, ISSN 0066-4170

Raupp, M. J., Holmes, J. J., Sadof, C., Shrewsbury, P., & Davidson, J. A. (2001). Effects of cover sprays and residual pesticides on scale insects and natural enemies in urban forests. *Journal of Arboriculture*, Vol. 27, No. 4, (July 2001), pp. 203-214, ISSN 0278-5226

Raupp, M. J., Webb, R., Szczepaniec, A., Booth, D., & Ahern, R. (2004). Incidence, abundance, and severity of mites on hemlocks following applications of imidacloprid. *Journal of Arboriculture*, Vol. 30, No. 2, (March 2004), pp. 108–13, ISSN 0278-5226

Raynolds, L. T. (2000). Re-embedding global agriculture: the international organic and fair trade movements. *Agriculture and Human Values*, Vol. 17, No. 3, (September 2007), pp. 297–309, ISSN 0889-048X

Rebek, E. J., & Sadof, C. S. (2003). Effects of pesticide applications on the euonymus scale (Homoptera: Diaspididae) and its parasitoid, *Encarsia citrina* (Hymenoptera: Aphelinidae). *Journal of Economic Entomology*, Vol. 96, No. 2, (April 2003), pp. 446-452, ISSN 0022-0493

Rebek, E. J., Sadof, C. S., & Hanks, L. M. (2005). Manipulating the abundance of natural enemies in ornamental landscapes with floral resource plants. *Biological Control*, Vol. 33, No. 2, (May 2005), pp. 203-216, ISSN 1049-9644

Rebek, E. J., Sadof, C. S., & Hanks, L. M. (2006). Influence of floral resource plants on control of an armored scale pest by the parasitoid *Encarsia citrina* (Craw.) (Hymenoptera: Aphelinidae). *Biological Control*, Vol. 37, No. 3, (June 2006), pp. 320-328, ISSN 1049-9644

Rekika, D., Stewart, K. A., Boivin, G., & Jenni, S. (2008). Reduction of insect damage in radish with floating row covers. *International Journal of Vegetable Science*, Vol. 14, No. 2, pp. 177-193, ISSN 1931-5260

Ridgway, R. L. (Ed.) (1998). *Mass-Reared Natural Enemies: Application, Regulation, and Needs*. Thomas Say Publications in Entomology: Proceedings. Entomological Society of America, ISBN 0938522663, Lanham, Maryland, USA

Rieux, R., Simon, S., & Defrance, H. (1999). Role of hedgerows and ground cover management on arthropod populations in pear orchards. *Agriculture, Ecosystems and Environment*, Vol. 73, No. 2, (April 1999), pp. 119-127, ISSN 0167-8809

Rhainds, M., Kovach, J., Dosa, E. L., & English-Loeb, G., (2001). Impact of reflective mulch on yield of strawberry plants and incidence of damage by tarnished plant bug (Heteroptera: Miridae). *Journal of Economic Entomology*, Vol. 94, No. 6, (December 2001), pp. 1477-1484, ISSN 0022-0493

Roland, J., & Taylor, P. (1997). Insect parasitoid species respond to forest structure at different spatial scales. *Nature*, Vol. 386, No. 6626, (17 April 1997), pp. 710-713, ISSN 0028-0836

Royer, T. A., Mulder, P. G., & Cuperus, G. W. (1999). Renaming (redefining) integrated pest management: fumble, pass, or play? *American Entomologist*, Vol. 45, No. 3, pp. 136-139, ISSN 1046-2821

Sadof, C. S., O'Neil, R. J., Heraux, F. M., & Wiedenmann, R. N. (2004). Reducing insecticide use in home gardens: effects of training and volunteer research on adoption of biological control. *HortTechnology*, Vol. 14, No. 1, (January-March 2004), pp. 149-154, ISSN 1063-0198

Sanford, K. H. (1967). The influence of spray programs on the fauna of apple orchards in Nova Scotia. XVII. Effects on some predaceous mites. *Canadian Entomologist*, Vol. 99, pp. 197-201, ISSN 0008-347X

Sclar, D. C., Gerace, D., & Cranshaw, W. S. (1998). Observations of population increase and injury by spider mites (Acari: Tetranychidae) on ornamental plants treated with imidacloprid. *Journal of Economic Entomology*, Vol. 91, No. 1, (February 1998), pp. 250-255, ISSN 0022-0493

Shelton, A. M., Zhao, J. Z., & Roush, R. T. (2002). Economic, ecological, food safety, and social consequences of the deployment of Bt transgenic plants. *Annual Review of Entomology*, Vol. 47, pp. 845-881, ISSN 0066-4170

Simmons, A. T., & Gurr, G. M. (2005). Trichomes of *Lycopersicon* species and their hybrids: effects on pests and natural enemies. *Agricultural and Forest Entomology*, Vol. 7, No. 4, (November 2004), pp. 265-276, ISSN 1461-9555

Smith, H. S. (1919). On some phases of insect control by the biological method. *Journal of Economic Entomology*, Vol. 12, No. 4, (August 1919), pp. 288-292, ISSN 0022-0493

Smith, S. F., & Krischik, V. A. (1999). Effects of systemic imidacloprid on *Coleomegilla maculata* (Coleoptera: Coccinellidae). *Environmental Entomology*, Vol. 28, No. 6, (December 1999), pp. 1189-95, ISSN 0046-225X

Snodgrass, G. L., & Stadelbacher, E. A. (1989). Effect of different grass and legume combinations on spider (Araneae) and ground beetle (Coleoptera: Carabidae) populations in roadside habitats in the Mississippi Delta. *Environmental Entomology*, Vol. 18, No. 4, (August 1989), pp. 575-581, ISSN 0046-225X

Spencer, J. L., Hibbard, B. E., Moeser, J., & Onstad, D. W. (2009). Behaviour and ecology of the western corn rootworm (*Diabrotica virgifera virgifera* LeConte). *Agricultural and Forest Entomology*, Vol. 11, No. 1, (February 2009), pp. 9-27, ISSN 1461-9555

Spencer, J. L., & Levine, E. (2008). Resistance to Crop Rotation, In: *Insect Resistance Management: Biology, Economics and Prediction*, D. W. Onstad (Ed.), pp. 153-183, Academic Press, ISBN 9780123738585, New York, New York, USA

Stern, V. M., Smith, R. F., van den Bosch, R., & Hagen, K. S. (1959). The integration of chemical and biological control of the spotted alfalfa aphid. I. The integrated control concept. *Hilgardia*, Vol. 29, pp. 81-101, ISSN 0073-2230

Szczepaniec, A., Creary, S. F., Laskowski, K. L., Nyrop, J. P., & Raupp, M. J. (2011). Neonicotinoid insecticide imidacloprid causes outbreaks of spider mites on elm trees in urban landscapes. *PLoS ONE*, Vol. 6, No. 5, e20018. doi:10.1371/journal.pone.0020018, ISSN 1932-6203

Theriault, F., Sequin, P., & Stewart, K. A. (2009). Incidence of *Pieris rapae* in organic broccoli grown with living mulches under floating row cover. *International Journal of Vegetable Science*, Vol. 15, No. 3, pp. 218-225, ISSN 1931-5260

Tooker, J. F., & Hanks, L. M. (2000). Influence of plant community structure on natural enemies of pine needle scale (Homoptera: Diaspididae) in urban landscapes, *Environmental Entomology*, Vol. 29, No. 6, (December 2000), pp. 1305-1311, ISSN 0046-225X

United States Department of Agriculture, Animal and Plant Health Inspection Service (8 July, 2011). Biological Control Program, 03.08.11, Available from http://www.aphis.usda.gov/plant_health/plant_pest_info/biocontrol

van Driesche, R. G., & Bellows, T. S. (1996). *Biological Control.* Chapman & Hall, ISBN 0412028611, New York, New York, USA

van Driesche, R. G., & Heinz, K.M. (2004). An Overview of Biological Control in Protected Culture, In: *Biocontrol in Protected Culture*, K. M. Heinz, R. G. Van Driesche, & M. Parrella (Eds.), pp. 1-24, Ball Publishing, ISBN 1883052394, Batavia, IL, USA

van Haren, R. J. F., Steenhuis, M. M., Sabelis, M. W., & de Ponti, O. M. B. (1987). Tomato stem trichomes and dispersal success of *Phytoseiulus persimilis* relative to its prey *Tetranychus urticae. Experimental and Applied Acarology*, Vol. 3, pp. 115–121, ISSN 0168-8162

van Lenteren, J. C. (1980). Evaluation of control capabilities of natural enemies: does art have to become science? *Netherlands Journal of Zoology*, Vol. 30, pp. 369-381, ISSN 0028-2960

van Lenteren, J. C., Hua, L. Z., Kamerman, J. W., & Xu, R. (1995). The parasite-host relationship between *Encarsia formosa* (Hymenoptera: Aphelinidae) and *Trialeurodes vaporariorum* (Homoptera: Aleyrodidae). XXVI. Leaf hairs reduce the capacity of *Encarsia* to control greenhouse whitefly on cucumber. *Journal of Applied Entomology*, Vol. 119, No. 1-5, (January/December 1995), pp. 553–559, ISSN 0931-2048

Vincent, C., Hallman, G., Panneton, B, & Fleurat-Lessard, F. (2003). Management of agricultural insects with physical control methods. *Annual Review of Entomology*, Vol. 48, pp. 261-281, ISSN 0066-4170

Vincent, C., Weintraub, P., & Hallman, G. (2009). Physical Control of Insect Pests, In: *Encyclopedia of Insects* (2nd edition), V. H. Resh & R. T. Cardé (Eds.), pp. 794-798, Academic Press, ISBN 9780123741448, San Diego, California, USA

Wäckers, F. L. (2004). Assessing the suitability of flowering herbs as parasitoid food sources: flower attractiveness and nectar accessibility. *Biological Control*, Vol. 29, No. 3, (March 2004), pp. 307-314, ISSN 1049-9644

Ware, G. W., & Whitacre, D. M. (2004). *The Pesticide Book* (6th edition), MeisterPro, Willoughby, Ohio, USA

Weber, D. C., Ferro, D. N., Buonaccorsi, J., & Hazzard, R. V. (1994). Disrupting spring colonization of Colorado potato beetle to nonrotated potato fields. *Entomologia Experimentalis et Applicata*, Vol. 73, No. 1, (October 1994), pp. 39–50, ISSN 0013-8703

Wong, S. W., & Chapman, R. B. (1979). Toxicity of synthetic pyrethroid insecticides to predaceous phytoseiid mites and their prey. *Australian Journal of Agricultural Research*, Vol. 30, No. 3, pp. 497-501, ISSN 0004-9409

Zalom, F. (1997). IPM Practices for Reducing Insecticide Use in U.S. Fruit Crops, In: *Techniques for Reducing Pesticide Use: Economic and Environmental Benefits*, D. Pimental (Ed.), pp. 317-342, John Wiley and Sons, ISBN 0471968382, Chichester, West Sussex, UK

9

Factors Affecting Performance of Soil Termiticides

Beverly A. Wiltz
USDA-ARS
New Orleans, LA
USA

1. Introduction

Although baits have increased in popularity in recent years, the application of liquid termiticide to soil remains the most widely used method for protecting structures against subterranean termites (Anonymous, 2008). In addition to fast acting, repellent toxicants such as bifenthrin and other pyrethroids that act as barriers to termite movement, non-repellent, slower acting compounds including fipronil, imidacloprid and thiamethoxam are now among the preferred soil treatments. Delayed toxicity can provide opportunity for horizontal transfer of the active ingredient, potentially reducing termite activity (Remmen & Su, 2005; Shelton & Grace, 2003). While there is some evidence of colony suppression or elimination following perimeter treatments with imidacloprid (Parman & Vargo, 2010), other studies have shown that a reduction in activity occurs over only a small portion of a colony's foraging range, making it unlikely that soil treatments affect the overall termite population (Osbrink et al., 2005; Rust & Saran, 2006; Saran & Rust, 2007; Su, 2005). This limited potential for transfer emphasizes the importance of bioavailability of termiticides in soil over an extended period of time.

Failure of soil termiticide treatments is often related to factors other than the active ingredient (Su & Scheffrahn, 1990b). Efficacy and longevity of soil treatments varies greatly with application rate, soil properties (Gold et al., 1996; Su & Scheffrahn, 1990b), termite pressure (Jones, 1990), and application technique (Forschler, 1994; Su et al., 1995). Factors influencing the performance of soil termiticides can be grouped into those that determine toxicity, bioavailability, or persistence. Each of these factors is affected by properties of the termiticide and soil (Gold et al., 1996; Spomer et al., 2009; Wiltz, 2010). Although some generalizations can be made about relative toxicity of different termiticides, susceptibility differences occur among species and colonies (Beal & Smith, 1971; Osbrink & Lax, 2002). Termiticide rate and application technique influence both bioavailability and long-term persistence (Peterson, 2010). Termite population pressure and satellite nests can reduce availability of the toxicant. Finally, other environmental factors such as moisture, temperature, and microbial communities affect termiticide degradation (Baskaran et al. 1999, Saran & Kamble 2008).

2. Soil termiticides

Long-term studies evaluating chemicals as potential termiticides were initiated in the 1920's and 1930's (Randall & Doody, 1934), but it was not until after World War II that the cyclodienes, a class of chemical compounds identified as highly effective termiticides, became commercially available (Ware, 2000). Pre-construction soil treatments with cyclodienes became the standard method of subterranean termite prevention from the late 1940s until 1988 (Lewis, 1980; Su & Scheffrahn, 1990b). The cyclodienes, particularly chlordane, were extremely efficacious and stable in soil, often protecting structures from subterranean termite infestation for several decades (Grace et al., 1993; Lenz et al., 1990; Su & Scheffrahn, 1990b).

Because of their residual longevity, questions were raised about the environmental impact of these chemicals (Lewis, 1980; Su & Scheffrahn, 1990a; Wood & Pierce, 1991). Chlordane and related chemicals were banned in most of the world in the 1970's and 1980's (Ware, 2000). However, they constitute a major environmental problem because of their high toxicity, persistence in the environment, and ability to bioaccumulate in the food chain and because they are still being used in certain countries for agricultural and public health purposes (Itawa et al., 1993; Ntow, 2005; Xue et al., 2006).

Following the loss of chlordane as a soil termiticide, the only termiticides available for use as soil barrier treatments were chlorpyrifos (an organophosphate) and several pyrethroids. The residual activity of chlorpyrifos was significantly shorter than that of the cyclodienes (Grace et al., 1993; Lenz et al., 1990). As a result of the Food Quality Protection Act of 1996, the U. S. Environmental Protection Agency (EPA) revised its risk assessment of chlorpyrifos and, in 2000, the use of chlorpyrifos as a soil termiticide was canceled (EPA, 2000).

With the loss of chlorpyrifos, pyrethroids were the primary weapon available for subterranean termite prevention. The pyrethroids are more persistent than chlorpyrifos, but less stable in the soil than the cyclodienes (Lenz et al., 1990; Pawson & Gold, 1996; Su & Scheffrahn, 1990b). Soil barriers composed of pyrethroids are more likely to fail than barriers composed of cyclodienes or chlorpyrifos (Forschler, 1994; Kard, 1999; Lenz et al., 1990; Su & Scheffrahn, 1990b; Su et al., 1993) because pyrethroids are repellant to subterranean termites (Rust & Smith, 1993; Su & Scheffrahn, 1990b; Su et al., 1993).

Beginning in 2000, several new nonrepellant soil termiticides appeared on the market: fipronil, a phenyl pyrazole (Aventis Corp., 2001), imidacloprid, a chloronicotinyl (Bayer Corp., 2000), and chlorfenapyr, a pyrrole (BASF Corp., 2001). Nonrepellant termiticides are an improvement over the pyrethroids because subterranean termites cannot detect gaps in the treatment and use them to gain access to structures (Potter & Hillery, 2001). Subterranean termites are unable to detect the termiticide and do not avoid soil that has been treated with them (Kuriachan & Gold, 1998). Chlorantraniliprole is a new termiticide belonging to the anthranilic diamide class of insecticides. It targets a unique receptor site, the ryanodine receptor, causing the release of stored calcium, resulting in loss of muscle control, cessation of feeding, and eventually death of the termite (Cordova et al., 2006). Unlike other soil termiticides, chlorantraniliprole has no known health effects to humans and no personal protective equipment is required for application (Dupont, 2010). Also being developed for subterranean termite control is indoxacarb, an oxadiazine proinsecticide that is metabolically activated after entering the insect (Spomer et al. 2009; Wing et al., 2000).

A large amount of the variability in effectiveness of different soil treatments can be attributed to the termiticide itself. In a study evaluating *Coptotermes formosanus* mortality on treated soils, bifenthrin performed better than fipronil or chlorfenapyr (Wiltz 2010). Bifenthrin was also found to have the highest activity against *Reticulitermes hesperus* when compared with other pyrethroids (Smith & Rust, 1990).

Saran and Rust (2007) found that *R. hesperus* tunneled through untreated sand and stopped near the interface of fipronil treated sand. There was little tunneling in the treated sand, but termites tunneled close enough to obtain a lethal dose of fipronil. To some extent, *C. formosanus* and *Reticulitermes flavipes* penetrated sand treated with 0 - 64ppm fipronil, indicating non-repellency, but complete penetration of the treated sand was prevented by high mortality (≥88% for *C. formosanus* and ≥89% for *R. flavipes* after 7 d) (Remmen & Su, 2005). While several studies conducted in small laboratory arenas have found high mortality in fipronil treatments, extended foraging arena assays demonstrated that fipronil barriers can split termite populations, with high mortality occurring close to the treatment site, but little mortality at distances >5 m (Su, 2005).

Although imidacloprid is slow to induce mortality, mobility impairment occurs within hours of exposure (Thorne & Breisch, 2001). Imidacloprid is non-repellent (Remmen & Su, 2005), but this combination of delayed mortality and rapid mobility impairment results in limited movement of termites into treated barriers and limited mortality after 7d in close proximity to imidacloprid-treated sand.

Several studies have demonstrated differences in degradation rates among insecticides. Baker and Bellamy (2006) found that of the termiticides tested, the organophosphate, chlorpyrifos, degraded the quickest, while chloronicotinyls and pyrethroids degraded at slower rates. Horwood (2007) measured termiticide residues in a weathered sand: loam mixture, finding that bifenthrin and chlorfenapyr were more persistent than chlorpyrifos, fipronil, and imidacloprid. Horwood (2007) found that after 15 months, chlorpyrifos and fipronil concentrations at lower depths were little changed from the time of treatment, but there was a major reduction in imidacloprid concentration at all depths.

3. Soil-termiticide interactions

Because soil consists of a heterogeneous mixture of mineral and organic particles, it is difficult to predict the influence of soil type on termiticides. When soil conditions fall outside an optimum range, termiticides can be immobilized or adsorbed by the soil or altered chemically to an inactive form.

Laboratory studies have found interactions between soil and termiticide properties. Effects of clay (Henderson et al. 1998; Smith & Rust 1993) and organic carbon (Felsot & Lew 1989; Forschler & Townsend, 1996; Gold et al., 1996; Spomer et al., 2009) content on bioavailability to termites differ with termiticide. Termiticide effectiveness diminishes over time, especially on soils that pose bioavailability problems (Gold et al., 1996; Su et al., 1993; Tamashiro et al., 1987).

Variation in soil properties, such as pH, clay and organic matter content, soil moisture, and electrolyte concentration, influence the adsorption and desorption characteristics of termiticides to soils. Of equal importance are the physical and chemical properties of the toxicant, including concentration, pH, and solubility.

3.1 Mobility

Mobility is one of the most important factors in determining bioavailability and efficacy of a soil treatment. If a pesticide is too mobile, it fails to protect the structure, while increasing risk of groundwater contamination. However, if the chemical is too tightly bound to soil particles, bioavailability is limited. Mobility is affected by the pesticide's sorption, water solubility, and vapor pressure and by external influences that include soil properties, weather, topography, and vegetation. Sorption describes the attraction between a chemical and soil, vegetation, or other surfaces. However, the term most often refers to the binding of a chemical to soil particles. Sorption is defined as the attraction of an aqueous species to the surface of a solid (Alley, 1993). The sorbing species, usually an organic compound, is called the sorbate, and the solid, usually soil, to which the sorbate is attracted is known as the sorbent. This attraction results from some form of bonding between the chemical and adsorption receptor sites on the solid. Several mechanisms may operate in a particular situation, including ionic attraction, hydrophobic attraction, and hydrogen bonding. For pesticides that are weak acids or bases, sorption is influenced by soil pH.

Sorption is also influenced by soil moisture, organic matter content, and texture. Pesticides are more readily sorbed onto dry soil because water competes with pesticides for binding sites in moist soil. More sorption occurs in soils made largely of clay and organic matter. Organic matter and clay particles have small particle size, large surface area, and high surface charge. Sand particles provide less surface area for sorption, making pesticides more likely to move away from the point of application.

Several parameters are used to describe a pesticide's sorption behavior in soils. Table 1 contains sorption parameters for selected chemicals currently and previously used as soil termiticides.

Two related measures of a pesticide's sorption are the sorption coefficient (K_d) and the soil organic carbon coefficient (K_{OC}). These coefficients are ratios of adsorbed to dissolved pesticide for a specific soil (K_d) or for the organic carbon fraction of a soil (K_{OC}). These values are useful for broadly discriminating between leaching classes of pesticides, but actual adsorption depends on many factors, including soil moisture, temperature, soil pH, and type of organic matter (Wauchope et al., 2002).

Termiticide	K_{OC} (L/kg)	Log K_{ow}	H_2O Solubility (mg/L)	Henry's law constant (atmm^3/mol)	Reference
Bifenthrin	1.31×10^5 - 3.02×10^5	6.0	0.1 (25°C)	7.2×10^{-3}	Fecko (1999)
Chlordane	4.19 – 4.39	2.78	1.0×10^{-4}	1.3×10^{-3}	USEPA (1986)
Chlorantraniliprole	3.3×10^2 (average)	2.8	1.023 (20°C)	3.1×10^{-15}	USEPA (2008)
Cypermethrin	6.1×10^4 (average)	6.6	4×10^{-3} (20°C)	2.5×10^{-7}	Jones (1999)
Fipronil	3.8×10^3 - 1.2×10^4	4.01	2.4 (pH 5) 2.2 (pH 9)	3.7×10^{-5}	Connelly (2001)
Imidacloprid	1.3×10^2 - 3.1×10^2	0.57	514 (20°C, pH 7)	6.5×10^{-11}	Fossen (2006)
Indoxacarb	2.2×10^3 - 9.4×10^3	4.7	0.2 (25°C)	$< 6.1 \times 10^{-10}$	USEPA (2000)

Table 1. Soil sorption parameters of selected soil termiticides.

Because sorption coefficient values for the same pesticide vary widely with soil properties, reported values are not included in Table 1. However, K_d values are useful for comparing sorption of different chemicals to the same soil. The organic carbon sorption coefficient is a property of the pesticide and is independent of soil organic matter. The sorption coefficient and organic carbon sorption coefficient are related by the equation:

$$K_d = K_{OC} \, (\%O.C.) \tag{1}$$

Where O.C. is the percentage organic carbon the soil contains. This relationship shows that as the organic fraction of soil increases, the distribution coefficient, K_d, increases. For this relationship to hold true, the chemical must be non-ionic because soil pH affects sorption of ionic sorbates.

As Table 1 illustrates, K_{OC} values for a pesticide are not constant. Pesticide concentration affects adsorption (Kamble & Saran, 2005), but not to an extent that prevents comparison of relative mobility of different pesticides. For polar solutes, surfaces other than organic carbon can become important sorbents particularly when soils are low in organic matter (Cheung et al., 1979; Cox et al., 1998; Means et al., 1982).

Another useful measure of potential pesticide mobility is the octanol - water partition coefficient (K_{ow}). K_{ow} is a measure of the hydrophobicity of an organic compound. The more hydrophobic a compound, the less soluble it is, therefore the more likely it will adsorb to soil particles (Bedient et al, 1994). To evaluate hydrophobicity, the organic solvent octanol is used as a surrogate for organic matter. The octanol-water partition coefficient is the ratio of the concentration of a chemical in octanol and in water at equilibrium and at a specified temperature. K_{ow} is determined by adding a known amount of the pesticide to equal volumes of octanol and water. The coefficient is determined by calculating the concentration in the octanol phase compared to the concentration in the water phase. K_{ow} values vary by several orders of magnitude and may be reported as either K_{ow} or log K_{ow} values. The octanol-water partition coefficient is correlated with water solubility; therefore, the water solubility of a substance can be used to estimate its octanol-water partition coefficient.

Water solubility describes the amount of pesticide that will dissolve in a known amount of water. Highly soluble pesticides are more likely to be moved by runoff or leaching. As with sorption parameters, solubility values are useful as a means of comparison, but actual values will vary with field conditions. Solubility is affected by temperature, water pH, and the presence of other chemicals. The solubility of a compound tends to be inversely proportional to the amount of sorption that it can undergo.

In addition to being adsorbed to soil or transported by water, pesticides can be volatilized. Pesticide volatilization from moist soil is described by the Henry's law constant (K_h). K_h is defined as the concentration of pesticide in air divided by the concentration in water. K_h can be measured experimentally or estimated by dividing the saturation vapor pressure of the compound by its solubility (Suntio et al., 1988). Like other sorption parameters, K_h is temperature dependent, but values are useful for comparing volatility of different compounds. Pesticides with higher K_h are more likely to volatilize from moist soil. Because sorption affects the amount of pesticide in the soil water, the tendency to volatilize from moist soil depends on both the Henry's law constant and sorption coefficients.

3.2 Clay

Soil texture has a strong impact on termiticide performance, but effects differ with insecticide. In assays conducted with bifenthrin, chlorfenapyr, and fipronil, *C. formosanus* mortality was generally highest when clay content was low (Wiltz, 2010). Clay content of soil was significantly related to termite mortality across all termiticides, rates, and exposure times (Wiltz, 2010). Likewise, Osbrink and Lax (2002) found that *C. formosanus* workers experienced greater mortality in fipronil-treated sand than in treated potting soil or a mixture of soil and clay. Bobé et al. (1997) reported that for fipronil there was a significant decrease in adsorption coefficient as the soil clay content decreased, thereby increasing bioavailability. However, the opposite result occurs with some insecticides. Smith and Rust (1993) found that increased clay content increased the toxicity of certain pyrethroids, such as cypermethrin. The authors concluded that cypermethrin and clay apparently interacted creating a formulation similar to a wettable powder, which may have an increased affinity for the nonpolar termite integument. Gao et al. (1998) investigated the adsorption of seven pesticides and metabolites with different physiochemical properties, finding that adsorption was generally more effective on smaller and larger soil particles than on intermediate-sized particles.

3.3 pH

Effects of pH on adsorption and desorption vary with insecticide chemistry and interact with other soil properties. Low pH soils increase the adsorption of weakly acidic pesticides (Boivin et al., 2005; Carrizosa et al., 2000; Halfon et al., 1996,). Desorption of endosulfan was higher at both acidic and alkaline pH ranges compared to neutral pH (Kumar & Philip, 2006). The authors found that in clay soil, adsorption decreased drastically when the pH was reduced. In soil column experiments, deltamethrin was essentially immobile in three different soils. Kaufman et al. (1981) suggested that for nonacidic soils, the pH may be a primary factor affecting mobility of deltamethrin. In bioassays of treated soils against *C. formosanus*, there was an interaction between effects of soil pH and clay content on effectiveness of chlorfenapyr and fipronil (Wiltz, 2010).

3.4 Organic carbon

Partitioning of insecticides between soil organic matter and soil solution affects bioavailability (Felsot & Lew, 1989). Like clay, organic matter decreases adsorption of fipronil (Bobé et al., 1997). Mulrooney and Gerard (2007) applied fipronil to 3 soils and found that *R. flavipes* mortality decreased with increasing organic carbon. Although no soil effects were found when soils were treated at label rates, pyrethroids applied at low rates were less available in soils with high OC (Henderson et al., 1998). Soil OC has been shown to affect adsorption of several non-acidic pesticides, but have little or no effect on weakly acidic chemicals (Barriuso & Calvet, 1992; Boivin et al., 2005; Worrall et al., 2001).

4. Application technique

In addition to properties of the soil and chemicals, variations in application technique can influence availability, persistence, and impenetrability of toxins. Such variables include gaps in soil treatment, thickness of treated layer, and watering method. Within certain ranges of application rates, availability increases with rate; however, the opposite is true at other rates.

Initial concentration of termiticides in soil varies from several hundred to over one thousand micrograms per gram. Kard and McDaniel (1993) reported initial concentrations of 858± 990mg/g after application to a Mississippi soil, and Davis and Kamble (1992) reported initial concentrations of chlorpyrifos as high as 1500mg/g in a Nebraska loamy sand soil. Application rate affects both initial availability and degradation rate. Saran and Kamble (2008) reported an inverse relationship between the initial concentrations of bifenthrin, fipronil, and imidacloprid and their LT50 and LT90 values against *R. flavipes*. Greater bioavailability at higher concentrations may explain similar trends reported by Smith and Rust (1992), Forschler and Townsend (1996), and Ramakrishnan et al. (2000). At low rates, fipronil has low soil affinity, but adsorption increases with concentration (Bobé et al., 1997). Kamble and Saran (2005) found that at termiticide application rates of 0.06–0.125%, there is a reversal in the fipronil adsorption process, whereby there is a decrease in adsorption coefficient with an increase in concentration, resulting in an increase in bioavailability. Chlorpyrifos exhibited a lower degradation rate when applied at ≈1,000 μg/g soil than when applied at typical agricultural levels of 0.3–32 μg/g (Racke et al., 1994). When fipronil was applied at the labeled rate for locust control (8g AI per ha), 75% degraded within 3 d (Bobé et al., 1998). However, when applied at termiticidal rates (60-125 μg AI per g), fipronil did not show much degradation, and no metabolites were detected in residue analysis after 180 d (Saran and Kamble 2008). Gahlhoff and Koehler (2001) found that concentration and treatment thickness significantly affected both mortality and penetration by *R. flavipes* into imidacloprid-treated soil, with mortality remaining low after 7 d exposure to low concentrations, as well as affecting bioavailability, high termiticide concentrations may indirectly affect degradation by negatively impacting bacterial and fungal populations, resulting in prolonged inhibition of soil dehydrogenase and esterase activities (Felsot & Dzantor, 1995). Racke et al. (1996) examined hydrolysis of chlorpyrifos in 37 soils at agricultural application rates (10mg/g) and observed that in some alkaline soils hydrolysis constituted the major degradation pathway. However, they also noted that in several soils, with pH values in the range of 7.1 to 8.5, the hydrolytic reaction was inhibited at higher concentrations (1000mg/g).

Termites can circumvent soil treatments by using untreated gaps, building materials, or debris as bridges between the surrounding soil and structure (Forschler, 1994; Smith & Zungoli, 1995; Su & Scheffrahn, 1998). Subterranean termite foragers are able to detect and avoid repellant termiticides so areas treated with pyrethroids are rarely contacted. The subterranean termites' ability to detect chemical barriers allows termite foragers to follow the edge of the pyrethroid treated area until they find a gap in the treatment (Forschler, 1994; Rust & Smith, 1993; Su & Scheffrahn, 1990b; Su et al., 1982). Thus, gaps in pyrethroid applications may actually funnel foragers toward the structures they are intended to protect (Forschler, 1994; Kuriachan & Gold, 1998). The inevitability of gaps in soil termiticide barriers is a major limitation to the efficacy of repellant liquid termiticides (Forschler, 1994; Kuriachan & Gold, 1998). Gaps may exist in a soil termiticide treatment for a number of reasons. Pre-construction treatments often contain gaps due to imperfect initial application or physical disturbance of the soil after application (Koehler et al., 2000; Su & Scheffrahn, 1990a, 1998). When an existing structure becomes infested and requires a remedial termiticide application, it is difficult to create a continuous horizontal barrier of liquid termiticide beneath the structure (Su & Scheffrahn, 1990a, 1998; Koehler et al., 2000). Finally, all termiticides degrade over time. An ageing soil treatment, applied below the foundation before a structure was built, is inaccessible after construction and cannot be reapplied (Su & Scheffrahn, 1990a; Su, 1997; Koehler et al., 2000).

The total volume of pesticide suspension applied to soil affects penetration depth and concentration in the soil. In tests using imidacloprid and fipronil in five different soils, when equal amounts of pesticide were diluted in different volumes of water, the higher volume treatments penetrated further into the soil, but the more concentrated treatments deposited more pesticide in the top 1cm of soil (Peterson 2010). It is likely that the thicker barrier of lower active ingredient concentration would provide better protection, at least in the short term because it might be better able to withstand disturbances to the top 1 cm of soil. Additionally, termites are less able to tunnel through thicker barriers of lower active ingredient concentration than through thinner barriers of higher concentration (Smith et al. 2008). However, pesticide treatments with low initial concentrations degrade faster than those with higher initial concentrations (Bobé et al., 1998; Felsot & Dzantor, 1995; Saran & Kamble, 2008). In addition to total volume of liquid applied, initial thickness of the treated zone depends on soil and termiticide properties. Smith and Rust (1992) found that termiticidal amounts of chlordane and cypermethrin moved to soil depths of at least 7 cm, while chlorpyrifos moved to a depth of at least 30 cm.

5. Environmental factors

Both biotic and abiotic pathways have been found to be important for insecticide degradation and transformation in soils (Racke et al., 1996).

5.1 Moisture

Water can compete with pesticides for sorption sites on soil particles. Dry soils become more sorptive for both polar and non-polar chemicals (Chen et al., 2000). However, chemicals with low polarity are released when soil becomes wet (Harper et al., 1976). Repeated cycles of wetting and drying affect pesticide availability and degradation, but depend on properties of the chemical, soil, number of wetting and drying cycles, time since pesticide application, and time since wetting (Garcia-Valcarcel & Tadeo, 1999; Xia & Brandenburg 2000; Ying & Kookana, 2006; Peterson, 2007).

5.2 Temperature

Soil temperature affects termiticide bioavailability through its influence on solubility and adsorption. In addition to its effect on the physical and chemical properties of the pesticide, extreme temperatures affect the rate of microbial degradation, as described in the following section. Several studies have demonstrated that temperature affects adsorption of pesticides to soil, but that the nature of this effect varies among pesticides. Although most of the work on pesticide availability and degradation has been conducted in the temperate climates of North America and Europe, soil temperature is likely to play an important role in termiticide degradation in tropical regions. Khan et al. (1996) found that lindane adsorption to silty loam and silty clay loam soils increased with temperature. Likewise, Valverde-Garcia et al. (1988) found that higher temperatures increased the adsorption of the fungicide thiram and the organophosphate insecticide dimethoate to organic soils. Temperature may interact with pH, particularly in saturated soils. In aqueous solutions, fenamiphos, fipronil, and trifluralin degradation increased with temperature and pH (Ramesh & Balasubramanian, 1999). Other studies have demonstrated a reduction in pesticide adsorption at higher temperatures. Dios-Cancela et al. (1990) found that sorption of the herbicide cyanazine to peats decreased with increasing temperature.

5.3 Micromial degradation

Microbial degradation occurs when fungi, bacteria, and other soil microorganisms use pesticides as food or consume pesticides along with other substances. Activity of microbes is affected by soil organic matter and texture and is usually highest in warm, moist, well-aerated soils with a neutral pH. Because microbial degradation is mediated by enzymes, temperature is important in determining the rate degradation: the rate of most reactions catalyzed by enzymes tends to double for each 10°C increase in temperature between 10° and 45°C and is greatly reduced above and below these temperatures.

Naturally-occurring pesticide-degrading microorganisms may be relatively rare in pristine environments and non-exposed agricultural soils (Bartha, 1990). Some of the pesticide-degrading microbes that have been identified include *Arthrobacter*, *Brevibacterium*, *Clavibacter*, *Corynebacterium*, *Micromonospora*, *Mycobacterium*, *Nocardia*, *Nocardioides*, *Rhodococcus* and *Streptomyces* genera (De Schrijver and De Mot, 1999).

A review of earlier work on organophosphate and carbamate insecticide degradation was prepared by Laveglia and Dahm (1977). Although organophosphates are no longer used in many parts of the world, there have been several recent studies on their degradation by microbes. Li et al. (2007) reported the isolation of a bacterium, *Sphingomonas* sp., that degrades chlorpyrifos, parathion, parathion-methyl, fenitrothion and profenofos. However, several other studies have found little microbial degradation of chlorpyrifos. Goda et al. (2010) showed that the intact cells of *Pseudomonas putida* IS168 were able to degrade fenitrothion, diazinon and profenofos when present as sole carbon sources, but failed to grow on chlorpyrifos. Trichloropyridinol (TCP), one of the main chlorpyrifos metabolites, has antimicrobial properties (Cáceres et al., 2007; Feng et al., 1997; Racke et al., 1990), possibly accounting for the scarcity of chlorpyrifos-degrading microorganisms. Degradation of pyrethroids in soil has also been extensively studied (Gan et al., 2005; Jorhan & Kaufman, 1986; Kaufman et al., 1981; Lee et al., 2004; Lord et al., 1982). Most of these studies show that microorganisms play an important role in the degradation of pyrethroid compounds in soils and sediments.

6. Termite pressure and susceptibility

In studies evaluating termite tunneling through chlordane, chlorpyrifos, or permethrin treated soil, large groups of termites were able to tunnel farther than small groups (Beal & Smith, 1971; Jones, 1990). At low population density, different colonies of *C. formosanus* either totally avoided permethrin-treated soil or tunneled slightly (Jones, 1989, 1990). Jones (1990) found that while large groups of termites tunneled more than small groups in soils treated with chlordane, chlorpyrifos, or permethrin, group size had different effects on mortality in different soil treatments. Several experiments have demonstrated correlations between termite survival rates and population density (Lenz et al., 1984; Lenz, 2009; Santos et al., 2004). At population densities below 0.1 g termites / ml, Lenz et al. (1984) found that, in the absence of termiticide treatment, survival of *Coptotermes lacteus* (Froggatt) and *Nasutitermes exitiosus* (Hill) increased with population density.

Susceptibility differences occur among termite species and colonies. Most soil termiticide evaluations have included only one target species. However, in studies comparing responses of two or more species, there are frequently differences in susceptibility. *C. formosanus* penetrated soil treated with aldrin, chlordane, dieldrin, or heptachlor, while *R. virginicus* and *R. flavipes* failed to penetrate lower rates of the same chemicals and were killed more quickly than *C. formosanus* (Beal & Smith 1971). In a laboratory assay, chlorpyrifos, permethrin, cypermethrin, bifenthrin, isofenphos, lambda-cyhalothirn, and fenitrothion all provided equal barrier protection against *R. flavipes* (Su et al. 1993). However, in the same assay, *C. formosanus* generally tunneled deeper into sand treated with organophosphates than with pyrethroids. Penetration of sand treated with thiamethoxam or fipronil was similar for *C. formosanus* and *R. flavipes*, but thiamethoxam was more toxic to *C. formosanus* than to *R. flavipes* (Remmen & Su 2005). Osbrink and Lax (2002) evaluated seven insecticides against termites from colonies that had been previously been identified as either insecticide susceptible or tolerant, finding differences in substrate penetration and mortality among colonies and insecticides.

Termite traits other than population size or susceptibility to toxicants can increase the likelihood that soil treatments will fail to protect a structure. Aerial infestations account for a large percentage of structural infestations by *C. formosanus* (Su & Scheffrahn, 1990), making soil treatments ineffective. Additionally, *C. formosanus* colonies may seal off or avoid treated areas (Su et al. 1982) when repellent toxicants are used, but use gaps in the soil barrier to access the structure (Forschler, 1994).

7. Conclusion

Several long-term studies of termiticide persistence have been conducted. In USDA Forest Service trials, which have been conducted for the past 40 years, tests consist of treated soil plots covered by concrete slabs. Treatments are considered failures when termites penetrate >50% of field replicates to reach a wood block placed in a pipe running through the slab (Kard, 2003; Wagner, 2003). In these tests, longevity differed with geographic location and termiticide class (Mulrooney et al., 2007). Such studies have the advantage of being performed under natural soil weathering conditions for an extended period of time and provide a standard method of comparing termiticides. However, products are evaluated on a limited number of soils and it is impossible to tell if a lack of penetration into plots should be attributed to effectiveness of the termiticide or to the absence of termite pressure. To overcome this problem, other studies have included laboratory bioassays coupled with field termiticide persistence studies (Gold et al., 1996; Grace, 1991; Su et al., 1993). Unfortunately, most of these studies have evaluated relatively few soil types. Because performance is so dependent on a combination of temiticide and soil properties and weathering, more research is needed to evaluate new and existing products under a larger range of conditions. Soil termiticides have been extensively evaluated for toxicity, bioavailability, and degradation. However, reasons for termiticide failure are complex and often local in nature, indicating the need for more research and localized treatment recommendations regarding choice of toxicant, application technique, and treatment frequency.

8. References

Alley, W., (1993). *Regional Groundwater Quality*, Van Nostrand Reinhold, New York, New York, 1994.

Anonymous. (2008). State of the industry. Termites: Kicked to the "B" list? PCT October 10, 2008. http://www.pctonline.com/Article.aspx?article_id=37550.

Aventis Environmental Science. (2001). Termidor SC termiticide label. Montvale, NJ.

BASF Corporation. (2001). Phantom termiticide label. Triangle Park, NC.

Baker, P. B., and D. E. Bellamy. (2006). Field and laboratory evaluation of persistence and bioavailability of soil termiticides to desert subterranean termite *Heterotermes aureus* (Isoptera: Rhinotermitidae). *Journal of Economic Entomology*, 99: 1345-1353.

Barriuso, E. & Calvet, R. (1992). Soil type and herbicide adsorption. *International Journal of Environmental Analytical Chemistry*, 46: 117-128.

Bartha, R., (1990). Isolation of microorganisms that metabolize xenobiotic compounds. In: *Isolation of Biotechnological Organisms from Nature*, Labeda, D. P. (Ed.), pp. 283-307, McGraw-Hill Publishing Company, New York, USA.

Baskaran, S., Kookana, R. S. & Naidu, R. (1999). Degradation of bifenthrin, chlopyrifos, and imidacloprid in soil and bedding materials at termiticidal application rates. *Pesticide Science*, 55:1222-1228.

Bayer Corporation. (2000). Premise 75 WSP termiticide label. Kansas City, MO.

Beal, R. H., & Smith, V. K. (1971). Relative susceptibilities of *Coptotermes formosanus*, *Reticulitermes flavipes*, and *R. virginicus* to soil insecticides. *Journal of Economic Entomology*, 64: 472-475.

Bedient, P.H., Rifai, H. S. & Newell C. J. (1994). Ground Water Contamination: Transport and Remediation, Prentice Hall, Englewood Cliffs, NJ, 1994.

Bobé, A., Coste, C. M. & Copper, J -F. (1997). Factors influencing the adsorption of fipronil on soils. *Journal of Agricultural and Food Chemistry*, 45: 4861-4865.

Bobé, A., Cooper, J. -F., Coste, C. M. & Muller, M. -A. (1998). Behaviour of fipronil in soil under Sahelian plain field conditions. *Pesticide Science*, 52: 275-281.

Boivin, A., Cherrier, R. & Schiavon, M. (2005). A comparison of five pesticides adsorption and desorption processes in thirteen contrasting field soils. *Chemosphere*, 61: 668-676.

Cáceres T, He, W., Naidu, R. & Megharaj, M. (2007). Toxicity of chlorpyrifos and TCP alone and in combination to *Daphnia carinata*: the influence of microbial degradation in natural water. *Water Research*, 41: 4497-4503

Carrizosa, M. J., Calderon, M. J, Hermosin, M. C. & Cornejo, J. (2000). Organosmectites as sorbent and carrier of the herbicide bentazone. *Science of the Total Environment*, 247: 285-293.

Chen D., Rolston, D. E., and Yumaguchi, T. (2000). Calculating partition coefficients of organic vapors in unsaturated soils and clays. *Soil Science*, 165: 217-225.

Cheung, M. W., Mingelgrin, V. & Biggar, J. W. (1979). Equilibrium and kinetics of desorption of picloram and parathion in soils. *Journal of Agricultural and Food Chemistry*, 27: 1201-1206.

Connelly, P. (2001). Environmental fate of fipronil. Environmental Monitoring Branch. Department of Pesticide Regulation. California Environmental Protection Agency.

Cox, L. Koskinen W. C., Celis, R., Yeen, P. Y., Hermosin, M. C. & Cornejo, J. (1998). Sorption of imidacloprid on soil clay mineral and organic components. *Soil Science of America Journal*, 69: 911-915.

Davis, R. W. & Kamble, S. T. (1992). Distribution of sub-slab injected Dursban TC (chlorpyrifos) in a loamy sand soil when used for subterranean termite control, *Bulletin of Environmental Contamination and Toxicology*, 48: 585-591.

De Schrijver, A. & De Mot, R., (1999). Degradation of pesticides by actinomycetes. *Critical Reviews in Microbiology*, 25: 85-119.

Dios-Cancela, G., Romero, T. E. & Sanchez-Rasero, F. (1990). Adsorption of cyanazine on peat and montmorillonite clay surfaces. *Soil Science*, 150: 836-843

Dupont. (2010). Dupont Altriset MSDS.

Environmental Protection Agency. (2000). Chlorpyrifos revised risk assessment and agreement with registrants. - http://www.epa.gov/pesticides/op/chlorpyrifos/agreement.pdf.

Fecko, A. (1999). Environmental fate of bifenthrin. Environmental Monitoring and Pest Management Branch. California Department of Pesticide Regulation.

Felsot, A. S., & Lew, A. (1989). Factors affecting bioactivity of soil insecticides: relationships among uptake, desorption, and toxicity of carbofuran and terbufos. *Journal of Economic Entomology*, 82: 389-395.

Felsot, A. S., & Dzantor, E. K. (1995). Effect of alachlor concentration and an organic amendment on soil dehydrogenase activity and pesticide degradation rate. *Environmental Toxicology and Chemistry*, 14: 23-28.

Feng Y, Racke, K. D. & Bollag, J. M. (1997). Isolation and characterization of a chlorinated-pyridinol-degrading bacterium. *Applied Environmental Microbiology*, 63: 4096–4409

Forschler, B. T. (1994). Survivorship and tunneling activity of *Reticulitermes flavipes* (Kollar) (Isoptera: Rhinotermitidae) in response to termiticide soil barriers with and without gaps of untreated soil. *Journal of Entomological Science*, 29: 43–54.

Forschler, B. T. & Townsend, M. L. (1996). Mortality of Eastern subterranean termites (Isoptera: Rhinotermitidae) exposed to four soils treated with termiticides. *Journal of Economic Entomology*, 89: 678-681.

Gahlhoff, J. E. & Koehler, P. G. (2001). Penetration of the eastern subterranean termite into soil treated at various thicknesses and concentrations of Dursban TC and Premise 75. *Journal of Economic Entomology*, 94: 486-491.

Gan, J., Lee, S. J., Liu, W. P., Haver, D. L., Kabashima, J. N. (2005). Distribution and persistence of pyrethroids in runoff sediment. *Journal of Environmental Quality*, 36: 834-841.

Gao, J. P., Maguhn, J., Spitzauer, P. & Kettrup, A. (1998). Sorption of pesticides in the sediment of the Teufelsweiher pond (Southern Germany). I: Equilibrium assessments, effect of organic carbon content and pH. *Water Research*, 32: 1662-1672.

Garcia-Valcarcel, A. I. & Tadeo, J. L. (1999). Influence of soil moisture on sorption and degradation of hexazinone and simazine in soil. *Journal of Agricultural and Food Chemistry*, 47: 3895-3900.

Goda, S. K., Elsayed, I. E., Khodair, T. A., El-Sayed, W. & Mohamed, M. E. (2010). Screening for and isolation and identification of malathion-degrading bacteria: cloning and sequencing a gene that potentially encodes the malathion-degrading enzyme, carboxylestrase in soil bacteria. *Biodegradation*, 21: 903–913.

Gold, R. E., Howell, H. N., Pawson, B. M., Wright, M. S. & J. C. Lutz. (1996). Persistence and bioavailability of termiticides to subterranean termites (Isoptera: Rhinotermitidae) from five soil types in Texas. *Sociobiology,* 28: 337-363.

Grace, J. K. (1991). Response of eastern and Formosan subterranean termites (Isoptera, Rhinotermitidae) to borate dust and soil treatments. *Journal of Economic Entomology,* 84: 1753-1757.

Grace, J. K., Yates, J. R., Tamishiro, M. & Yamamoto, R. T. (1993). Persistence of organochlorine insecticides for Formosan subterranean termite (Isoptera: Rhinotermitidae) control in Hawaii. *Journal of Economic Entomology,* 86: 761-766.

Halfon, E., Galassi, S., Bruggemann, R. &. Provini, A. (1996). Selection of priority properties to assess environmental hazard of pesticides, *Chemosphere,* 33: 1543-1562.

Harper, L. A., White, A. W. Jr., Bruce, R. R., Thomas, A. W., and Leonard, R, A. (1976) Soil and microclimate effects on trifluralin volatilization. *Journal of Environmental Quality,* 5: 236-242.

Henderson, G., Walthall, P. M., Wiltz, B. A. Rivera-Monroy, V. H., Ganaway, D. R. & H. M. Selim. (1998). Analysis of soil properties in relation to termiticide performance in Louisiana. In Proceedings of the National Conference on Urban Entomology, San Diego, CA, April 26-28, 1998.

Horwood, M. A. (2007). Rapid degradation of termiticides under field conditions. *Australian Journal of Entomology,* 46: 75-78.

Itawa, H., Tanabe, S., Sakai, N., Tatsukawa, R., (1993). Distribution of persistent organochlorines in the oceanic air and surface seawater and the role of ocean on their global transport and fate. *Environmental Science & Technology,* 27: 1080-1098.

Jones, D. (1999). Environmental fate of cypermethrin. Environmental Monitoring and Pest Management Branch. California Department of Pesticide Regulation.

Jones S. C. (1989). Field evaluation of fenoxycarb as a bait toxicant for subterranean termite control. *Sociobiology,* 15: 33-41.

Jones, S. C. (1990). Effects of population density on tunneling by Formosan subterranean termite (Isoptera: Rhinotermitidae) through treated soil. *Journal of Economic Entomology,* 83: 875-878.

Jorhan, E. G. & Kaufman, D. D. (1986). Degradation of cis- and transpermethrin in flooded soil. *Journal of Agricultural and Food Chemistry,* 34: 880-884.

Kamble, S. T. & Saran, R. K. (2005). Effect of concentration on the adsorption of three termiticides in soil. *Bulletin of Environmental Contamination and Toxicology,* 75: 1077-1085.

Kard, B. (1999). Termiticides: the Gulfport report. *Pest Control,* 67: 42-46.

Kard, B. M. (2003). Integrated pest management of subterranean termites (Isoptera*). Journal of Entomological Science,* 38: 200-224.

Kard B. M. & McDaniel, C. A. (1993). Field evaluation of the persistence and efficacy of pesticides used for termite control, in Pesticides in Urban Environments - Fate and Significance, ed by Racke KD and Leslie AR, ACS Symposium Series 522, American Chemical Society: Washington DC. pp 42-62.

Kaufman, D. D., B. A. Russell, C. S. Helling, and A. Kayser. (1981). Movement of cypermethrin, decamethrin, permethrin and their degradation productions in soil. *Journal of Agricultural and Food Chemistry*, 29: 239-345.

Koehler, P.G., Su, N. -Y., Scheffrahn, R. H. & Oi, F.M. (2000). Baits to control subterranean termites: the Sentricon system, ENY 2000, Florida Cooperative Extension Service, Institute of Food and Agricultural Services.

Kumar, M., & Philip, L. (2006). Adsorption and desorption characteristics of hydrophobic pesticide endosulfan in four Indian soils. *Chemosphere*, 62: 1064-1077.

Kuriachan, I. & Gold, R. E. (1998). Evaluation of the ability of *Reticulitermes flavipes* Kollar, a subterranean termite (Isoptera: Rhinotermitidae), to differentiate between termiticide treated and untreated soils in laboratory tests. *Sociobiology*, 32: 151-166.

Lee, S, Gan, J. Y., Kim, J. S., Kabashima, J. N. & Crowley, D. E. (2004). Microbial transformation of pyrethroid insecticides in aqueous and sediment phases. *Environmental Toxicology & Chemistry* ,23: 1-6.

Lenz, M. (2009). Laboratory bioassays with subterranean termites (Isoptera) – the importance of termite biology. *Sociobiology*, 53: 573-595.

Lenz, M., Barrett, R. A. & Williams, E. R.. (1984). Implications for comparability of laboratory experiments revealed in studies on the effects of population density on vigor in *Coptotermes lacteus* (Froggatt) and *Nasutitermes exitious* (Hill) (Isoptera: Rhinotermitidae & Termitidae). *Bulletin of Entomological Research*. 74: 477-485.

Lenz, M, Watson, J.A L., Barrett, R.A. & Runko, S. (1990). The effectiveness of insecticidal soil barriers against subterranean termites in Australia. *Sociobiology*, 17: 9-35.

Lewis, D. L. (1980). Environmental and health aspects of termite control chemicals. Sociobiology. 5: 197-203.

Li, X., He, J., & Li, S. (2007). Isolation of a chlorpyrifos-degrading bacterium, *Sphingomonas* sp. strain Dsp-2, and cloning of the mpd gene. *Research in Microbiology* 158(2):14–143.

Lord, K. A., Mckinley, M. & Walker, N. (1982). Degradation of permethrin in soils. *Environmental Pollution*, 29 (2): 81-90.

Masters, G. M., (1991). Introduction to Environmental Engineering and Science, Prentice Hall, Englewood Cliffs, NJ.

Means, J. C., Wood, S. G. Hassett, J. J., & Banwart W. L. (1982). Sorption of amino-and carboxy-substituted polynuclead aromatic hydrocarbons by sediments and soils. *Environmental Science Technology*, 16: 93-98.

Mulrooney, J. E., & Gerard, P. D. (2007). Toxicity of fipronil in Mississippi soil types against *Reticulitermes flavipes* (Isoptera: Rhinotermitidae). *Sociobiology* 50: 63-70.

Mulrooney, J. E., Wagner, T. L., Shelton, T. G., Peterson, C. J. & Gerard, P. D. (2007). Historical review of termite activity at Forest Service termiticide test sites from 1971 to 2004. *Journal of Economic Entomology*, 100: 488-494.

Ntow, W. J. (2005). Pesticide residues in Volta Lake, Ghana, Lakes Reservoirs. *Research and Management*, 10: 243-248.

Osbrink, W. L. A., & Lax, A. R. (2002). Effect of tolerance to insecticides on substrate penetration by Formosan subterranean termites (Isoptera: Rhinotermitidae). *Journal of Economic Entomology*, 95: 989-1000.

Osbrink, W. L A., Cornelius, M. L. & Lax, A. R. (2005). Effect of imidacloprid soil treatments on occurrence of Formosan subterranean termites (Isoptera: Rhinotermitidae) in independent monitors. *Journal of Economic Entomology*, 98: 2160–2168.

Parman, V. & Vargo, E. L. (2010). Colony-Level Effects of Imidacloprid in Subterranean Termites (Isoptera: Rhinotermitidae). *Journal of Economic Entomology*, 103: 791-798.

Pawson, B. M. & R. E. Gold. (1996). Evaluation of baits for termites (Isoptera: Rhinotermitidae) in Texas. *Sociobiology*, 28: 485-491.

Peterson, C. J. (2007). Imidacloprid mobility and longevity in soil columns at a termiticidal application rate. *Pest Management Science*, 63: 1124-1132.

Peterson, C. J. (2010). Varying termiticide application rate and volume affect initial soil penetration. *Journal of Economic Entomology*. 103: 433-436.

Potter, M. H. & Hillery, A. E. (2001). Exterior-targeted liquid termiticides an alternative approach to managing subterranean termites (Isoptera: Rhinotermitidae) in buildings. *Sociobiology* 39: 373–405.

Racke, K. D., Fontaine, D. D., Yoder, R. N. & Miller, J. R. (1994). Chlorpyrifos degradation in soil at termiticidal application rates. *Pesticide Science* 42: 43-51.

Racke K. D., Laskowski D. A. & Schultz M. R. (1990). Resistance of chlorpyrifos to enhanced biodegradation in soil. *Journal of Agricultural and Food Chemistry*, 38: 1430–1436.

Racke K. D., Steele, K. P., Yoder, R. N., Dick, W.A. & Avidov E. (1996). Factors affecting the hydrolytic degradation of chlorpyrifos in soil, *Journal of Agricultural and Food Chemistry*, 44: 1582-1592.

Ramakrishnan, R., Suiter, D. R., Nakatsu C. H. & Bennett, G. W. (2000). Feeding inhibition and mortality in *Reticulitermes flavipes* (Isoptera: Rhinotermitidae) after exposure to imidacloprid-treated soils. *Journal of Economic Entomology*, 93: 422-428.

Randall, M. and Doody, T. C. (1934). Wood preservatives and protective treatments, pp. 463-476. In C.A. Kofoid [ed.] Termites and Termite Control. University of California Press, Berkeley, CA.

Remmen, L. N. & Su, N.-Y. (2005). Tunneling and mortality of eastern and Formosan subterranean termites (Isoptera: Rhinotermitidae) in sand treated with thiamethoxam or fipronil. *Journal of Economic Entomology*, 98: 906–910.

Rice, R. C., Jaynes, D. B. & Bowman, R. S. (1991). Preferential flow of solutes and herbicide under irrigated fields. *Transactions of the American Society of Agricultural Engineers*, 34: 914-918.

Ramesh, A. & Balasubramanian, M. (1999). Kinetics and Hydrolysis of Fenamiphos, Fipronil, and Trifluralin in Aqueous Buffer Solutions. *Journal of Agricultural and Food Chemistry*, 47: 3367-3371.

Rust, M. K. & Saran, R. K. (2006). The toxicity, repellency, and transfer of chlorfenapyr against western subterranean termites (Isoptera: Rhinotermitidae). *Journal of Economic Entomology*, 99: 864–872.

Rust, M. K. & Smith, J. L. (1993). Toxicity and repellency of components in formulated termiticides against western subterranean termites (Isoptera: Rhinotermitidae). *Journal of Economic Entomology*, 86: 1131-1135.

Santos, C. A., DeSouza, O. & Guedes, R. N. C. (2004). Social facilitation attenuating insecticide-driven stress in termites (Isoptera: Nasutitermitinae). *Sociobiology*, 44: 539-546.

Saran, R. K., & Kamble, S. T. (2008). Concentration-dependent degradation of three termiticides in soil under laboratory conditions and their bioavailability to eastern subterranean termites (Isoptera: Rhinotermitidae). *Journal of Economic Entomology*, 101: 1373-1383.

Saran, R. K. & Rust, M. K. (2007). The toxicity, uptake, and transfer efficiency of fipronil in western subterranean termites (Isoptera: Rhinotermitidae). *Journal of Economic Entomology*, 100: 495–508.

Shelton, T. G. & Grace, J. K. (2003). Effects of exposure duration on transfer of nonrepellent termiticides among workers of *Coptotermes formosanus* Shiraki (Isoptera: Rhinotermitidae*). *Journal of Economic Entomology*, 96: 456–460.

Smith, B. C. & Zungoli, P. A. (1995). Rigid board insulation in South Carolina: its impact on damage, inspection and control of termites (Isoptera: Rhinotermitidae). *Florida Entomologist*, 78: 507-515.

Smith, J. L. & Rust, M. K. (1990). Tunneling response and mortality of the western subterranean termite (Isoptera: Rhinotermitidae) to soil treated with termiticides. *Journal of Economic Entomology*, 83: 1395–1401.

Smith, J. L., & Rust, M. K. (1992). Activity of water-induced movement of termiticides in soil. *Journal of Economic Entomology*, 85: 430-434.

Smith, J. L. & Rust, M. K. (1993). Cellulose and clay in sand affects termiticide treatments. *Journal of Economic Entomology*, 86: 53-60.

Spomer, N. A., Kamble, S. T. & Siegfried, B. D. (2009). Bioavailability of chlorantraniliprole and indoxacarb to eastern subterranean termites (Isoptera: Rhinotermitidae) in various soils. *Journal of Economic Entomology*, 102: 1922-1927.

Su, N.-Y. (1997). Protecting Historic Buildings and structures from termites. Cultural Resource Management 20(7). National Park Service. http://crm.cr.nps.gov/issueindex.cfm.

Su, N. -Y. (2005). Response of the Formosan subterranean termites (Isoptera: Rhinotermitidae) to baits or nonrepellent termiticides in extended foraging arenas. *Journal of Economic Entomology*, 98: 2143–2152.

Su, N.-Y. & Scheffrahn, R. H. (1990a). Economically important termites in the United States and their control. *Sociobiology*, 17: 77-92.

Su, N.-Y. & Scheffrahn, R. H. (1990b). Comparison of eleven soil termiticides against the Formosan subterranean termite and the eastern subterranean termite (Isoptera: Rhinotermitidae). *Journal of Economic Entomology*, 83: 1918-1924.

Su, N.-Y. & Scheffrahn, R. H. (1998). A review of subterranean termite control practices and prospects for integrated pest management programs. *Integrated Pest Management Reviews*, 3: 1 -13.

Su, N.-Y., Scheffrahn, R. H. & Ban P. M. (1993). Barrier efficacy of pyrethroid and organophosphate formulations against subterranean termites (Isoptera: Rhinotermitidae). *Journal of Economic Entomology*, 86: 772-776.

Su, N. -Y., Tamashiro, M., Yates, J. R. & Haverty, M. I. (1982). Effect of behavior on the evaluation of insecticides for prevention of or remedial control of the Formosan subterranean termite. *Journal of Economic Entomology*, 75: 188-193.

Su, N. -Y., Wheeler, G. S. & Scheffrahn, R. H. (1995). Subterranean termite (Isoptera: Rhinotermitidae) penetration into sand treated at various thicknesses with termiticides. *Journal of Economic Entomology*, 88: 1690-1694.

Sunti, L. R., Shiu, W. Y., Mackay, D., Sieber, J. N. & Glotfelty, D. (1988). Critical review of Henry's law constants for pesticides. Reviews of Environmental Contamination & Toxicology. 103: 1–59.

Tamashiro, M., Yates, J. R. & Ebesu, R. H. (1987). The Formosan termite in Hawaii: problems and control, pp. 15-22. In Biology and control of the Formosan subterranean termite: M. Tamashiro and N. -Y. Su, eds., 67th Meeting of the Pacific Branch of the Entomological Society of America, Research Extension Series 083, University of Hawaii, Honolulu.

Thorne, B. L. & Breisch, N. L. (2001). Effects of sublethal exposure to imidacloprid on subsequent behavior of subterranean termite *Reticulitermes virginicus* (Isoptera: Rhinotermitidae). *Journal of Economic Entomology*, 94: 492–498.

USEPA. (1986). U.S. Environmental Protection Agency. Pesticide fact sheet: chlordane. U.S. Environmental Protection Agency. Washington, DC.

USEPA. (2001). U.S. Environmental Protection Agency. Pesticide fact sheet: chlorfenapyr. U.S. Environmental Protection Agency. Washington, DC.

USEPA. (2000). U.S. Environmental Protection Agency. Pesticide fact sheet. Indoxacarb. U.S. Environmental Protection Agency. Washington, DC.

USEPA. (2008). U.S. Environmental Protection Agency. (2008). Pesticide fact sheet: chlorantraniliprole. U.S. Environmental Protection Agency. Washington, DC.

Wagner, T. L. (2003). U.S. Forest Service termiticide tests. *Sociobiology*, 41: 131–141.

Ware, G.W. (2000). The Pesticide Book 5th Ed. Thompson Publications. Fresno, CA. 386 pages.

Wauchope, R. D., Yeh, S., Linders, J., Kloskowski, R., Tanaka, K., Rubin, B., Katayama, A., Kördel, W., Gerstl, Z., Lane, M. & Unsworth, J. B. (2002). Review: Pesticide soil sorption parameters: theory, measurement, uses, limitations, and reliability. *Pest Management Science*, 58: 419–445.

Wiltz, B. A. (2010). Laboratory Evaluation of Effects of Soil Properties on Termiticide Performance against Formosan Subterranean Termites (Isoptera: Rhinotermitidae). *Sociobiology*, 56 (3): 755-773.

Wing, K. D., Sacher, M., Kagaya, Y., Tsurubuchi, Y., Mulderig, L., Connair, M. & Schnee, M. (2000). Bioactivation and mode of action of the oxadiazine indoxacarb in insects. *Crop Protection*, 19: 537-545.

Wood, T. G. & Pierce, M. J. (1991). Termites in Africa: The environmental impact of control measures and damage to crops, trees, rangeland and rural buildings. *Sociobiology*, 19: 221-234.

Worrall, F., Fernandez-Perez, M., Johnson, A. C., Flores-Cesperedes, F. & Gonzalez-Pradas, E. (2001). Limitations on the role of incorporated organic matter in reducing pesticide leaching, *Journal of Contaminant Hydrology*, 49: 241–262.

Xia, Y. & Brandenburg, R. L. (2000). Effect of irrigation on the efficacy of insecticides for controlling two species of mole crickets (Orthoptera: Gryllotalpidae) on golf courses. *Journal of Economic Entomology*, 93: 852–857.

Xue, N., Zhang, D. & Xu, X., (2006). Organochlorinated pesticide multiresidues in surface sediments from Beijing Guanting reservoir. *Water Research*, 40: 183-194.

Ying, G. G. & Kookana, R. S. (2006). Persistence and movement of fipronil termiticide with under slab and trenching treatments. *Environmental Toxicology and Chemistry*, 25: 2045-2050.

Part 2

Further Applications

10

Non-Chemical Disinfestation of Food and Agricultural Commodities with Radiofrequency Power

Manuel C. Lagunas-Solar
University of California, Davis
RF Biocidics Inc., Vacaville, California
USA

1. Introduction

The presence of microbial and insect/mite pests in foods and agricultural commodities, particularly in fresh produce, dried foods, nuts, grains, seeds, nursery plants, ornamental flowers and in wood products (i.e. pallets), continues to be a major factor affecting their condition for safe distribution and use in local, regional and international markets. As a mean to reduce the potential of propagating non-indigenous pests, postharvest (mandatory) treatment modalities and quarantine barriers have been imposed to regulate transportation and distribution of many of these products worldwide. These regulations define strategies for the detection, control, or eradication techniques for controlling quarantine insect and mite pests.

Today, more than 6,500 nonnative species are already established in the United Sates and approximately 15% of these species are either economically or environmentally harmful (Pimentel, Lach, Zuñiga et al., 1999). Control or eradication practices for arthropod pests are mostly based on chemical pesticides, although host removal, adequate agricultural production practices, biological control agents, and sterile insect release are often techniques applied in place off or in conjunction with pesticides.

Among the most important quarantine plant pests, various exotic fruit flies have been identified in the USA as threats to more than 250 crops. On the other hand, the presence of moths in stored products represents important and unacceptable risks to many growing and expanding agricultural regions worldwide. If detected, affected commodities must be processed with effective control or eradication techniques. If unattended, losses in product's quality represent unacceptable economic losses.

Chemical pesticides, waxes, coatings, thermal treatments (heated air; hot water immersion), modified atmospheres, cold storage (refrigeration), and irradiation are some of the processes that have helped industry meet current challenges and demands. Lately, however, new consumer preferences, trends and regulatory interventions have increased the needs for minimally processed foods with low or no residual chemicals. This new trend requires that less invasive or chemical-free alternatives become available to replace or minimize the use of pesticides. Furthermore, recent concerns associated with potential terrorist threats using microbial contaminants or other pests, have increased the need to develop alternatives to assure the safety of the food supply while minimizing economical risks associated with production and export agriculture. These combined challenges are now familiar to affected governments as well as to industry and regulators worldwide.

Historically, and with a few exceptions, pesticides have provided an ample spectrum of effective techniques to control pests and there is a continual industry trend to maintain and improve their use. However, this practice and its effects and limitations have partially fueled the emergence of organic agriculture. This in turn has prompted conventional agriculture to review its practices, its traditional processes, and to investigate new types of pesticides as well as to develop new disinfestation techniques. The incorporation of fluorine in agrochemicals to enhance stability and bioavailability is the latest attempt to increase their effectiveness while reducing their secondary impact (Jeschke, 2004). Nevertheless, their invasiveness and persistence in all environs surrounding agricultural practices continues to be resisted by consumers and by increased limiting regulations.

Past and even present industry reliance on methyl bromide fumigation for quarantine pest controls is the best and most recent example of the changing attitude that exists today with respect to invasive chemical processes. The existing ban and the new restrictions on production levels have forced agriculture to look for new and better alternatives. Fumigation, vacuum techniques and controlled atmospheres (CA) for insect (quarantine) control are marginally successful and restricted to long-storage commodities (i.e. grains, nut products, raisins) (Bond, 2007; Calderon, 1990). For perishable fresh commodities, these techniques have failed to provide the required and timely disinfestation level. Nevertheless, while somewhat successful, the needed long processing times (days or weeks) increases cost and is inadequate to fit with the logistics of marketing fresh agricultural products.

The use of low-level doses of ionizing radiation (i.e. food irradiation) is another effective and approved technique providing an alternative to disinfestation and disinfection of many commodities (Urbain, 1986). However, while technically useful and approved for certain applications, this approach prompts many public concerns and is usually and effectively resisted. Furthermore, because irradiation facilities require a high capital investment to install and operate in order to remain economically viable, it also forces the irradiation industry to operate as major centralized facilities located near high productivity agricultural areas. The seasonal nature of agriculture, however, forces the irradiation industry to meet the peak demands with excess processing capacity and to broaden off-season applications (i.e. disinfection of medical supplies) to remain viable. Consequently, the handling and distribution of to-be-treated food and agricultural commodities imposes new and severe logistical and cost adjustments to the user community. As a result, few if any agricultural export areas rely on irradiation facilities and those operating represent a small and stagnant resource for insect control.

Despite the above limitations, ionizing radiation also provide means to sterilize insects that once released in specific areas can reduce the impact of local/regional infestations.

As of today, with the exception of food irradiation, few attempts to fulfill the need for new alternatives to pesticides have been investigated using single or combined physical processes. If effective, these processes are inherently safer, eliminating the risks associated with the presence of pesticides in products and ultimately easing the current concerns with disposal issues, worker safety, and environmental impacts. Non-chemical or residue-free alternatives also provide opportunities to yield products with attributes closer to their natural sensory and nutritional properties. Furthermore, because physical processes are solely based on the use of energy, they are naturally free of residues and therefore can serve the needs of both conventional and organic agriculture.

Since 2002, research at the University of California, Davis established the use of RF power for disinfestation as well as for many novel sanitation and preservation purposes for a variety of food, non-food and agricultural commodities. Since then, RF processing has been established as a novel methodology able to provide new alternatives for chemical-free disinfestation, disinfection and enzyme deactivation effects on various commodities (Lagunas-Solar, 2003; Lagunas-Solar, Zeng & Essert, 2003; Lagunas-Solar, Zeng, Essert et al. 2005a; Lagunas-Solar, Cullor, Zeng, et al. 2005b; Lagunas-Solar, Zeng, Essert et al. 2006a). RF disinfestation, in particular, was proven as an effective, rapid, and a reliable chemical-free alternative to pesticides and capable of large-scale processing.

Radiofrequency waves using designated, single frequencies are approved for industrial, scientific and medical uses by national (US Federal Communication Commission, FCC) and international organizations. Currently, limited but increasing commercial use in all these areas to heat-treat and dry a variety of commodities is underway. Radiofrequency power provides well-controlled, volumetric (internal) and rapid heating of a diverse variety of food and non-food commodities. Appropriate food and non-food products to be processed and heated with RF power are generally known as dielectrics (poor electric conductors) and include pests, microbes, foods and non-food agricultural commodities such as soil, packaging and wood (pallets) products.

Dielectric properties are directly related to the material's chemical (molecular) composition and due to the presence and relative abundance of dipoles like water and/or induced dipoles like proteins, lipids, and carbohydrates. Therefore, the material's ability to absorb RF power and convert it to thermal power resides at the molecular level. Because molecules are well distributed and organized within and on the surface of dielectric materials, the effect of absorbing RF power occurs throughout its volume and to a lesser extent on its surface (lower concentrations) where temperatures are slightly lower than its internal volume (< 1ºC). For this reason, RF processing is said to be a volumetric process, comparable to microwave heating, but in contrast with any other conventional surface thermal process known today. By comparison, the volumetric nature of RF processing provides with unique opportunities to reduce the needed thermal load (i.e. temperature over time) required for an intended effect as heat losses by radiation are larger at the surface. This volumetric property applies equally to arthropod and microbial pests as well as to the host commodity and its package.

The RF disinfestation process is rapid (seconds to minutes) and proven effective when reaches lethal thermal levels (50-60ºC). These levels are sufficient to provide thermal loads able to irreversibly disrupt essential and common metabolic pathways and to affect all biological stages of arthropod (and other) pests. Furthermore, as the interaction of RF photons with molecules is frequency dependent, at specific frequencies insect pests exhibit a higher heating rate than the host commodity allowing a somewhat selective heating process to be realized. This selective process minimizes processing time and lowers the overall thermal load applied to the commodity thus decreasing the potential for any adverse effects on its quality attributes.

The fundamental physical concepts and the rationale behind the RF disinfestation process, including the interactive energy-transfer and conversion mechanisms (RF to thermal power) with arthropod pests are explained below.

2. Physics of RF power

2.1 RF photons and the electromagnetic spectrum

Radiofrequency photons belong to the electromagnetic spectrum of radiant energy. The electromagnetic spectrum covers a very large range of wave photons with frequencies ranging from 10^6 to 10^{20} Hz (1 Hz = 1 cycle/sec) and wavelengths from 10^3 to 10^{-12} m. As shown below in Figure 1, this range covers radiowaves (~10^6 to 10^{10} Hz), microwaves (~10^{10} to 10^{12} Hz), infrared, visible and ultraviolet radiation (~10^{12} to 10^{16} Hz) and soft, hard X rays and gamma rays (10^{16} to 10^{20} Hz).

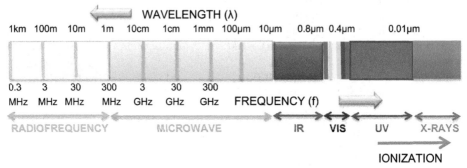

Fig. 1. Electromagnetic spectrum (simplified).

Radiofrequency power is, however, a small segment of the radiowaves region with an arbitrarily defined range of frequencies between ~ 1 MHz (300 m wavelengths) to 300 MHz (1 m wavelengths). In the defined frequency range, the RF photon energy is in the 6.6×10^{-28} to 6.6×10^{-26} J/photon (or 4.1×10^{-9} to 4.1×10^{-7} eV/photon). Therefore, RF processing involves photons of very low energy and long wavelength and therefore absorbing dipole or induced dipole molecules can only experience excitation effects (i.e. vibrational and rotational) but will not lose electrons to cause ionization or the formation of free radicals.[1]

Radiofrequency waves are produced by rapid electrical oscillations and generally are able to penetrate deep into various materials, but are reflected by electric conductors and by the ionized layers in the upper atmosphere. Like all other photons in the electromagnetic spectrum, RF photons consists of electric and magnetic waves oscillating at right angles to the direction of propagation (i.e. transverse waves) and moving through space at the speed of light (c = 2.998×10^8 m/sec). The combination of electric and magnetic fields originates an electromagnetic field.

The relationship between the RF photon energy and its frequency is given by Einstein's classical expression as:

$$E = hf \tag{1}$$

where: E is the photon energy (Joules);

---

[1] Chemical bond energies are in the range of 1 to 10 eV per bond. Therefore, RF photons (1 to 100MHz) carry one billionths to 100 millionths less energy than is required to break a single bond. Free radicals are extremely reactive (short lived) chemical species capable of inducing chemical reactions. Their formation is associated exclusively with sources of ionizing radiation (> 1 eV/photon).

h is the Planck's constant (6.626×10^{-34} Joules sec or 4.136×10^{-15} eV sec); and
f is the photon frequency (Hz or cycles/sec).
This expression indicates that all photons in the electromagnetic spectrum come as discrete quantities named "quanta" and moving at the speed of light. It also indicates that photon energy is always a multiple of Planck's constant times its frequency (cycles/sec).
Because frequency (f in Hz) and wavelength (λ in m) of an electromagnetic wave are related to the speed of light as

$$c = f\lambda \tag{2}$$

formula 1 can also be expressed as

$$E = hc / \lambda \tag{3}$$

indicating that photon energy E is inversely proportional to its wavelength λ.

2.2 Interactions of RF photons with matter

Biological materials - including foods, microbes, arthropods and many agricultural products, are non-magnetic in nature, therefore, only the electric field component of an electromagnetic wave is able to interact and strongly affect the polar and induced polar molecules in the product.
In the presence of an oscillating electric field (changing polarity at a set frequency), the interactive mechanisms of the electric field with RF active molecules (i.e. dielectrics or poor electric conductors) include: (1) reorientation of permanent dipoles (i.e. water); (2) inducing dipoles by polarization of bound charges (proteins, carbohydrates, lipids); and (3) forcing the drift (displacement) of electronic and ionic conduction charges (mineral nutrients) (Klauenberg & Miklavcic, 2000).
The above interactive mechanisms only act at the molecular level and thus the effects of RF processing is based solely on the material's chemical composition in which permanent dipoles (i.e. water) play a major role while other lower concentration non-polar or weakly polar molecules are activated in proportion to the magnitude of the electric field. Initially, and without an electric field, polar and non-polar molecules in any material are randomly oriented due to thermal excitation, which forces their multi-directional movement and spatial distribution.
When an electric field is applied, dipole (polar) molecules tend to re-orient and become aligned according to the direction of the electric field in a phenomenon known as "orientation polarization". Still, orientation is opposed by thermal excitation and therefore, the net orientation effect is proportional to the intensity of the electric field once it overcomes the random distribution of the active molecules in the RF field.
In non-polar molecules, the electric forces separate positive and negative charges a small distance thus inducing temporal dipoles. This type of induced dipole exists only when the electric field is present and occurs via electronic (displacement of electrons) or atomic (displacement of charged atoms) mechanisms known collectively as "distortion polarization".
In both cases with orientation or distortion polarization, the charges in dipoles or in induced dipoles do not cancel and, therefore, new internal electric fields are formed. Distortion polarization is temperature dependent while orientation polarization is inversely

proportional to temperature as RF active molecules must overcome the randomness from thermal excitation.

Furthermore, all polarization effects can only operate up to a limiting frequency after which if frequency increases, orientation polarization effects tend to disappear as the inertial effect of permanent polar molecules prevent reversal of their direction of motion and thus their inertial movement (i.e. momentum) cannot be overcome. The RF process is thus frequency dependent and can be optimized at certain selective frequencies matching the dielectric properties of a material (Lagunas-Solar, Zeng & Essert, 2003).

In arthropod pests, as in all biological systems, water (free and bound) and to a lesser extend proteins, lipids, carbohydrates are the major chemical constituents while mineral nutrients are at trace levels. Water is a natural permanent dipole but its degree of freedom depends on its chemical environ with free (unbound) water being the most active dipole to interact with oscillating electric fields. Bound water, on the other hand, because of its binding (coordination) with other molecules, may still be active but is somewhat restricted to respond to electric field oscillations. Proteins, including enzymes, lipids and carbohydrates are polarizable under a voltage difference and therefore become temporal induced dipoles able to experience electric field interactions and be actively involved in generating thermal energy within the material. Inorganic ions (i.e. mineral nutrients) are always charged and can be displaced by the electric fields and generate small electric currents which converts to heat through resistance (Ohm's law). Overall, although at different levels, all constituents may be actively re-oriented or displaced generating thermal energy by combination of the above different interactive mechanisms.

Although most permanent and induced dipoles are not free to drift, displacements of conduction charges or free ions under the influence of an electric field is a classical phenomenon known as ionic conductivity. Conduction effects (J_c in Amperes/m^2) are related directly to both conductivity (σ in Siemens/m^2) and the net electric field E (Amperes/Siemens) (Lea & Burke, 1998).

2.3 Mechanisms of RF heating

The ability to induce polarization effects in a material by an applied electric field and the creation of new, transient electric fields and currents within the material is characterized by a quantity noted as ε and called "dielectric constant" or "permittivity" (Klauenberg & Miklavcic, 2000). Therefore, the dielectric constant measures how easily a material is polarized to store electric energy.

However, dielectric constants are measured in relation to vacuum or air (ε_o = 1.00000 and 1.00054, respectively) as they represent the ability of a material to store electric energy (i.e. capacitance) at a given voltage as compared to vacuum or air. Therefore, relative dielectric constants for a material are given by

$$\varepsilon' = |\hat{E}_a| / |\hat{E}| \qquad (4)$$

where ε' is the relative dielectric constant and $\hat{E}_a$ and $\hat{E}$ are the applied and the net electric field strengths (vectors), respectively.

In real practice, the ratio by which each mechanism intervenes in storing electric energy is accompanied by effective dissipation losses due to thermal excitation, inertial motions and due to the different binding forces in lattices or accompanying the RF active chemicals. These losses force molecules to lag behind the frequency of the oscillating electric field or restrict drifting and thus resist movements of electric currents. These types of losses are represented by a relative complex dielectric constant ($\mathcal{E}^*$) which is given by the expression (Metaxas & Meredith, 1983):

$$\mathcal{E}^* = \mathcal{E}' - j\,\mathcal{E}'' \tag{5}$$

In this expression, $\mathcal{E}''$ is a measure of the dissipation losses per cycle and is known as the "dielectric loss factor" and j is the imaginary unit. The dielectric loss factor measures the ability of a material to convert electric energy to thermal energy purely based on polarization effects (i.e. no resistance heating) and is always positive and much smaller than the relative complex dielectric constant ($\mathcal{E}^*$) (Mudgett, 1986; Nyfors & Vainikainen, 1989). Both relative complex dielectric constant and dielectric loss factors are related to the absolute dielectric constant in vacuum ($\mathcal{E}_o$ = 8.85 x 10^{-12} F/m). For clarity, the use of the word "relative" is omitted from this point and therefore $\mathcal{E}^*$, $\mathcal{E}'$ and $\mathcal{E}''$ will be known simply as complex dielectric constant, dielectric constant and dielectric loss factor, respectively.

While most products have small dielectric loss factors, it increases rapidly with temperature but only slightly with pressure. However, these factors can vary drastically with operating frequencies but are independent of the applied electric-field magnitude.

Finally, dielectric constants are the factor by which a dielectric material increases the capacitance of a parallel-plate RF system (i.e. RF cavity, see section 3.1 below) in relation to its capacitance in vacuum or air under the same electric field conditions. Examples of $\mathcal{E}^*$ values for selected materials are given in Table 1, below (Clarke, 2006). Worth noting is that $\mathcal{E}^*$ values for codling moth (71.5; 84.5) and Mexican fruit fly (90; 141) are exceptionally high and similar to water and much larger than values for some host materials (i.e. nuts). Thus, RF disinfestation applications with nuts or similar products, selective (higher) heating of insects - as compared with heating of the host commodity, can be realized and is advantageous for effective insect control while lowering overall thermal loads applied to the host commodity. This phenomenon is further explained in section 3.5.2, below.

Material (Moisture %)	Temperature (°C)	Frequency	$\mathcal{E}^*$
Water	100	> 1MHz	80
Ethanol	20	10 MHz	24
Sand (dry)	20	1	2.5
Walnut (0%)	20	10 MHz	2.0
Walnut (17%)	20 (60)	10 (27) MHz	5.0 (4.9)
Almonds (5%)	20 (60)	27 MHz	5.9 (6.0)
Codling moth	20 (60)	27 MHz	71.5 (84.5)
Mexican fruit fly	20 (60)	27 MHz	90 (141)
Douglass fir (11%)	15	1 MHz/10 MHz	3.2
Compressed	15		4.3
Paper Fiber	20	1 MHz	4.5
Polyethylene (non polar)	20	50 Hz/1 GHz	2.3
Polycarbonate (polar)	20	1 MHz	3.0

(*) From: National Physical Laboratory (www.kayelaby.npl.co.uk/general_physics/2_6/2_6_5.html) and Wang et al., (2003)

Table 1. Values of complex dielectric constants for selected materials (*).

2.4 RF power dissipation as thermal power

The ability of molecules within a material to store electric energy from an operating RF system is the first step towards an effective heating process able to induce a desirable biological effect (i.e. disinfestation). As indicated above, this is expressed by the complex dielectric constant which combines dielectric properties defined by molecular composition. Therefore, the conversion of RF power into thermal power is directly related to the polarization and ionic conduction mechanisms described above. However, the fractional contribution of each interactive mechanism is determined by the frequency (Hz) of the oscillating electric field.

At low frequencies, all dipole molecules (permanent and induced) have sufficient time to follow and adjust to the reversal cycles of the oscillating electric fields. In this case, no or negligible energy dissipation losses occur due to orientation polarization effects. Under this condition, the dielectric constant is at its maximum value and the dielectric material is capable of storing a maximum energy from the applied electric field. The RF heating is then mostly due to combined polarization and ionic conduction effects.

As frequency increases, dipoles gradually lose their ability to fully adjust to the oscillations in polarity of the electric field and polarization effects lag behind and contribute less to the total polarization. To minimize this lagging, the electric field transfers its energy to the dipoles forcing them to respond faster. However, this electric-field forced adjustment reaches a limit at which no further corrections occur. At this point, lags in dipolar polarization become larger forcing the dielectric constant to fall in value while dielectric loss factors increases. Under this scenario, RF heating depends less on polarization effects but more on ionic conduction effects (displacements or drifting of charged molecules and ions) leading to resistance heating. This variation in mechanism is therefore highly influenced by commodity temperature.

The total RF power dissipation into a sample is derived from the fundamental laws of electromagnetism. For a steady-state sinusoidal electric field, the time-average of RF thermal power dissipation per unit volume of the sample is given by:

$$P_v = 2\pi\varepsilon_o f \varepsilon''_{eff} E^2_{rms} \qquad (6)$$

where P_v is in watt per cubic meter (W/m³; or [Joules/sec]/m³); f is the frequency of the oscillating electric field (Hz); ε'' is the dielectric loss factor and E_{rms} is the root-mean-square of the applied electric field in Volts per meter (V/m) (Metaxas & Meredith, 1983).

The total amount of heat Q (Joules) needed for a mass m (kg) of a dielectric material to increase its temperature from an initial value (T_i) to a final temperature (T_f) (i.e. $\Delta T = T_f - T_i$) is given by the classical expression

$$Q = mC_p(T_f - T_i) \qquad (7)$$

where C_p is the specific heat of the material (Joules/kg°C).

Power per unit volume in formula 6 can be combined with the energy required as given in formula 7 to provide a combined formula (formula 8) leading to RF throughput as determined with RF processing parameters:

$$mC_p(T_f - T_i)/Vt = 2\pi\varepsilon_o f \varepsilon''_{eff} E^2_{rms} \qquad (8)$$

which can be expressed as

$$\Delta T/t = (2\pi\mathcal{E}_o f\mathcal{E}''_{eff}E^2_{rms})/d \tag{9}$$

where the ratio $\Delta T/t$ is the rate of heating expressed as a function of processing parameters (f, E) and the other factors are associated with the product properties ($\mathcal{E}''$ and d), where d (kg/m³) is its density.

2.5 Temperature distribution and depth of penetration in RF processing

Reaching all pests in a volume of material to be disinfested is an important feature of RF disinfestation as the process must be effective over large volumes of material to assure reliable control with adequate throughputs. Temperature distribution and depth of penetration are thus important aspects that need to be considered for RF disinfestation of large volumes of commodities.

In standard volumes of boxed or palletized materials processed with a parallel-plate capacitor (see section 3, below), the intensity of the electric field is largely unaffected by the load and it contributes to similar energy absorption throughout the material. In addition, depth of penetration is an important added factor.

An electromagnetic wave incident on the surface of a dielectric material can either be reflected (i.e. reflected wave) or be transmitted into the material (i.e. transmitted wave).

In good dielectrics (including its package), a great fraction of the wave energy is transmitted but is gradually attenuated as it is converted to thermal energy. The extent (length) of the wave transmitted into the material is known as "penetration depth" (D_p) and is arbitrarily defined as the distance from the surface to the point (plane) at which its energy is reduced to e^{-1} (1/2.71 or 37%).

Because the effective loss tangent (tan δ_{eff} = $\mathcal{E}''_{eff}$/ $\mathcal{E}'$); the penetration depth can be approximated by

$$D_p = \frac{c}{2\pi f(\varepsilon')^{1/2}\tan\delta_{eff}} = \frac{c(\varepsilon')^{1/2}}{2\pi f\varepsilon''_{eff}} \tag{10}$$

Penetration depth (D_p in meters) is therefore proportional to the dielectric constant ($\mathcal{E}'$) and inversely proportional to the dielectric loss factor ($\mathcal{E}''_{eff}$) and to the frequency of oscillation of the electric field. In general, at frequencies below 100 MHz, penetration depth is of the order of meters unless the dielectric loss factors are exceedingly high (Metaxas & Meredith, 1983). Despite its penetration, however, the energy distribution and thus the thermal profiles of the RF heated material must be taken into account when the process's efficacy requires a pre-defined or a narrow temperature range.

For disinfestation applications, however, the threshold temperature to assure lethal effects in all insects and mites at any biological stage is rather small (50-60°C) and requires a short time (< 1 min). This allows the use of RF disinfestation in large-volume containers (i.e. pallets at 2 x 2 x 2.2 m high) as material handling techniques can also be applied to improve temperature homogeneity to narrower ranges (but assuring reaching a threshold thermal load) as some limitations are expected by penetration depth factors.

However, as explained above, changes in the dielectric behavior of the load due to increased temperature (i.e. increased dielectric loss factors) induce rapid and significant changes in the fraction of the electric energy being absorbed and converted to thermal power. Unattended, these factors could lead to severe localized, uneven heating of the packaged commodity with potential loss of quality. Therefore, process controls need to be focused into maintaining adequate RF power densities to be applied and by controlling product temperatures during the process. Besides, product geometry, package material and its geometry, and air gaps within the material clearly contribute further to different power densities being generated in their volumes and thus temperature variations are to be expected. As lethal thermal loads in insects and mites are low (50-60°C; ~ 1 min), process effectiveness is assured by reaching the relatively low lethal thermal loads needed. This occurs at levels below than those affecting quality in the host commodity and is due to the higher metabolic complexity of arthropod pests as compared with the much simpler metabolism of the host food (i.e. insects in grains) or agricultural commodity (i.e. insects in pallets).

Heat transfer and temperature distribution across a material in RF disinfestation is critical to assure effectiveness and both phenomena have been studied intensively (Giles, Moore & Bounds, 1970). In the absence of any significant mass transfer (e.g. evaporation), the temperature distribution in the medium obeys the heat diffusion law and is given by:

$$\frac{\partial T}{\partial t} = \alpha_T \nabla^2 T + \frac{\delta P}{dc_p} \tag{11}$$

where α_T is the thermal diffusivity of the medium (m²/s), T is temperature, δP is the localized power density (W/m³), ∇ is the Laplacian operator, d is density and c_p the specific heat (Metaxas & Meredith, 1983). The thermal diffusivity measures the ability of a material to conduct thermal energy relative to its ability to store thermal energy. Materials of large a_T will respond quickly to changes in their thermal environment, while materials of small a_T will respond more sluggishly and take more time to reach a new temperature equilibrium condition.

In RF processing, materials are usually heterogeneous, and therefore a_T plays an important role because different parts absorb RF power at different rates. For homogenous materials, a_T is less important in temperature distribution, and δP can be approximated as P_v (formula 6) for the temperature analysis. However, thermal diffusivity (or thermal mass effect) of insects and mites is not known despite the many reported studies on thermoregulation of common habitats with its surroundings. However, the rapid heating of insects with RF power (Nelson & Charity, 1972) suggests an appreciable value of thermal diffusivity for insects.

Finally, due to its direct heating effects at molecular levels, RF heating is independent of temperature differences and heat transfer coefficients, although both these factors will influence the subsequent dynamic distribution of thermal energy within the volume of the material (Hill et al., 1969).

3. Principles of RF processing

3.1 The RF cavity – A parallel-plate capacitor

In order to best realize and apply the above mechanisms in a controlled and safe RF disinfestation process, a parallel-plate capacitor is used with materials to be treated placed in between and named the "load". The process can be performed either statically (batch mode) or continuously (conveyorized mode). This type of capacitor is known as a "RF cavity" and is shown schematically in Figure 2, below.

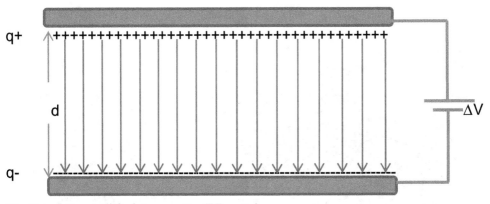

Fig. 2. A static parallel-plate capacitor (RF cavity).

The RF cavity operates with equally charged plates (top positive and bottom negative) formed when a voltage difference is applied. Electric field lines (red) are directed towards the negative (ground) plate and are equally spaced and parallel to each other. Transverse waves (not shown) are perpendicular to the electric field. When activated, however, by placing a material (load) in between, the electric field geometry is changed as field lines are distorted (i.e. fringe effects especially at low frequencies) due to the load and its package, and its intensity is decreased because of new charges created in the load. The presence of air gaps in between and on top of the packaged dielectric load also contributes to field distortion and localization effects. Therefore, an active RF cavity needs to be properly designed and managed in order to minimize the above effects and maintain field homogeneity and thus treat with adequate uniformity.

A schematic of the major features of a RF power system is shown in Figure 3, below, while a version of an operating commercial-scale prototype is shown in Figure 4.

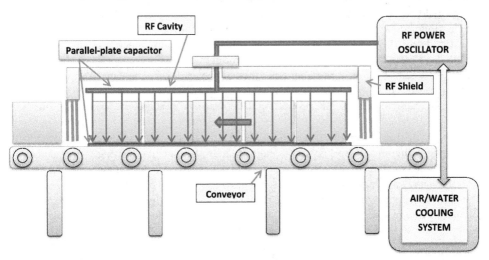

Fig. 3. Schematics of a conveyorized RF processing system.

Fig. 4. RF system for batch processing (13.15 MHz, 10 kW). Designed by UC Davis & RF Biocidics Inc.

In this latter system, the RF cavity is shielded in all directions with a metallic enclosure (shown in light blue) so as to prevent propagation or reflections of RF waves outside its boundaries and thus eliminate the potential to expose workers, the surrounding environs or interfere with other radiowaves. This basic configuration, singly or in modules, is able to operate and meet the conditions to generate and delivery RF energy safely and efficiently to food and agricultural commodities at commercial-scale levels.

The parallel-plate configuration shown in Figure 2 (above) is said to be in a static condition in which no material (other than air or vacuum) is placed in between and therefore the electric field lines are equally spaced and parallel to each other while the overall electric field is uniform except at the edges. However, when a product (load) (i.e. a dielectric) is introduced and the electric field is rapidly oscillated (changing electric polarities at every cycle) with a certain frequency, the dielectric product (load) is now capable of absorbing RF energy by a combination of the above mentioned molecular mechanisms and convert it to thermal power.

The main characteristic of RF processing (RF heating) is therefore, based on the high frequency alternating oscillating electric fields interacting with the dielectric medium (dipoles and induced dipoles) in between the plates and generating thermal energy (heat). RF heating is therefore, also known as "high frequency capacitive heating" (Piyasena et al., 2003), although as the medium in between the plates is also a dielectric material, the process is often referred as "high frequency dielectric heating" (Zhao et al., 2000).

The generation of thermal energy is due to the ability of the applied oscillating electric field to polarize and re-orient internal electric fields of charges formed in the load (material). The rotating electric field exerts torques on permanent and induced dipoles to force them into flip-flop motions. During the rapid cycling, friction and heat is generated between polarized molecules (permanent or induced dipoles) and their neighbors including lattice losses as they move. The higher the frequency of oscillations the greater is the energy available or created to be converted to heat. However, due to lattice limitations, when the frequency is at the maximum equilibrium between rotation and inertial restrictions, it is said to be at a "Debye resonance" at which there is maximum conversion to heat. If operating frequency is beyond the ability of the molecules to react due to inertial motion, the process loose overall energy-conversion efficiency. This suggests that specific materials, due to their own unique chemical composition will present an optimal frequency at which to operate with maximum energy-use efficiency. In materials with complex or different composition (i.e. pest and host) is therefore possible to establish selective RF heating effects and establish a process with minimal energy input to the lesser dielectric component (Lagunas-Solar et al. 2006).

In addition to polarization mechanisms, a dielectric material can also be heated by the resistance to direct ionic conduction or drift mechanism as given by Ohm's law and that states that the current (I in Amperes) through a conductor between two points is directly proportional to the voltage difference (V in volts) across the two points and inversely proportional to the resistance (R in Ohms). The heating level through these mechanisms depends on the electric conductivity of the material which is generally low as dielectric (i.e. poor conductor) properties prevail.

Finally, because these mechanisms occur with equal intensity between the RF cavities (i.e. same electric field intensity) and are only dependent on the material's chemical composition, RF heating is in principle homogeneous and a volumetric (internal) method in contrast with all other surface heating methods known today. However, at a microscopic scale within biological materials, some differences do occur due to variations in chemical composition and moisture levels. These differences allow for the enhancement of the rate of heating with distinct materials and are the basis for selective RF heating effects (Zimmerman, Pilwat & Riemann, 1974).

3.2 Advantages of RF disinfestation

For disinfestation purposes, RF power provides a unique mean to heat an arthropod pest (small mass or volume) inside a host commodity (large mass or volume) volumetrically (internally) and with penetrating RF waves. This behavior is opposite to the use of conventional surface-heat methods such as infrared, dry and wet steam, or hot water where the host's surface becomes a physical barrier to the applied thermal energy. In all latter cases, the distribution of the applied heat to reach the entire volume depends on heat-transport mechanisms and time. In addition, heat is only applied at its surface. Furthermore, under these conditions, many commodities experience undesirable changes that lower product value. In contrast, because of its penetration, RF waves are effective in reaching deeply internalized pests such as eggs and larva deposited in internal cavities by borer insects, a situation in which the effectiveness of fumigants is restricted by the presence of air-locks impeding penetration of fumigants.

Radiofrequency processing is volumetric heating and its energy transfers directly to the product without the need of intermediate transfer mechanism such as conduction, radiation, or convection. This allows RF energy to be transfer to the load much faster and more effectively. The amount of input energy can be controlled by reducing the input power or switching the system on and off in order to achieve precisely the final temperature. These characteristics allow the RF process to be operated within low and high thermal boundaries, called "thermal windows". Thermal windows for RF disinfestation as compared with other biological effects (i.e. pasteurization and enzyme deactivation) are given in Figure 5, below.

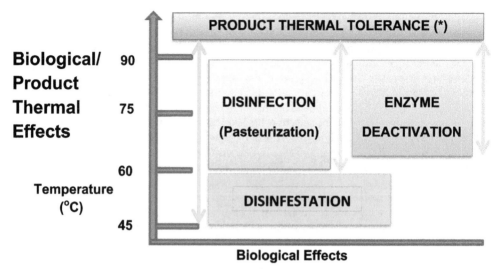

Biological Effects

Fig. 5. Thermal Windows (colored arrows) for RF processing effects.

A thermal window represents the differential thermal sensitivity between living organisms (highly-heat sensitive) and the more heat-tolerant properties of agricultural products. Therefore, operating within a product's thermal window minimizes the impact on the host commodity. This is a critical advantage of RF processing over any conventional (surface) heating method as disinfestation effects can be well controlled because of the high sensitivity of arthropod pests to thermal energy and the higher heat tolerance of most affected foods and agricultural commodities.

By comparison with conventional heating processes, overheating the surface is very common because energy is first applied to the surface and then is conducted to its interior. Because energy loss from the surface (by radiation and/or convection) is unavoidable, significant and fast, these processes often require additional heat input on the surface in order to produce internal temperatures high enough to achieve a uniform biological effect. The host, however, received higher and usually damaging thermal loads on its surface.

As a result, the upper boundary of the thermal window (especially for the surface) is frequently exceeded causing unacceptable changes in the physiological, sensory, and quality of foods. In RF disinfestation, the surface temperature is usually lower than the internal temperature due to the heat loss from surface radiation and due to evaporation. This can be effectively prevented during RF processing by reducing evaporation (e.g. high humidity inside the chamber), by adding moisture before processing and by providing good radiation reflectors in the RF cavity design.

3.3 RF and microwave processing

Frequently, microwave heating is confused with RF heating. While fundamentally similar, microwave heating (also an energy source in the electromagnetic spectrum) is operated at 915 MHz (λ = 0.3 m) and 2,450 MHz (λ = 0.1 m), that is with higher frequency and shorter waves than RF. For most commercial scales (i.e. large amounts) of foods and agricultural products, microwave heating is not adequate also has many disadvantages in aspects of heating homogeneity, energy penetration, and energy-use efficiency. First, it does not produce homogeneous heating because of the limited penetration of the shorter wavelength and the complex non-uniform standing wave patterns. The penetration depth of microwave is in the order of 5 cm to 10 cm for bodies with high water content, and may be higher (in several tens of centimeters) for other drier materials (Orfeuil, 1987). In addition, the electric field inside the microwave oven is not uniform due to the nature of standing waves. In fact, the enclosed electric field and power density vary with the location and the sample's shape and size. Non-uniform electric field patterns and variable power densities often lead to local (or uneven) heating in the material. Besides, the power density in microwave heating are much higher than in RF heating (due to much higher operational frequencies) and is associated with non-uniform electric fields. Therefore microwave heating normally causes local hot spots to the commodity.

In contrast, the RF process is operated at frequencies much lower than conventional microwaves hence the penetration of RF energy is greater, usually higher than 1 m and even several tens of meters at low frequencies (Orfeuil, 1987). Furthermore, the electric fields generated between two parallel plates are very uniform; therefore, RF transversal waves interact and heat the material more homogeneously (Wig *et al.*, 1999; Mitcham *et al.*, 2004).

3.4 Comparison with conventional disinfestation technologies
Today, conventional or emerging alternatives face several restrictions or their use is associated with many safety concerns many of which prompted the development of RF disinfestation as well. The contributing factors from the industry perspective are summarized below.

3.4.1 Chemical pesticides issues and concerns
Methyl bromide fumigation was for many decades the preferred treatment applied to many stored food commodities. It was used worldwide to meet quarantine and phytosanitary restrictions and quality requirements as mandated by global agriculture markets. Current alternative methods used to control insects in grains include the use of insecticides (e.g. Malathion), fumigants (e.g. phosphine, carbon dioxide) and temperature treatment (Bond, 2007).

Malathion (American Cyanamid Co., USA) is one of the safest organophosphate insecticides. Nevertheless, existing regulations demands that the treated grains should not be sold for at least 7 days nor should be eaten within 60 days after treatment to avoid potential toxic effects from residues left.

Phosphine gas is very toxic to human therefore its application requires strict controls, even though there is no residue left to the treated grains.

Other pesticides in use include Chloropicrin, 1,3-dichloropropene, Telone/Vapam, sulfaryl fluoride and hydrogen cyanide.

However, all pesticides available and those mentioned in particular are of global concern due to the potential for causing detrimental effects on animals, air, water and soil as well as potentially impacting public health and workers safety.

Conventional carbon dioxide fumigation of grains usually referred as modified atmospheres requires a lengthy treatment (i.e. days to weeks) therefore its cost is high as well as its impact on the logistics of product distribution to markets.

3.4.2 Conventional heat processing

Conventional high-temperature treatments of grains, such as hot air or hot water immersion and dry or wet steam are usually less effective to internally hidden eggs or pupae inside grain kernels. As adequate lethal temperature for insect pests need to be applied throughout the volume of the commodity, surface overheating and diminishing quality attributes usually occurs due to slow dynamics of heat transport from the outside to the core of grain kernels. Overheating also leads to the deterioration of grain quality and viability.

Because of the above, there is a clear need to develop and establish better, less or non-invasive alternatives to disinfect grains and other commodities to overcome safety concerns associated to invasive methods (leaving residues). Highly desirable is the long-stated need to reduce risks to consumers, workers and the environment as indicated by international organizations (UNEP 1998; WMO 2003).

3.5 Mechanisms of RF Disinfestation

As RF disinfestation is to initiate energy-transfer mechanisms at the molecular level, there are two possible mechanisms for the inactivation/control of insect and mites using RF power: thermal and non-thermal effects. The thermal effect of RF power is essential to the destruction of microorganisms and many studies have proven its validity (Goldblith & Wang, 1967; Fujikawa et al., 1992; Kozempel, Annous, Cook et al., 1998). The energy absorption from RF power can raise the temperature of contaminant organisms high and fast enough to induce irreversible (i.e. non repairable) biochemical damage to cells such as the denaturation of enzymes, proteins, DNA, RNA, or of other vital cellular components, as well as disruption of cell membranes (Heddleson & Doores, 1994). Reports of potential non-thermal effects (effects unrelated to heat stress) with higher-frequency dielectric heating (basically at microwave frequencies) are still controversial. While some researchers have announced these effects (Burton, 1949; Olsen, 1965; Fung & Cunningham, 1980; Cross & Fung, 1982), other researches have concluded there is little or no non-thermal effect on cells (Goldblith & Wang, 1967; Carroll & Lopez, 1969; Rosenberg & Bögl, 1987; Knutson *et al.*, 1987). However, using high-peak power RF technologies capable of delivering ultra-short pulses with very high instant power (>MW/pulse) remain as a potentially successful approach for disinfestation and in particular for fresh produce and other high-thermally sensitive commodities (Lagunas-Solar, Zeng & Essert, 2003).

3.5.1 RF disinfestation thermal effects

The cell is the fundamental unit of all living matter. Living cells (prokaryotes and eukaryotes) are basically composed of high-molecular-weight polymeric compounds (macromolecules) such as proteins, DNA, RNA, polysaccharides, lipids, and storage materials such as fats, glycogen, polyhydroxybutyrate, etc. (Madigan, Martinko & Parker, 2000). These macromolecules are only functional in the proper three-dimensional structures. The structural property is affected by thermal energy and is especially important for enzymes as they are very effective biological catalysts and involved in most of cellular reactions (Shuler and Kargi, 1992).

Because RF power generates heat at the molecular level, RF energy can effectively increase the kinetic energy of molecules and make these molecules vibrate more rapidly and violently. These molecular vibrations, up to a point, are strong enough to disrupt weak intermolecular forces, such as hydrogen bonds, salt bridges, disulfide bonds, and non-polar hydrophobic interactions in secondary, tertiary and quaternary structures of macromolecules and denature their normal biological order and function. The most essential thermal damage that leads to cell death is the denaturation of enzymes, especially some critical enzymes responsible for DNA and RNA replications in the cell (Roti Roti, 1982). Thermal energy or heat can cause non-repairable denaturation of DNA, RNA, and sometimes create structural DNA lesions (sections of DNA contain elementary damage sites) that cause the loss of cellular genetic information (Ward, 1985).

Heat also transfers its energy to make molecules more energetic which leads to weaker hydrogen bonds and hydrophobic interactions sustaining the cell membrane, and eventually causes its disruption or collapse. The disruption of cell membrane leads to uncontrollable material exchange between the cell and its environment, which causes the cell to lose its optimum microenvironment required for its metabolisms and the cell dies eventually (Bowler & Fuller, 1987). Heat can also destroy storage materials in cells such as lipids, fats, and carbohydrates by oxidation.

Thermal energy from RF power can increase insect body temperature high enough to be lethal and destroy them (disinfestation) by causing cellular damages (i.e. cell death or dysfunctional) or body dehydration. The thermal death due to cellular damages of this multi-cellular organism is not usually the consequence of massive cell death per unit time, but it may due to the loss or disruption of cells in a certain critical tissues (Denlinger & Yocum, 1998).

Differences in species and developmental stages are also likely to influence the site of lethal thermal wounding. The more complex the biological system, the more susceptible it is to high thermal stress. Therefore, it is expected that macromolecules (e.g. proteins, DNA, RNA, lipid, fat, etc.) are more resistant to thermal stress than cellular organelles (e.g. mitochondria, nucleus, Golgi complex, etc.), cellular organelles are more resistant than cells, cells are more resistant than tissues, and tissues are more resistant than the whole organism (Ushakov, 1964; Prosser, 1986). Hence for a multi-cellular organism, lethal wounding may be inflicted from cellular damages of an organization with a high level of complexity.

The above concepts explains the prevalence of the concept "living dead" in the insect control, which means organisms are still alive but will not survive and reproduce due to cellular thermal injuries (Bowler, 1963; Chen, Lee & Denlinger, 1990). Therefore, as insect's biology is more complex than unicellular organisms (e.g. bacteria, fungi), they are expectedly more susceptible to thermal stress.

High temperature can also be lethal to insects by causing dehydration and promoting desiccation. Above a certain temperature, the critical transition temperature, the rates that insects lose water from their bodies increase dramatically (Yoder & Denlinger, 1991). Critical transition temperature values commonly range from 30 to 60°C for different species and developmental stages (Hadley, 1994). Most insects contain about 60 to 70% water in their body weights, and many can tolerate a loss of 20 to 30% of water for brief periods. The loss of water will increase the osmotic stress and concurrently increase the solute concentration within the body, presumably leading to irreversible cell damages. This also increases RF-induced ionic conduction effects in insects thus enhancing thermal energy production and thermal stress favoring lethality.

3.5.2 RF selective heating of insects

The main mechanism of disinfestation in RF selective heating is also thermal stress (i.e. heat). In the selective RF heating, a proper operating frequency is selected so that the effective dielectric loss factor (ε'') of the target material is close to its maximum value and the load (material) can be heated fast. Because different materials have different dielectric properties (i.e. dielectric constant [$\mathcal{E}'$] and effective dielectric loss factor [$\mathcal{E}''$]) - both of which depend on the composition and frequency, they interact and convert RF energy into heat at different rates at the same frequency.

This leads to the potential that different materials in the same load can have different heating rates, depending on the values of their effective loss factor ($\mathcal{E}''$) at that frequency. If an appropriate frequency is chosen so that contaminant organisms (e.g. arthropods, arachnids) can absorb RF energy faster than host material, those organisms can be heated much faster than other components in the same load (Lagunas-Solar et al., 2006; Lagunas-Solar et al., 2008).   As a result, insects/mites are destroyed by heat while the host commodity is unaffected. This treatment is proposed for somewhat thermally resistance fresh products (i.e. tomatoes, avocados, apples, grapes, and broccoli) which can tolerate some low thermal loads but sufficiently high to be effective for disinfestation applications using a controlled- thermal RF treatment.

While theoretically applicable to selective RF heating of microorganisms, their small size prevents adequate absorption of the penetrating RF energy waves and thus there is no evidence today for the availability of this selective mechanism for microorganisms.

Finally, the above and other technological and consumer factors prompted the investigation on the use of RF power for disinfestation of various commodities by several authors (Ikediala et al., 2000; Wang et al., 2003; Mitcham et al., 2004; Wang et al., 2007a; Wang et al., 2007b). Results and conclusions of all these studies corroborated the advantages of RF disinfestation over available techniques and also helped identify remaining challenges (Prakash & Rao, 2002).

4. Case study: RF disinfestation of rough (paddy) rice

During long-term storage, insects can cause considerable damage to grains (and to other products, i.e. nuts), with weight and nutritional losses reducing yields and quality which reduces market values. Furthermore, deterioration of grains intended for seedling purposes may cause further losses in quality and viability (germination) thus affecting future yields in crop production.

Under current storage (bulk) conditions over long periods of time, the presence of even a few viable colonies of insect pests may result in the emergence of much larger populations as the storage conditions are favorable to insect reproduction and propagation due to the abundant presence of nutrients and lack of antagonistic organisms. In rough (paddy) rice, two major insects Angoumois grain moths (Sitotroga cerealella [Oliver]) and lesser grain borers (Rhyzopertha dominica [F.]) represent major threats as primary grain insects whose larvae feed entirely inside the kernel of the grain and eat from inside becoming more tolerant to fumigation as diffusion of gas into kernels is severely restricted or blocked by the presence of air locks (pockets). Therefore, infestation with primary insects are critically more damaging to stored grains than secondary insects that eat grains from outside and are more easily controlled with conventional fumigation or heat treatments.

As explained above (see section 3.5.2), selective heating of arthropod pests is feasible via a differential heating mechanism based upon the higher ionic conductivity in pests (see Table 1). Therefore, all biological stages of arthropod pests do heat faster than the host commodity leading to their effective biological inactivation (Wang et al., 2003). As shown in table 1 (above), insects such as codling moths and Mexican fruit flies have large dielectric constants ($\mathcal{E}^*$ 71.5-84.5 and 90 to 141; respectively at 27 MHz). Therefore, when treated with RF power these pests can absorb a larger proportion of the available RF energy delivered. By comparison, the host commodity is expected to have complex dielectric constants in the range of 3 to 6 for low-moisture foods (nuts, seeds, grains) or up to 60 to 70 for high-moisture foods (i.e. fruits) although considerable higher dielectric loss factors (>200) for insects have been reported under the same processing conditions (frequency) (Ikediala et al., 2000). The difference in dielectric properties between insects and host generates lower thermal effects on the commodity (Kunze, 1979).

As arthropods (arachnids as well) have similar chemical composition, selective heating effects have been demonstrated with ants, aphids, beetles, borers, bugs, fruit flies, moths, thrips, mites and arachnids confirming the validity of the selective heating process in different food hosts as well as soil and wood products (unpublished results).

Therefore, disinfestation appears as an effective RF application that can heat arthropod pests rapidly (45 to 65°C; 3-4 min) inducing lethal conditions that are well tolerated by a large variety of foods. As proven in various laboratory-scale experimentations, this approach is being developed for commercial-scale applications with RF systems designed and engineered for full optimization.

4.1 Experimental results of RF disinfestation of rough (paddy) rice

A full control of all life cycles of Angoumois grain moths (*Sitotroga cerealella* [Oliver]) and lesser grain borers *Rhyzopertha dominica* [F.]), in laboratory-scale experimentation with rough (paddy) rice as host was reported (Lagunas-Solar et al., 2008).

Samples of rough rice (13.5% and 11.0% moisture) were obtained from Pacific International Rice Mills Inc., (Woodland, CA) and were infested in separate batches (~ 10 kg each) with adult grain moth *Sitotroga cerealella* (13.5% batch) and with both *Sitotroga cerealella* and adult lesser-grain borer *Rhyzopertha dominica* (11.0% batch).

After approximately one month at 28-30°C (35-40% relative humidity) both colonies were well established showing abundant populations of all biological stages. RF disinfestation was conducted at the University of California, Davis using several processing conditions with 500 W of RF power at 20.3 MHz (Lagunas-Solar et al., 2008). Samples were treated at the same temperature (60°C) but with different times (5 and 30 min; 1 and 2 h) so as to vary thermal loads (temperature x time) delivered. Effectiveness of the RF disinfestation process was determined by assaying the emergence of adult insects found over ~40 days of periodic observations. However, as no adult insects survived any of the initial treatments, adult emergence was assumed to be due to the presence of surviving eggs, larva and/or pupas. Results from replicates in triplicate (control and treated) are shown in Figures 6 and 7, below.

The response of grain moth *Sitotroga cerealella* and lesser-grain borer *Rhyzopertha dominica* to the same RF processing conditions were different indicating that other parameters need to be considered in establishing an optimized process.

As expected, Sitotroga cerealella was found to be more sensitive to the RF disinfestation process as these insects are normally on the outside surface of the grain. While disinfestation effects were observed at all conditions (Figure 6), some adult emergence (~ 16%) was observed in the 60°C/5 min samples after ~40-day incubation and observation period. This was attributed to the partial survival of eggs at different eclosion stages prompting a delayed emergence of adult insects. In all other treatments (60°C/30min; 60°C/1h; 60°C/2h) the thermal loads were sufficiently high to cause a full control of all stages of *Sitotroga cerealella*, as no adult emerged in the treated samples.

Rhyzopertha dominica showed higher tolerance under the same processing conditions.

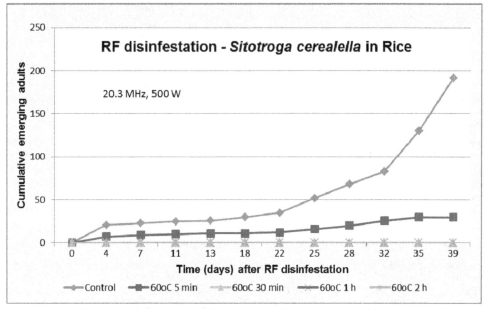

Fig. 6. RF disinfestation of *Sitotroga cereallela* grain moths in rough (paddy) rice.

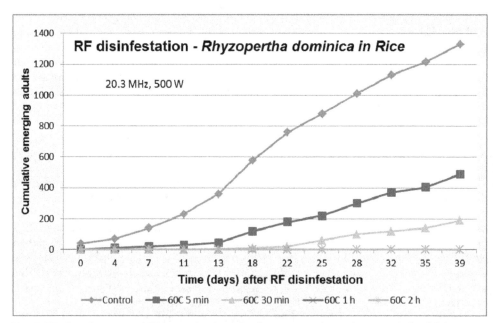

Fig. 7. RF disinfestation of *Rhyzopertha dominica* (lesser grain borers) in rough (paddy) rice.

As compared with controls (1330 adults/40 days); in the 60°C/5min batch 490 adults/40 days were observed for ~37% emergence (~63% control). As thermal load was increased, the 60°C/30 min batch showed only 190 adults/40 days (~14% emergence; 86% control). With either 60°C/1h and 60°C/2h processes, a 100% control were observed as no adults emerged during the 40-day observation period. It was concluded that the ability of *Rhyzopertha dominica* to bore into grains and deposit eggs from which larva emerged, provided additional barriers and protection due to the internalized condition of the pest.

As stated earlier, RF disinfestation is applicable under similar conditions to all arthropod pests as the interactive mechanisms utilized operate with similar molecules present in all arthropods and thus is independent from the biological speciation, developmental stages, or behavioral patterns. Optimization of the RF disinfestation process is also straightforward as thermal loads required for full control (i.e. 60°C/1h for both insects) can be achieved rapidly by increasing RF power. As only 500 W were used in previous experimentation, an operation at 10 kW would only require a 5-min processing time. Other commercial-scale conditions with increased RF power (i.e. 25-50 kW) are also possible and available (www.rfbiocidics.com) for processing larger throughputs (> 2-4 tons/h) while taking full advantage of this emerging chemical-free alternative.

4.2 Quality attributes of host commodity

The application of RF power in disinfestation applications should also consider the potential effects of the applied thermal load to the host commodity. Therefore, the potential for changes of quality attributes in RF treated rough (paddy) rice was also studied and the results are summarized in Table 2, below. These measurements were conducted using standard commercial laboratory tests and indicated no adverse effects.

Rough (paddy) rice samples				
Quality attributes (%)	Controls	Batch 1 (50°C)	Batch 2 (60°C)	Batch 3 (70°C)
Moisture	13.5 ± 0.1	13.5 ± 0.1	13.5 ± 0.1	13.5 ± 0.1
Whole kernel	79.3 ± 1.1	81.1 ± 7.9	78.3 ± 0.1	77.9 ± 0.8
Total rice	68.1 ± 0.3	68.3 ± 0.1	68.2 ± 0.1	68.0 ± 0.1
Dockage	16.9 ± 4.8	11.7 ± 1.0	12.4 ± 1.6	13.2 ± 1.7
Brown rice	81.1 ± 0.4	81.1 ± 0.4	81.3 ± 0.2	81.3 ± 0.1
Whiteness	44.2 ± 0.2	44.1 ± 0.2	44.2 ± 0.2	44.3 ± 0.3
* Mean values and standard deviations for triplicate measurements with 1-kg samples. Data courtesy of California Rice Association.				

Table 2. Quality attributes of RF disinfested rough (paddy) rice*

5. RF process economics

Commercial application of RF disinfestation is already taking place on various commodities and is combined with simultaneous disinfection (pasteurization) and enzyme deactivation effects.[2] This combination of desirable and simultaneous sanitation effects is unique to RF processing and is only dependent on the applied thermal loads (see Figure 5, above). In

---

[2] RF Biocidics Inc., Vacaville, CA 95688 USA (www.rfbiocidics.com)

addition, and because of its penetration, RF power is capable of processing commodities in its package (boxes, bags) thus avoiding recontamination and facilitating logistics of operation. Therefore, due to its chemical-free nature and to the combined effects, RF processing offers many advantages over single-effect technologies including those based upon applications of conventional surface-heat sources (i.e. dry or wet heat, vapor, steam, etc.). Despite its unique advantages and multiple controlling effects, the current commercial application of RF disinfestation is priced competitively in comparison with the cost of using chemical pesticides or any other physical process.

As a new and emerging option for sanitation of foods and agricultural commodities, RF operating facilities are being established to operate at or near high agriculture production areas or near key distribution centers and facilities in which RF processing can be part of the overall chain of production and distribution for local, regional and overseas markets.

6. Conclusions

The application of RF power to disinfestation provides a rapid and effective chemical-free alternative capable of replacing the use of chemical and biological pesticides and as alternative to other conventional heating processes during post-harvest management of various foods and agricultural commodities. As a physical, electricity-based process, its operation is based on well-known, designed and engineered systems capable of safe and large-scale applications. Disinfestation efficacy requires reaching a relatively low thermal-load level as RF is a volumetric heating process with interactions and heating effects starting at the molecular level and somewhat selectively. It can be readily applied to arthropods and arachnids with equal effectiveness using thermal loads well below the threshold for impacting host's quality.

The RF process - with similar and even higher thermal loads, has been demonstrated at a commercial scale for various different commodities including nut products (no effects on free fatty acids, peroxide values), other grains (Quinoa, edible seeds (Chia, pumpkins, sunflower), spices (paprika, cumin, cardamom, nutmeg, coriander, etc.) and flours (brown rice, oat, wheat, flaxseed). Therefore, RF disinfestation is an emerging process with broad applications to many potentially infested commodities and can even be extended to disinfest some heat-tolerant fruits and vegetables as the required thermal load is low and the RF disinfestation process is rapid.

Furthermore, additional energy-use savings can be realized as less RF energy would be needed to control insects (a very small load) as compared with the larger mass (load) represented by the commodity. It is postulated that this approach would result in significant operational cost reductions for RF-based disinfestation applications of a variety of foods such as grains, nut products, flours, beans, spices, and agricultural commodities such as wood products (pallets), soil and soil amendments, and tobacco.

As the needs for non-chemical (residue-free), non-thermal technologies for disinfestation (and disinfection as well) continues to be a goal in production agriculture, a new non-thermal, residue-free process named metabolic stress has recently emerged and is soon to initiate commercialization (Lagunas-Solar, Essert, Piña et al., 2006b; Lagunas-Solar & Essert, 2011). Metabolic stress, singly or in combination with RF processing, is expected to overcome some of the limitations of RF disinfestation and be able to treat commercial levels of thermally-sensitive commodities in particular fresh fruits and vegetables.[3]

---

[3] RF Biocidics Inc., (www.rfbiocidics.com)

Finally, because of its nature, RF disinfestation can be applied to conventionally- and organically-produced food commodities.

7. Acknowledgements

The author wishes to acknowledge the scientific and technical assistance from Mr. Timothy Essert, Mrs. Cecilia Piña U., Drs. Nolan Zeng and Tin Truong from the University of California, Davis. The author also wishes to acknowledge the research support, encouragement and vision of Dr. Robert G. Flocchini, former director of Crocker Nuclear Laboratory, University of California, Davis. Sponsorship from Allied Minds Inc. Boston, MA, Ishida Ltd. Kyoto, Japan, Hortifrut S.A., Santiago, Chile, during the various stages of this research is also greatly appreciated.

8. References

Bond, E.J. 2007. Manual for fumigation for insect control. FAO Plant Production and Protection, Paper 54. FAO. www.fao.org/docrep/x5042E00.htm

Bowler, K. 1963. A Study of the Factors Involved in Acclimatization to Temperature and Death at High Temperatures in *Astacus pallipes*; II. Experiments at the Tissue Level. Journal of Cellular and Comparative Physiology, 62: 133-146.

Bowler, K. and Fuller, B.J. 1987. Temperature and Animal Cells. Cambridge, England.

Burton, H. 1949. A Survey of Literature on Bacterial Effects of Short Electromagnetic Waves. Shinfield, England. National Institute for Research in Dairying Shinfield. N.I.R.D. Paper No, 1041.

Calderon, M. 1990. Food Preservation by Modified Atmospheres. CRC Press (www.crcpress.com/product/isbn/97808493365690)

Carroll, D.E. and Lopez, A. 1969. Lethality of Radio Frequency Energy Upon Microorganisms in Liquid, Buffered, and Alcoholic Food System. Journal of Food Science, 34(4): 320-324.

Chen, C.P., Lee, R.E., and Denlinger, D.L. 1990. A Comparison of the Response of Tropical and Temperate Flies (Diptera: Sarcohagidae) to Cold and Heat Stress. Journal of Comparative Physiology B: Biochemical, Systematic, and Environmental Physiology, 160 (5): 543-547.

Clarke, R.N. 2006. Dielectric Properties of Materials. Kaye & Laby: Tables of Physical & Chemical Constants. National Physical Laboratory, © Crown Copy Right 2006. http://www.kayelaby.npl.co.uk/general_physics/2_6/2_6_5.html

Cross, G.A. and Fung, D.Y.C. 1982. The Effect of Microwaves on Nutrient Value of Foods. Critical Reviews in Food Science and Nutrition, 16: 355-381.

Denlinger, D.L. and Yocum, G.D. 1998. Physiology of Heat Sensitivity (Editors: G.J. Hallman and D.L. Denlinger). Westview Press, Colorado, USA.

Fröhlich, H. 1958. Theory of Dielectrics, 2nd Edition. Oxford University Press. London.

Fujikawa, H., Ushioda, H., and Kudo, Y. 1992. Kinetics of *Escherichia coli* Destruction by Microwave Irradiation. Applied and Environmental Microbiology, 58(3): 920-924.

Fung, D.Y.C and Cunningham, E.E. 1980. Effect of Microwaves on Microorganisms in Foods. Journal of Food Protection, 43(8): 641-650.

Giles, P.G., Moore, E.E., and Bounds, L. 1970. Investigation of Heat Penetration of Food Sample at Various Frequencies. Journal of Microwave Power, 5(1): 40-48.

Goldblith, S.A. and Wang, D.I.C. 1967. Effect of Microwaves on *Escherichia coli* and *Bacillus subtilis*. Applied Microbiology, 15(6): 1371-1375.

Hadley, N.R. 1994. Water Relations of Terrestial Arthropods. Academic Press, San Diego, California, USA.

Heddleson, R.A. and Doores, S. 1994. Factors Affecting Microwave Heating of Foods and Microwave Induced Destruction of Foodborne Pathogens – A Review. Journal of Food Protection, 57 (11): 1025-1037.

Hill, N., Vaughan, W.E., Price, A.H., and Davies, M. 1969. Dielectric Properties and Molecular Behavior. Von Nostrand, New York, USA.

Jeschke, P. 2004. The unique role of fluorine in the design of active ingredients for modern crop production. ChemBioChem 5: 570-589.

Kasevich, R. S. 1998. Understand the Potential of Radiofrequency Energy. Chemical Engineering Progress, 94(1): 75-81.

Klauenberg, B.J. and Miklavcic, D. 2000. Radio Frequency Radiation Dosimetry and Its Relationship to the Biological Effects of Electromagnetic Fields. Kluwer Academic Publishers, Dordrecht, The Netherlands.

Knutson, K.M., Marth, E.H., and Wagner, M.K. 1987. Microwave Heating of Food. Lebensm Wiss Technology, 20: 101-110.

Kozempel, M.F., Annous, B.A., Cook, R.D., Scullen, O.J., and Whiting, R.C. 1998. Inactivation of Microorganisms with Microwaves at Reduced Temperatures. Journal of Food Protection, 61(5): 582-585.

Kunze, O.R. 1979. Fissuring of The Rice After Heated Air Drying. Transactions of the American Society of Agricultural Engineers, 22(5): 1197-1207.

Lagunas-Solar, M.C. 2003. Development of New Physical Methods as Alternative Clean Technologies for Production Agriculture in the XXI Century. Agro-Ciencia 19(1): 57-73.

Lagunas-Solar, M. C., Zeng, N.X., and Essert, T.K. (Inventors). 2003. Method for Inhibiting Pathogenic and Spoilage Organisms in Products. University of California, assignee. US Patent No. 6,638,475.

Lagunas-Solar, M.C., Zeng, N.X., Essert, T.K., Truong, T.D., Piña, C., Cullor, J.S., Smith, W.L., and Larraín, R. 2005a. Disinfection of Fishmeal with Radiofrequency Heating for Improved Quality and Energy Efficiency. Journal of the Science of Food and Agriculture, 85 (13): 2273-2280.

Lagunas-Solar, M.C., Cullor, J.S., Zeng, N.X., Truong, T.D., Essert, T. K., Smith, W.L., and Pina, C. 2005b. Disinfection of Dairy and Animal Farm Wastewater with Radiofrequency Power. Journal of Dairy Science, 88 (11): 4120-4131.

Lagunas-Solar, M.C., Zeng, N.X., Essert, T.K., Truong, T.D., and Piña, C. 2006a. Radiofrequency Power Disinfects and Disinfests Food, Soils and Wastewater. California Agriculture, 60(4): 192-199.

Lagunas-Solar, M.C., Essert, T.K., Pina U, C., Zeng, N.X., and Truong, T.D. 2006b. Metabolic Stress Disinfection and Disinfestation (MSDD): A New, Non-thermal, Residue-free Process for Fresh Agricultural Products. Journal of the Science of Food and Agriculture, 86: 1814-1825.

Lagunas-Solar, M.C., Pan, Z., Zeng, N.X., Truong, T.D., Khir, R., and Amaratunga, K.S.P. 2008. Application of Radiofrequency Power for Non-chemical Disinfestation of

Rough Rice with Full Retention of Quality Attributes. Applied Engineering in Agriculture 23(5): 647-654.

Lagunas-Solar, M.C. and Essert, T.K. (Inventors). 2011. Disinfection and Disinfestation of Foods, Perishables and Other Commodities. University of California, assignee. US Patent No. 7,975,427.

Lea, S.M. and Burke, J.R. 1998. Physics: The Nature of Things. Brooks/Cole Publishing Company, Pacific Grove, California, USA.

Metaxas, R. and Meredith, R.J. 1983. Industrial Microwave Heating. Peter Peregrinus Ltd., London, UK.

Mitcham, E.J, Veltman, R.H., Feng, X., de Castro, E., Johnson, J.A., Simpson, T.L., Biasi, W.V., Wang, S., and Tang, J. 2004. Application of Radio Frequency Treatments to Control Insects in In-Shell Walnuts. Postharvest Biology and Technology, 33: 93-100.

Mudgett, R.E. 1986. Electrical Properties of Foods. M.A. Rao, and S.S.H. Rizvi (Eds.). Engineering Properties of Foods. Marcel Dekker, New York, 329-390.

Nelson, S.O. and Charity, L.F. 1972. Frequency Dependence of Energy Absorption by Insects and Grains. Transactions of the American Society of Agricultural Engineers, 15(6): 1099-1102.

Nyfors, E. and Vainikainen, P. 1989. Industrial Microwave Sensors. Artech House, Norwood.

Olsen, C.M. 1965. Microwaves Inhibit Bread Mold. Food Engineering, 37(7): 51-53.

Orefeuil, M. 1987. Electric Process Heating: Technologies/Equipment/Applications. Battelle Press, Columbus, Ohio, USA.

Pimentel, D., Lach, L., Zuniga, R., and Morrison, D. 1999. Environmental and Economic Costs Associated with Non-indigenous Species in the United States. Report. College of Agriculture and Life Sciences, Cornell University.

Piyasena P., Dussault, C., Koutchma, T., Ramaswamy, H.S., and Awuah, G.B. 2003. Radio Frequency Heating of Foods: Principles, Applications and Related Properties-A Review. Critical Reviews in Food Science and Nutrition, 43(6): 587-606.

Prakash, A. and Rao, J. 2002. Integrated Management of Rice Storage Insects. The Central Rice Research Institute (CRRI), Cuttack, India. http://www.crriicar.org/

Prosser, C.L. 1986. Adaptational Biology: Molecules to Organisms. Wiley, New York, USA.

Rosenberg, U. and Bögl, W. 1987. Microwave Pasteurization, Sterilization, Blanching, and Pest Control in the Food Industry. Food Technology, 41(6): 92-99.

Roti Roti, J.L. 1982. Heat-induced Cell Death and Radiosensitization: Molecular Mechanisms. Proceedings of the Third International Symposium: Cancer Therapy by Hyperthermia, Drugs and Radiation: 3-10.

Shuler, M.L. and Kargi, F. 1992. Bioprocess Engineering Basic Concepts. Prentice Hall, Inc., Englewood Cliffs, New Jersey, USA.

Urbain W.M. 1986. Food Irradiation, Academic Press, Orlando FL.

United Nations Environmental Program (UNEP). 1998. Montreal Protocol on Substances that Deplete the Ozone Layer.

USDA Federal Grain Inspection Service. 1994. Rice Inspection Handbook. Washington D.C.: USDA Agricultural Marketing Service.

Ushakov, B. 1964. Thermostability of Cells and Proteins of Poikilotherms and Its Significance in Speciation. Physiological Reviews, 44: 518-559.

Wang S., Tang J., Johnson J.A., Mitcham E., Hansen J.D., Hallman G., Drake S.R., and Wang
    Y. 2003. Dielectric Properties of Fruits and Insect Pests as related to Radiofrequency
    and Microwave Treatments. Byosystems Engineering 85 (2): 201-212.
Wang S., Monzon M., Johnson J.A., Mitcham E.J., and Tang J. 2007a. Industrial-scale radio
    frequency treatments for insect control in walnuts. I: Heating uniformity and
    energy efficiency. Postharvest Biology and Technology 45: 240-246.
Wang S., Monzon M., Johnson J.A., Mitcham E.J., and Tang J. 2007b. Industrial-scale radio
    frequency treatments for insect control in walnuts. II: Insect mortality and product
    quality. Postharvest Biology and Technology 45: 247-253.
Ward, J.F. 1985. Biochemistry of DNA Lesions. Radiation Research-Supplement, 104: S103-
    S111.
Wig, T., Tang, J., Younce, F., Hallberg, L., Sunne, C.P., and Koral, T. 1999. Radio Frequency
    Sterilization of Military Group Rations. AIChE Annual Meeting.
World Meteorological Organization (WMO). 2003. Scientific Assessment of Ozone
    Depletion: 2002 Global Ozone Research Monitoring Project Report # 47, Geneva,
    Switzerland, 498pp.
Yoder, J.A., and Denlinger, D.L. 1991. Water Balance in Flesh Fly Pupae and Water Vapor
    Absorption Associated with Diapause. The Journal of Experimental Biology, 157:
    273-286.
Zhao, Y., Flugstad, B., Kolbe, E., Park, J.W., and Wells, J.H. 2000. Using Capacitive (Radio
    Frequency) Dielectric Heating in Food Processing and Preservation – A Review.
    Journal of Food Process Engineering, 23: 25-55.
Zimmermann, U., Pilwat, G., and Riemann, F. 1974. Dielectric Breakdown of Cell
    Membranes. Biophysical Journal, 14(11): 881-889.

11

Proteomic Profiling of *Escherichia coli* in Response to Carbamate Pesticide - Methomyl

Amritha G. Kulkarni and B. B. Kaliwal

P.G. Department of Studies in Biotechnology and Microbiology
Karnatak University,
India

1. Introduction

Since decades, there has been mounting concern regarding the adverse health effects of environmental contaminants in general and carbamate in particular. Methomyl is a carbamate and widely used throughout the world since it is effective as "contact insecticide" as well as "systemic insecticide" for fruits and vegetables and is well known established cholinesterase inhibitor [1]. Methomyl has been classified as a pesticide of category-I toxicity [2]. Methomyl is a metabolite of thiodicarb and acetimidate is suspected oncogen, which is a metabolite in animal tissues [3]. It has been classified by the WHO, EPA (Environmental Protection Agency, USA), and EC (European Commission) as a very toxic and hazardous pesticide [4]. Methomyl is highly soluble in water and can therefore, easily cause ground water contamination in agricultural areas [5]. Bonatti *et al.,* [6] have shown genotoxic effects of methomyl in *in vitro* studies. Methomyl is potent genotoxic and is capable of inducing structural and numerical chromosomal aberration in mammalian cells [7].

Prokaryotic cells respond to environmental or chemical stress by inducing specific sites of proteins characteristic to each stress [8]. Studies on stress response and survival strategies of enteric bacteria have evolved a range of complex mechanisms, which use different regulatory structures and genetic components for their survival and virulence [9]. The stress protein induced in response to four different pesticides viz. cypermethrin, zeta-cypermethrin, carbofuran and bifenthrin were analyzed by protein profiling of *Escherichia coli* by Asghar *et al.,* [10]. Mechanisms of cellular adaptation and compensation against different kinds of toxic metals have been proposed. However, the molecular mechanisms and underlying responses of cells against various pesticides are not yet completely understood [11].

Proteomics is a technique used to investigate whole proteins expressed by an organism, tissue or a cell at a specific time point under defined environmental conditions. Nowadays, proteomics has been used for many research purposes e.g. disease diagnosis, drug target and biomarkers of pollutants [12,13]. Proteomics, transcriptomics and metabolomics are powerful tools for acquiring information on gene/protein function and regulatory networks[28]. Using proteomics, one can determine protein expression profiles related to research for both microbial isolates and communities. Proteomics provides a global view of the protein complement of biological systems and, in combination with other omics technologies, has an important role in helping uncover the mechanisms of these cellular processes and thereby advance the development of environmental biotechnologies[29].

The polyacrylamide gel electrophoresis has been used extensively for the separation of proteins in yeast, bacteria and higher organisms with the successful separation of whole cell extracts or specific proteins under selected conditions. This is an excellent method to attempt a global depiction of the cells protein profile. Thus, this technique is being extensively used to determine the *in vivo* amount of protein, its rate of synthesis, and rate its rate of degradation [13]. SDS-PAGE is an important molecular technique used for the identification of whole cell proteins and it has the advantage of being fairly simple and rapid to perform [14]. Therefore, the present investigation was undertaken to study the proteomic profiling of *Escherichia coli* on dose and durational exposure to methomyl by gel electrophoresis.

2. Materials and methods

2.1 Preparation of stock solution of methomyl
The sample of methomyl (Lannate ®) used in the experiment was supplied by E.I. Dupont India Pvt. Ltd., Haryana obtained. The stock solution of 1 M of methomyl was prepared and further diluted to give different required molar concentrations.

2.2 Maintenance and propagation of culture
The organism *Escherichia coli* was procured from NCL, Pune and the bacteria was maintained at 4°C on nutrient agar formulated by Lapage and Shelton [15] and sub cultured very fortnight.

2.3 Medium used for the study
Synthetic sewage medium (S-medium) formulated by Babich and Stotzky [16] was used as the medium for toxicity testing.

2.4 Preparation of inoculum for free cells
Pre-inoculum was prepared by inoculating a loopful of bacteria from the overnight incubated nutrient agar slant cultures on a 100 ml sterilized synthetic sewage medium and incubated for 18-24 hours at 37°C under static conditions depending on the exponential phases of bacteria under test.

2.5 Experimental procedures
Free cells: Five ml of the pre-inoculum was inoculated to 250 ml Erlenmeyer's flask containing 100 ml of sterilized S-medium amended with different molar concentrations of heavy metals. The flasks were incubated at 37°C for 96 hours under shaking conditions at 120 rpm on a rotary shaker (REMI – CIS-24). At regular intervals sample was taken out from each flask aseptically for analysis.

2.6 Isolation of protein
The bacterial cell pellet was dissolved in 100μl of lysis buffer and incubated at 37°c for 15 min. the tubes were centrifuged and the supernatant was used as protein sample. PAGE according to Laemmli [17] analyzed these protein samples.

3. Results and discussion

The present investigation was attempted to elucidate the protein profiling in *Escherichia coli* cells that were exposed to different concentrations of methomyl ranging from 10^{-7} M to 10^{-3}

M of methomyl for a period of 96 hrs and at regular intervals of 24 hrs, the proteins induced were analyzed. The protein expression was observed at 29, 45, 48, 55, 63, 92 and 114 kDa at 24 hrs (Fig. 1). On exposure to methomyl for 48 hrs the bands were observed at 29, 45, 48, 55, 63, 92 and 114 kDa (Fig. 2). The methomyl treated for 72 hrs showed expression at 29, 39, 45, 66 and 92 kDa (Fig. 3) and for 96 hrs the expressions was observed at 29, 35, 39, 45, 55, 63, and 92 kDa (Fig. 4) respectively. The expression of proteins were more conspicuous in our result which was obligatory, since the free *Escherichia coli* cells possess antioxidant enzymes, which are induced in response to the stress and are directly exposed to methomyl [18].

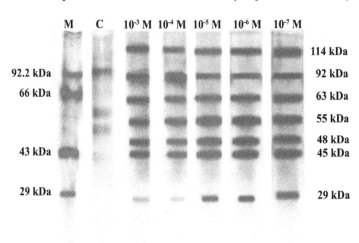

Fig. 1. Protein profile of *Esherichia coli* induced by methomyl for 24 hours.

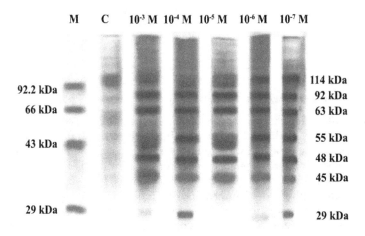

Fig. 2. Protein profile of *Esherichia coli* induced by methomyl for 48 hours.

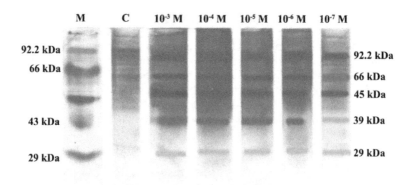

Fig. 3. Protein profile of *Esherichia coli* induced by methomyl for 72 hours.

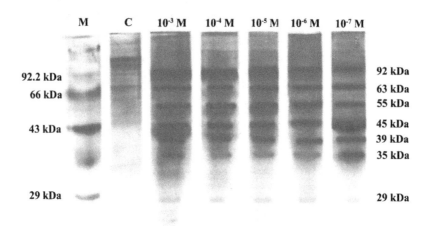

Fig. 4. Protein profile of *Esherichia coli* induced by methomyl for 96 hours.

The protein profiles were compared with the dose and duration of exposure of methomyl in *Escherichia coli* and the results revealed that the intensity of the proteins expressed increased with an increase in the dose and duration of exposure of methomyl when compared with those of the corresponding parameters of the control, indicating that the pesticide methomyl induces stress. Our results agreed with the observations made by Asghar *et al.,* [10] who analyzed the stress proteins of *Escherichia coli* induced in response to the pesticides cypermethrin, zeta-cypermethrin, carbofuran and bifenthrin.

The over expressions of some of the proteins observed in the present study at 29 and 45 kDa at all the dose and duration of exposure could be due to the fact that prokaryotic cells respond to environmental or chemical stress by inducing specific sets of proteins characteristic to each stress. It has been reported that the proteins in each set of their coding genes constitute a stimulon, such as heat shock, SOS response and oxidation stress. In some other cases, proteins, which are associated with one stimulon, can be induced during other stresses, such as various heat shock proteins in *Escherichia coli*. These proteins are also synthesized when the cells are exposed to different physical and chemical stress. In some stimulons, exposure to non-lethal levels of a stress agent can confer protection against subsequent exposure to lethal levels of the same stress agent [19]. Similarly, in the present study, the proteins expressed at 29 and 45 kDa could be unique or could be observed in the protein profiling of other micro-organisms exposed to various physical or chemical stress.

It has been suggested that the analysis of many proteins produced during the transition into stationery phase and under stress conditions demonstrated that a number of novel proteins were induced in common to each stress and could be the reason for cross protection in bacterial cells. It is necessary to investigate the synthesis of these proteins during different stress conditions [20]. Similarly it has been mentioned that when organisms or cells are exposed to low levels of certain harmful physical and chemical agents, the organisms acquire an induced tolerance against the adverse effects [7]. Hence, in the present study the high molecular weight proteins of 114 kDa at 24 and 48 hrs respectively observed in all the doses of exposure in comparison to their corresponding controls may be ascertained to the protein selective proteolytic degradation that appears to be rather significant in homeostasis maintaining and metabolism regulation in the cell [21]. It has been reported that along with short-lived regulatory proteins, the polypeptide chains with disrupted or changed structures are selectively hydrolyzed. Such defects might arise from inaccuracy during protein biosynthesis, chemical or physical damage [22] and moreover, the extracts of *Escherichia coli* have been shown to degrade rapidly the damaged enzyme, but not the native protein, and several preliminary reports have appeared concerning the *Escherichia coli* protease that may be responsible for selective degradation of the modified proteins [23].

Although it has been reported that the starvation for individual nutrients and other stress induce a unique and individual profile of protein expression, some proteins are common to different starvation and stress factors in *Escherichia coli*. However, the proteins of one stimulon do not respond coordinately to all the starvation and stress treatments and relatively few of the starvation- inducible proteins have been found to overlap with those induced by stress. This suggests that despite the regulation of a few specific proteins being interconnected, there are major difference in the regulatory pathways controlling the expression of starvation and different stress proteins [24]. Studies in the micro-organisms have provided evidence for increased longevity, cell division rate and survival when exposed to stress [25]. Similarly in the present study, the types of stress patterns observed with the dose and duration of exposure of methomyl were identical which agreed with the earlier reports [10] that the stress proteins produced in response to two different classes of pesticides showed that the same stress patterns were obtained for different substituent chemical groups within the same class and two different classes, indicating that the gene or set of genes responsible for stress expressions were the same irrespective of the class or nature of substituent's on the pesticide.

Further, an increase in the intensity in protein expression observed in the present study may be due to the fact that the major protein modification is observed due to stress, loss of catalytic activity, amino acid modification, carbonyl group formation, increase in acidity, decrease in thermal stability, change in viscosity, fluorescence, fragmentation, formation of protein protein crosslink's, s-s bridges and increased susceptibility to proteolysis [3]. It has been revealed that the secretion of extra cellular proteins, including toxins and cellular effectors, is one of the key contributing factors in a bacterium's ability to thrive in diverse environments [26]. Hence, the present study indicates that the protein expressions are dose and duration dependent. It has been suggested that there are many protein synthesized in common with many stress in *Escherichia coli* and some of these proteins may play a major role in the stability of the cells under different stresses. The fact that specific patterns of proteins are expressed for a particular stress has led to the use of stress proteins to monitor environmental samples for the presence of particular pollutants [27]. It has been suggested that the analysis of such stress proteins will aid in the development of more sensitive techniques for the pollutant analysis. The unique proteins could be purified and raised to enable quick detection, which could be used as biomarkers of xenobiotics in the environment [11].

4. Conclusions

The present study indicated the molecular weights of the various stress proteins induced in response to the dose and durational exposure of methomyl. Further, it indicates that the stress protein analysis is a promising alternative and more sensitive method for measuring toxic effects on the organisms at sub lethal levels. The study suggests that the proteomic profiling is a sensitive tool for environmental stress diagnosis, and that the stress proteins could be used as biomarkers for environmental pollution identification. The specific patterns of the proteins that are expressed in response to the stress induced by methomyl could be used to monitor the environmental samples for the presence of such pollutants. Although the application of gene and protein expression analysis to ecotoxicology is still at an early stage, this holistic approach seems to have several potentials in different fields of ecological risk assessment. It can be concluded that such extensive work on proteomics can be performed in understanding the proteomic/genomic response and tolerance of the microorganisms to the extreme environment.

5. Acknowledgements

The authors are grateful to the Post Graduate Department of Studies in Microbiology and Biotechnology, Karnatak University Dharwad for providing the necessary facilities.

6. References

[1] Barakat .K.K., Effect of Certain Insecticides on the Stabilization And Lysis of Human and Fish Erythocyte *Research Journal of Agriculture and Biological Sciences* ; 1(2), 195-199 (2005)

[2] Tamimi M., Qourzal S., Assabbane A., Chovelon J. M., Ferronato C., Emmelin C., Ait-Ichou Y., Photocatalytic degradation of pesticide methomyl Determination of the reaction pathway and identification of intermediate Products. *Photochem. Photobiol. Sci;* 5, 477-48 (2006)

[3] Stadtman E. R., protein oxidation and ageing. *Science* 257, 1220-1224 (1992)

[4] Bonatti.S., Bolognesi. C., Degan. P., Abbondandolo. A., Genotoxic effect of the carbamate insecticide methomyl. In vitro studies with pure compound and the technical formulation " Lannate 25". *Environmental and molecular mutagenesis*; vol 23, p.306-311. (2006)

[5] Laura L., Eerd. V., Hoagland. R. E., Zablotowicz. R M.., Hall C. J., Pesticide metabolism in plants and micro-organisms. *Weed science;* 51, 472-495 (2003)

[6] Benjamin I.J., McMillan D. R., Stress (heat shock) proteins: molecular chaperones in cardiovascular biology and disease. *Circ Res;* 83,117–132 (1998)

[7] Flahaut S., Hartke A., Giard J. Nystrom, T., Olsgon, R.M., and Kjelleberg, S. Survival, stress resistance, and alteration in protein expression in the marine *Vibrio* sp. strain S14 during starvation for different individual nutrients. *App. Environ. Microbiol.* 58, 55-65 (1992).

[8] Ronan, O.T., Marjan J., Smeulders, Marian, C., Blokpoel, Emily, J. Kay, Kathryn Lougheed, and Huw, D. Williams., A two-component regulator of universal stress protein expression and adaptation to oxygen starvation in *Mycobacterium smegmatis. J. Bacteriol.* 185 (5): 1543-1554. (2003)

[9] Kappke. J., da Silva E.R.,.Schelin H.R., Pashchuk S.A. and de Oliveira. A.., Evaluation of *Escherichia coli* cells damages induced by ultraviolet and proton beam radiation., *Brazilian journal of Physics,* 35, 3B. (2005).

[10] Asghar. M. N., Ashfaq. M., Ahmad. Z., Khan I. U., 2-D PAGE analysis of pesticide-induced stress proteins of E. coli. *Anal Bioanal Chem.* 384: 946–950 (2006).

[11] Patcharee. Isarankura-Na-Ayudhya., Virapong Prachayasittikul., Proteomic profiling of *Escherichia coli* in response to heavy metal stress. *European Journal of Scientific Research.* Vol.25. No 4. Pp 679-688 (2009).

[12] Mahashi Nakayama., Kyoko Ishizawa., Jiro Nakajima., Akiko Kawamura and Takako Umino: Cellular Protein Profile of Halobacteriurn Halophilic halobium, Archaea. *Memoirs of Osaka Kyoiku University, Ser. III,* Vol. 45, No. 1 pp. 81-91 (1996).

[13] Khemika Lomthaisong., Kanchanit Boonmaleerat and Aphinya Wongpia Proteomic study of recombinant *Escherichia coli* expressing *Beauveria bassiana* Chitinase Gene, *Chiang Mai J. Sci.* 35(2) : 324-330. (2008)

[14] Leisner J.J., Millan H. H., Huss and Larsen L. M., Production of histamine and tyramine by Lactic acid bactria isolated from vacuum packed sugar-salted fish. *J. Appl. Bacteriol,* 76. 417-423 (1994).

[15] Lapage S. P. and Shelton J. E., In Methods in Microbiology, (ed. Norris J. R.and Ribbons D. W.), academic Press. New York, N Y. pp, 1,3A (1970).

[16] Babich and Stotzky.. Reduction in the toxicity of cadmium to micro-organisms by clay minerals. *Appl. Environ. Microbiol.,* 33, 696-705. (1977).

[17] Laemmli, U. K. : Cleavage of structural proteins during the assembly of the head of bacteriophage T4, *Nature* 227: 680-685 (1970).

[18] Kulkarni A. G. and Kaliwal B. B., Studies on methomyl induced stress in free and immobilized *Escherichia coli. Proceedings of ISBT* 419-423 (2008).

[19] Nystrom T., Olsgon R.M. and Kjelleberg S., Survival, stress resistance and alteration in protein expression in the marine *Vibrio* sp. Strain S14 during starvation for different individual nutrients. *Appl environ. Microbial.* 58. 55-65. (1992).

[20] Jamshid Raheb., Shamim Naghdi1 and Ken P. Flint., The Effect of Starvation Stress on the Protein Profiles in *Flexibacter chinensis. Iranian Biomedical Journal* 12 (2): 67-75. (2008).

[21] Beckwith and Strauch., Periplasmic protease mutants of *Escherichia coli, World intellectual property organization,* 5. 819-821 (1988).

[22] Vasilyeva.O.V., Potapenko. N.A., Ovchinnikova.T.V: Limited proteolysis of *Escherichia coli* ATP-Dependent protease ion., *vestnik Moskovskogo Universtiteta, Khimiya.*,41., 6. (2000).

[23] Young.S.Lee., Sang.C.Park., Alfred.L.Goldberg., and Chin.Ha.Chung., Protease so from *Escherichia coli* preferentially degradesOxidatively damaged Glutathione synthetase.m, *the Journal of Biological Chemistry.*,263(14)., 6643-6646 (1988).

[24] Lambert N. H., Abshire k., Blanmenhorn D. and Slonczewski J. L. proteins induced in *Escherichia coli* by benzoic acid. *J. Bacteriol.* 179. 7595-7599 (1997).

[25] Smith S.J., Barbee A.S., Exercise stress response as an adaptive tolerance strategie. *Environ Health Persp* 106. 325-330. (1998).

[26] Werner G. and Stephen L. *Current opinion in Microbiology,* 9, 123-126. (2006).

[27] Sanders H.M., Martin L.S., Stress proteins as biomarkers of contaminant exposure in archived environmental samples. *Science Total Environ.*139/140. 459-470 (1993).

[28] Phelps, T.J., Palumbo, A.V., Beliaev, A.S., Metabolomics and microarrays for improved understanding of phenotypic characteristics controlled by both genomics and environmental constraints., *Curr Opin Biotechnol*, 13, 20–24 (2002).

[29] Carla, M. R., Lacerda and Kenneth F., Reardon Environmental proteomics: applications of proteome profiling in environmental microbiology and biotechnology., *briefings in functional genomics and proteomics.*, pp 1-13(2009).

12

Ameliorative Effect of Vitamin E on Sensorimotor and Cognitive Changes Induced by Chronic Chlorpyrifos Exposure in Wistar Rats

Suleiman F. Ambali[1], Joseph O. Ayo[1], Muftau Shittu[1],
Mohammed U. Kawu[1] and Suleiman O. Salami[2]
[1]Department of Veterinary Physiology and Pharmacology
[2]Department of Veterinary Anatomy
Ahmadu Bello University, Zaria,
Nigeria

1. Introduction

The use of pesticides is inevitable in contemporary world because of their role in the improvement of food production through increase in crop yields and quality, reduction of farm labour requirements hence lowering cost of production, and improving public health through control of vector and vector-borne diseases (Weiss et al., 2004). Despite all these benefits, pesticides constitute menace to the health of man, animals and even the environment. This is because they are poorly selective and are toxic to non-target species, including humans. The segments of the population that are at the greatest risk of exposure are those that are occupationally exposed, such as agricultural workers. Despite the strict measures put in place concerning its commercialization and use, pesticides sales has increased in recent years (Carlock et al., 1999). The World Health Organization (WHO) estimated that about 3 million cases of acute intoxication and 220,000 deaths are attributable to pesticides each year with majority of these cases occurring in less developed countries (He, 2000; Clegg & van Gemert, 1999), particularly in Africa, Asia, Central America, and South America (Pancetti et al., 2007). Although many pesticides cause neurotoxicity, insecticides are the most acutely neurotoxic to humans and other non-target species compared to other pesticides (Costa et al., 2008). Association between acute exposure to pesticides and neurotoxicity is well known (Lotti, 2000) but the potential effects of chronic low-level exposure are less well established (Alavanja et al., 2004; Ambali et al., 2010a; Ambali & Aliyu, 2012).

Organophosphate (OP) compounds are one of the most widely used constituting about 50% global insecticide use (Casida & Quistad, 2004). Studies in humans showed neurological, cognitive and psychomotor impairments following cumulative exposure to OPs and organochlorines in people from agricultural communities, without history of acute poisoning (Kamel & Hoppin 2004; Kamel et al., 2007). Neurobehavioural changes following low-dose OP exposure have been reported in sheep farmers (Stephens et al., 1995), greenhouse workers (Bazylewicz-Walczak et al. 1999), tree-fruit workers (Fiedler et al., 1997), and farm workers (Kamel et al., 2003). These studies have found deficits in measures of sustained attention, information processing, motor speed and coordination. The principal mode of insecticidal action of OPs relates to phosphorylation and subsequent inactivation of the esteratic sites of the acetylcholinesterase (AChE) enzyme. The classical role of AChE is to hydrolyze the neurotransmitter acetylcholine (ACh), effectively clearing it from the neuronal synapse and terminating impulse conduction (Farag et al., 2010). Inactivation of AChE results in the accumulation of ACh in the neuronal synapses in the central and peripheral nervous system, thereby overstimulating the nicotinic, muscarinic and central cholinergic receptors with consequent neurotoxicity. Thus, the acute neurotoxic effect of OP results in muscarinic, nicotinic and central cholinergic symptoms (Abou-Donia, 1992). However, toxicity has been reported at doses below the threshold required for inhibition of AChE (Pope, 1999; Slotkin, 2004, 2005) prompting search for other mechanisms. The induction of oxidative stress as one of the other molecular mechanisms involved in OP-induced neurotoxicity has received tremendous attention in recent years (Gultekin et al., 2007; Prendergast et al., 2007; El-Hossary et al., 2009; Ambali et al., 2010a, Ambali & Ayo, 2011a, 2011b; Ambali & Aliyu, 2012). Indeed, the enhanced production of reactive oxygen species (ROS) by pesticides has been used to explain the multiple types of responses associated with its toxic exposure (Bagchi et al., 1995; Verma et al., 2007).

Chlorpyrifos (O,O-diethyl-O-[3,5,6-trichloro-2-pyridyl] phosphorothioate) is a chlorinated OP insecticide that exhibit a broad spectrum of activity against arthropod pests of plants, animals, and humans, and has wide applications in both agricultural and commercial pest control (Rack, 1993). It is one of the most widely used insecticides and is applied about 20 million times per year in US to houses and lawns (Kingston et al., 1999) with 82% of adults having detectable levels of the 3,5,6-trichloro-2-pyridinol, the metabolite of CPF in their urine (Hill et al., 1995). However, the United States Environmental Protection Agency in 2000 placed ban on some its residential uses in 2000 because of the danger posed to children's health. However, CPF is still widely used as its residues have been detected in citrus fruits in some parts of the world (Iwasaki et al., 2007). Studies have shown that CPF induces neurobehavioural alterations following acute (Cañadas et al., 2005; Ambali et al., 2010a, Ambali & Aliyu, 2012) and repeated low-dose (Stamper et al., 1988; Sanchez-Santed et al., 2004; Ambali & Ayo, 2011a, 2011b) exposure. Similarly, CPF is a developmental neurotoxicant (Qiao et al., 2003; Dietrich et al., 2005; Colborn, 2006; Slotkin et al., 2006;) impairing children mental and behavioral health (Lizardi et al., 2008). Although, CPF like the other OP compounds phosphorylates and subsequently inactivate AChE, neurobehavioural and cognitive deficits have however been observed following repeated low-dose CPF exposure that cannot be attributed to the usual AChE inhibition and muscarinic receptor binding (Pope et al., 1992; Chakraborti et al., 1993; Saulsbury et al., 2009). Earlier studies have shown the involvement of oxidative stress in the neurotoxicity induced by CPF exposure (Gultekin et al., 2007; Ambali et al., 2010a; Ambali & Aliyu, 2012; Ambali and Ayo, 2011a, 2011b).

Oxidative stress, defined as a disruption of the prooxidant-antioxidant balance in favor of the former causes damage to the body tissue (Sies, 1991). Oxidative stress results from an increase in ROS, impairment of antioxidant defense system or insufficient capacity to repair oxidative damage (Halliwell, 1994; Aly et al., 2010). Damage induced by ROS which alters cellular macromolecules such as membrane lipids, DNA, and proteins results in impaired cell functions through changes in intracellular calcium or pH, and consequently leads to cell death (Kehrer, 1993; Sally et al., 2003). The body is however endowed with cellular defence systems to combat the menace posed by the oxidants to the body. These defensive systems are accomplished by the activities of both the enzymatic and non-enzymatic antioxidants which mitigate the toxic effect of oxidants. However, under increased ROS production, the antioxidant cellular defensive systems are overwhelmed, resulting in oxidative stress. Under this type of condition, exogenous supplementation of antioxidants becomes imperative to minimise tissue damage.

Vitamin E is nature's major lipid soluble chain breaking antioxidant that protects biological membranes and lipoproteins from oxidative stress (Osfor et al., 2010). The main biological function of vitamin E is its direct influence on cellular responses to oxidative stress through modulation of signal transduction pathway (Hsu & Guo, 2002). Vitamin E primarily scavenges peroxyl radicals and is a major inhibitor of the free radical chain reaction of lipid peroxidation (Maxwell, 1995; Halliwell & Gutteridge, 1999). We have earlier demonstrated the mitigating effect of vitamin E on short-term neurobehavioural changes induced by acute CPF exposure (Ambali & Aliyu, 2012). The present study was therefore aimed at evaluating the ameliorative effect of vitamin E on sensorimotor and cognitive changes induced by chronic CPF exposure in Wistar rats.

2. Materials and methods

2.1 Experimental animals and housing
Twenty 10 week old male Wistar rats (104±4.2) used for this study were obtained from the Laboratory Animal House of the Department of Veterinary Physiology and Pharmacology, Ahmadu Bello University, Zaria, Nigeria. The animals were housed in plastic cages and allowed to acclimatize for at least two weeks in the laboratory prior to the commencement of the experiment. They were fed on standard rat pellets and water was provided *ad libitum*.

2.2 Chemicals
Commercial grade CPF (20% EC, Termicot®, Sabero Organics, Gujarat limited, India), was prepared by reconstituting in soya oil (Grand Cereals and Oil Mills Ltd., Jos, Nigeria) to make 10% stock solution. Vitamin E (100 mg/capsule; Pharco Pharmaceuticals, Egypt) was reconstituted in soya oil (100% v/v) prior to daily use.

2.3 Animal treatment schedule
The rats were weighed and then assigned at random into 4 groups of 5 rats in each group. Group I (S/oil) served as the control and was given only soya oil (2mL/kg b.w.) while group II (VE) was dosed with vitamin E [75 mg/kg b.w. (Ambali et al., 2010b)]. Group III (CPF) was administered with CPF only [10.6 mg/kg b.w. ~1/8th LD_{50} of 85 mg/kg b.w., as determined by Ambali (2009)]. Group IV (VE+CPF) was pretreated with vitamin E (75 mg/kg b.w.), and then dosed with CPF (10.6 mg/kg b.w.), 30 min later. The regimens were administered once daily by oral gavage for a period of 17 weeks. During this period, the animals were monitored for clinical signs and death. Furthermore, at various intervals during the study period, the animals were evaluated for neurobehavioural parameters measuring motor coordination, neuromuscular coordination, and motor strength, efficiency of locomotion, learning and memory using the appropriate neurobehavioural devices. In order to avoid bias, the neurobehavioural parameters were evaluated by two trained observers blinded to the treatment schedules. At the end of the dosing period,

each of the animals was sacrificed by jugular venesection and the brain dissected, removed and evaluated for the levels of oxidative stress parameters and AChE inhibition. The experiment was conducted with the permission of the Animals Research Ethics Committee of the Ahmadu Bello University, Zaria, Nigeria and in accordance with the National Institutes of Health Guide for Care and Use of Laboratory Animals (Publication No. 85-23, revised 1985).

2.4 Evaluation of the effect of treatments on motor coordination

The assessment of motor coordination was performed using the beam walk performance task as described in an earlier study (Ambali et al., 2010a) on day 0, weeks 8 and 16. Briefly, each of the rats was allowed to walk across a wooden black beam of 106-cm length, beginning at 17.2 cm width and ending at 1.0-cm width. Periodic widths were marked on the side of the apparatus. On each side of the narrowing beam, there was a 1.8-cm step-down to a 3.0-cm area where subjects may step if necessary. As the subject walked across from the 17.2 cm to the 1.0 cm width, the width at which they stepped down was recorded by one rater on each side, and this was repeated twice during each trial session.

2.5 Evaluation of the effect of treatments on motor strength

The forepaw grip time was used to evaluate the motor strength of the rats, as described by Abou-Donia et al. (2001). This was conducted by having each of the rats hung down from a 5 mm diameter wooden dowel gripped with both forepaws. The time spent by each rat before releasing their grips was recorded in seconds. This parameter was evaluated on day 0, weeks 8 and 16.

2.6 Effect of treatments on neuromuscular coordination

The effect of treatments on neuromuscular coordination was assessed using the performance on incline plane as was described earlier (Ambali et al., 2010a). Briefly, each rat was placed on an apparatus made with an angled rough wooden plank with thick foam pad at its bottom end. The plank was first raised to an inclination of 35°, and thereafter gradually increased stepwise by 5° until the subject could no longer stay and be situated horizontally on the plank for 3s, without sliding down. Angles were measured and marked on the apparatus beforehand, and were obtained by propping the plank on a vertical bar with several notches. The test was performed with the head of the rat first facing left and then right hand side of the experimenter. The highest angle at which each rat stayed and stood horizontally, and facing each direction was recorded. Two trials were performed at 2 min apart for each animal. This procedure was carried out on each animal from all the groups on day 0, weeks 8 and 16 of the study.

2.7 Evaluation of the effect of treatments on efficiency of locomotion

The ladder walk was used to assess the efficiency of locomotion as described by Ambali and Aliyu (2012). Briefly, each rat was encouraged to walk across a black wooden ladder (106 cm x17 cm) with 0.8-cm diameter rungs, and 2.5-cm spaces between them. The number of times the rat missed a rung was counted by one rater on each side. The performance on ladder walk was evaluated on Day 0, weeks 3, 7 and 11. Two trials were performed for each testing session.

2.8 Assessment of the effect of treatments on learning

The effect of treatments on learning task in rats was assessed 48h to the final termination of the study in week 17 using the step-down inhibitory avoidance learning task as described by Zhu et al. (2001). The apparatus used was an acrylic chamber 40 x 25 x 25 cm consisting of a floor made of parallel 2-mm-caliber stainless steel bars spaced 1 cm apart. An electric shock was delivered through the floor bars. A 2.5-cm-high, 8 x 25 cm wooden platform was placed on the left extreme of the chamber. Each rat was gently placed on the platform. Upon stepping down, the rat immediately received a single 1.5 amp foot shock through the floor bars. If the animal did not return to the platform, the foot shock was repeated every 5s. A rat was considered to have learned the avoidance task if it remained on the platform for more than 2 min. The number of foot shocks was recorded as an index of learning acquisition.

2.9 Assessment of the effect of treatments on short-term memory

Short-term memory was assessed in individual rat from each group using the step-down avoidance inhibitory task as described by Zhu et al. (2001) 24h after the assessment of learning. The apparatus used was the same used earlier for the assessment of learning. In this test, each rat was again placed gently on the platform and the time an animal remained on the platform was recorded as an index of memory retention. Staying on the platform for 2 min was counted as maximum memory retention (ceiling response).

2.10 Brain tissue preparation

The whole brain tissue was carefully dissected and a known weight of the brain sample from each animal was homogenized in a known volume of ice cold phosphate buffer to obtain a 10% homogenate. This was then centrifuged at 3000 × g for 10 min to obtain the supernatant. The supernatant was then used to assess the levels of protein, malonaldehyde (MDA), superoxide dismutase (SOD), catalase (CAT) and AChE in the brain sample.

2.11 Effect of treatments on brain lipoperoxidation

The level of thiobarbituric acid reactive substance, malonaldehyde (MDA) as an index of lipid peroxidation was evaluated on the brain sample using the method of Draper & Hadley (1990) as modified (Freitas et al., 2005). The principle of the method was based on spectrophotometric measurement of the colour developed during reaction of thiobarbituric acid (TBA) with malonadehyde (MDA). The MDA concentration in each sample was calculated by the absorbance coefficient of MDA-TBA complex $1.56 \times 10^5/cm/M$ and expressed as nmol/mg of tissue protein. The concentration of protein in the brain homogenates was evaluated using the Lowry method (Lowry et al., 1951).

2.12 Evaluation of the effect of treatments on brain superoxide dismutase activity

Superoxide dismutase activity was evaluated using NWLSS™ superoxide dismutase activity assay kit (Northwest Life Science Specialities, Vancouver, WA 98662) as stated by the manufacturer and was expressed as mMol/mg tissue protein.

2.13 Evaluation of the effect of treatments on brain catalse activity

Catalase activity was evaluated using NWLSS™ catalase activity assay kit (Northwest Life Science Specialities, LLC, Vancouver, WA 98662) as stated by the manufacturer and was expressed as mMol/mg tissue protein.

2.14 Evaluation of the effect of treatments on brain acetylcholinesterase activity

Acetylcholinesterase activity was evaluated using the method of Ellman et al. (1961) with acetylthiocholine iodide as a substrate. Briefly, the whole brain of each animal was homogenized in a cold (0–4 °C) 20 mM phosphate buffer saline (PBS) incubated with 0.01M 5,5-dithio-bis(2-nitrobenzoic acid) in 0.1 M PBS, pH 7.0. Incubations were allowed to proceed at room temperature for 10 min. Then, acetylthiocholine iodide (0.075 M in 0.1 M PBS, pH 8.0) was added to each tube, and absorbance at 412 nm was measured continuously for 30 min using a UV spectrophotometer (T80+ UV/VIS spectrometer®, PG Instruments Ltd, Liicestershire, LE 175BE, United Kingdom). AChE activity was expressed as IU/g tissue.

2.15 Statistical analysis

Data were expressed as mean ± standard error of mean. Data obtained from the sensorimotor assessment were analyzed using repeated one-way analysis of variance followed by Tukey's posthoc test. The cognitive and biochemical parameters were analyzed using one-way analysis of variance followed by Tukey's posthoc test. Values of $P < 0.05$ were considered significant.

3. Results

3.1 Effect of treatments on clinical signs

There was no clinical manifestation recorded in the S/oil, VE and VE+CPF groups, while lacrimation, congested ocular mucous membranes and intermittent tremors were observed in the CPF group.

3.2 Effect of treatments on beam walk performance

There was no significant change (P>0.05) in the dynamics of beam walk performance in the S/oil group throughout the period of the study. There was a progressive decrease in the width at which VE group slipped off the beam (increase in beam walk length) throughout the study period. Although no significant change (P>0.05) was recorded in week 8 compared to day 0 or week 16, a significant decrease (P<0.05) in the width at which the VE group slipped off the beam in week 16 compared to that of day 0. There was a significant increase (P<0.01) in the width of slip off the beam (decrease in beam walk length) in the CPF group at weeks 8 and 16 when compared to that of day 0, and between week 16 and that recorded in week 8. There was no significant change (P>0.05) in the width at which VE+CPF group slipped off the beam at week 8 when compared to that recorded on day 0 or week 16 but a significant increase (P<0.01) was recorded at week 16 compared to that of day 0.

There was no significant change (P>0.05) in the width at which animals in all the groups slipped off the beam at day 0. At week 8, there was a significant increase (P<0.01) in the width at which the CPF group slipped off the beam compared to that of S/oil, VE or VE+CPF group. Similarly, there was a significant increase (P>0.05) in the width of slip in the VE+CPF group compare to that of VE group but no significant change (P>0.05) in the S/oil group compared to that of VE or VE+CPF group. At week 16, there was a significant increase (P<0.01) in the width of slip off the beam in the CPF group compared to the other groups but no significant change (P>0.05) in the S/oil group when compared to that of VE or VE+CPF group, and between VE group and that recorded in theVE+CPF group (Fig. 1).

Ameliorative Effect of Vitamin E on Sensorimotor and Cognitive Changes Induced by Chronic Chlorpyrifos Exposure in Wistar Rats

239

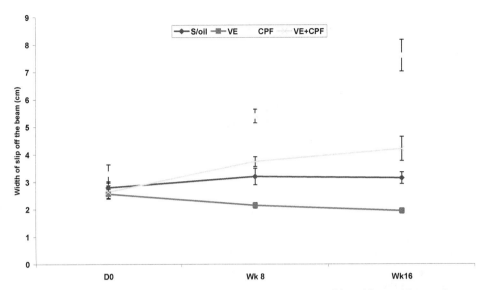

Fig. 1. Effect of chronic administration of soya oil, vitamin E and/or chlorpyrifos on the dynamic of beam walk performance in Wistar rats.

3.3 Effect of treatments on grip time

There was no significant change (P>0.05) in the grip time in the S/oil and VE groups throughout the study period. There was a significant increase (P<0.01) in the grip time of CPF and VE+CPF groups at day 0 compared to that of week 8 or 16, but not between week 8 and that of week 16. At day 0, there was no significant change (P>0.05) in the grip time of rats in between the groups. At week 8, there was a significant decrease (P<0.01) in the grip time of CPF group compared to that in the S/oil and VE groups, but not that of VE+CPF group. There was a significant decrease (P<0.05) in the grip time in the VE+CPF group compared to that in S/oil or VE group. There was no significant change (P>0.05) in the grip time in the VE group compared to that in S/oil group. At week 16, there was a significant decrease (P<0.01) in the grip time in the CPF group compared to that in S/oil or VE group but no significant change (P<0.05) compared to that in VE+CPF group. There was no significant change (P>0.05) in the grip time in the VE+CPF group compared to that in S/oil or VE group. Similarly, there was no significant change (P>0.05) in the grip time of S/oil group compared to that in VE group (Fig. 2).

3.4 Effect of treatments on incline plane performance

There was no significant change (P>0.05) in the angle at which the S/oil and VE groups slipped off the incline plane throughout the study period. There was a significant decrease (P<0.05) in the angle at which the CPF group slipped off the incline plane at weeks 8 and 16, respectively, compared to that of day 0 but no significant change (P>0.05) at week 8 relative to that recorded in week 16. There was a significant decrease (P<0.01) in the angle at which VE+CPF group slipped off the incline plane at week 16 compared to that of day 0 but no significant change (P>0.05) at week 8 relative to that recorded in day 0 or week 16.

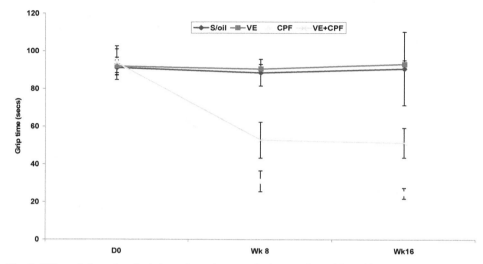

Fig. 2. Effect of chronic administration of soya oil, vitamin E and/or chlorpyrifos on the dynamic of grip time in Wistar rats.

At day 0, there was no significant change (P>0.05) in the angle of slip off the incline plane in between the groups. At week 8, there was a significant decrease in the angle of slip off the incline plane in the CPF group relative to that recorded in S/oil (P<0.05), VE (P<0.01) or VE+CPF group. No significant change (P>0.05) in the angle of slip in the VE+CPF group relative to that in S/oil or VE group, and between VE group and that of S/oil group. At week 16, there was a significant decrease in the angle of slip off the incline plane in the CPF group relative to that in S/oil (P<0.05) or VE (P<0.01) group. Although not significant, there was a 6.3% increase in the angle of slip off the incline plane in the VE+CPF group relative to that in CPF group. There was no significant change (P>0.05) in the angle of slip off the plane in the S/oil group compared to that in VE or VE+CPF group (Fig. 3).

3.5 Effect of treatments on ladderwalk performance
There was no significant change (P>0.05) in the dynamics of the number of missed rungs in the S/oil, VE and VE+CPF groups throughout the study period. There was a significant decrease (P<0.01) in the number of missed rungs in the CPF group at day 0 compared to that in week 8 or 16 but no significant change at week 8 compared to that of week 16.

There was no significant change (P>0.05) in the number of missed rungs in between the groups at day 0. At week 8, there was a significant decrease (P<0.01) in the number of missed rungs in the CPF group compared to that in S/oil or VE group. Although not significant (P>0.05), the mean number of missed rungs in the VE+CPF group was 26% higher relative to that recorded in the CPF group. There was a significant decrease (P<0.01) in the number of missed rungs in the VE+CPF group compared to that in S/oil or VE group. There was no significant change (P>0.05) in the number of missed rungs in the VE group compared to that in S/oil group. At week 16, there was a significant decrease (P<0.01) in the number of missed rungs in the CPF group compared to the VE group but no significant change (P>0.05) when compared to that recorded in S/oil or VE+CPF group. There was no significant change (P>0.05) in the VE+CPF group compared to that in S/oil or VE group.

Ameliorative Effect of Vitamin E on Sensorimotor and Cognitive Changes Induced by Chronic Chlorpyrifos
Exposure in Wistar Rats

241

Similarly, there was no significant change (P>0.05) in the number of missed rungs in the VE
group compared to that in the S/oil group (Fig. 4).

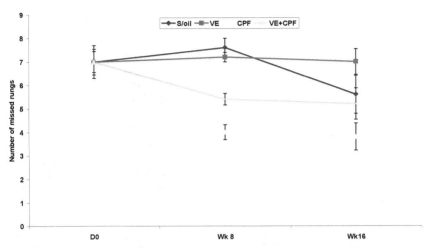

Fig. 3. Effect of chronic administration of soya oil, vitamin E and/or chlorpyrifos on the
dynamics of locomotion efficiency in Wistar rats.

3.6 Effect of treatments on learning acquisition

There was a significant increase (P<0.01) in the number of footshocks applied to the CPF
group relative to that recorded in the S/oil, VE or VE+CPF group. There was no significant
change (P>0.05) in the number of footshocks in the VE+CPF group relative to that in S/oil or
VE group (Fig. 5).

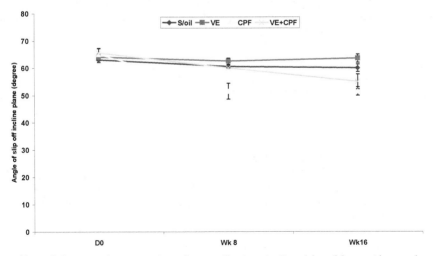

Fig. 4. Effect of chronic administration of soya oil, vitamin E and/or chlorpyrifos on the
dynamics of incline plane performance in Wistar rats.

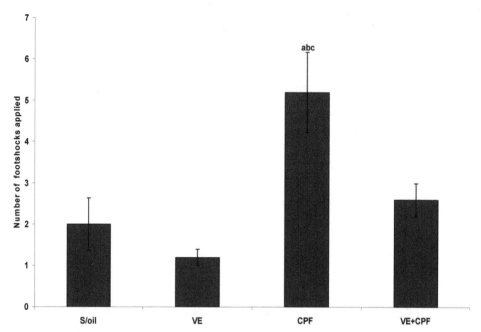

Fig. 5. Effect of chronic administration of soya oil, vitamin E and/or chlorpyrifos on the learning task in Wistar rats. [abc]P<0.01versus S/oil, VE and VE+CPF groups, respectively.

3.7 Effect of treatments on short-term memory
A significant decrease (P<0.01) in the duration of stay on platform (latency on platform) was recorded in the CPF group compared to that in the S/oil, VE or VE+CPF group. There was no significant change (P>0.05) in the duration of stay on the platform in the VE+CPF group compared to that in the S/oil or VE group (Fig. 6).

3.8 Effect of treatments on brain malonaldehyde concentration
A significant increase (P<0.01) in MDA concentration was recorded in the CPF group relative to that in the S/oil, VE or VE+CPF group. There was no significant change (P>0.05) in the brain MDA concentration in the VE+CPF group compared to that in S/oil or VE group, nor between VE and S/oil groups (Fig. 7).

3.9 Effect of treatments on brain superoxide dismutase activity
There was a significant decrease (P<0.01) in SOD activity in the CPF group relative to the S/oil, VE or VE+CPF group. No significant change (P>0.05) was recorded in SOD activity in the VE+CPF group relative to that in S/oil or VE group, nor between VE and that recorded in the S/oil group (Fig. 8).

3.10 Effect of treatments on brain catalase activity
A significant decrease (P<0.01) in brain CAT activity was recorded in the CPF group relative that in the S/oil, VE or VE+CPF group. The CAT activity in the VE+CPF group did not

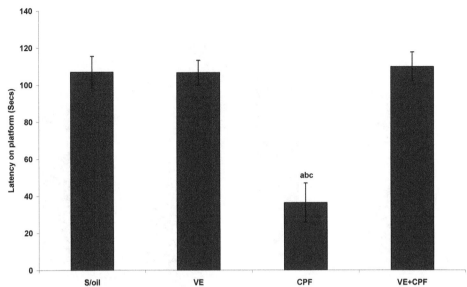

Fig. 6. Effect of chronic administration of soya oil, vitamin E and/or chlorpyrifos on short-
term memory in Wistar rats. abcP<0.01versus S/oil, VE and VE+CPF groups, respectively.

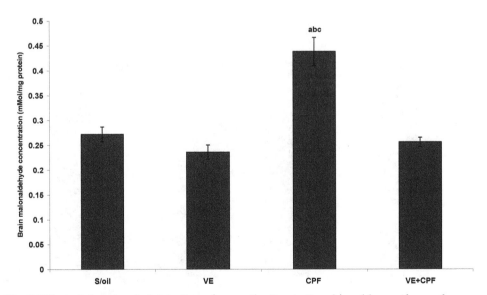

Fig. 7. Effect of chronic administration of soya oil, vitamin E and/or chlorpyrifos on the
brain malonaldehyde concentration in Wistar rats. abcP<0.01versus S/oil, VE and VE+CPF
groups, respectively.

differ significantly (P>0.05) when compared to that in the S/oil or VE group, and between VE and that recorded in the S/oil group (Fig. 9).

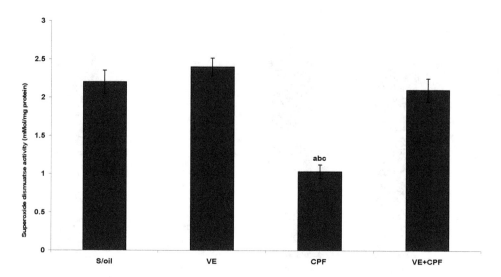

Fig. 8. Effect of chronic administration of soya oil, vitamin E and/or chlorpyrifos on the superoxide dismutase activity in Wistar rats. [abc]P<0.01versus S/oil, VE and VE+CPF groups, respectively.

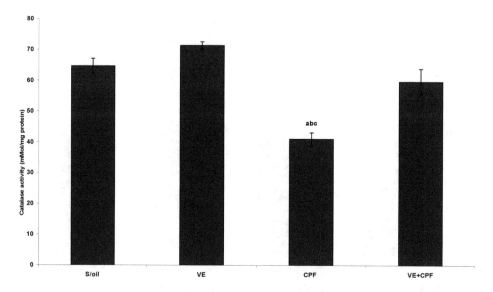

Fig. 9. Effect of chronic administration of soya oil, vitamin E and/or chlorpyrifos on the catalase activity in Wistar rats. [abc]P<0.01versus S/oil, VE and VE+CPF groups, respectively

Ameliorative Effect of Vitamin E on Sensorimotor and Cognitive Changes Induced by Chronic Chlorpyrifos
Exposure in Wistar Rats

245

3.11 Effect of treatments on brain acetylcholinesterase activity

There was a significant decrease in brain AChE activity in the CPF group compared to that in the S/oil ($P<0.01$), VE ($P<0.01$) or VE+CPF ($P<0.05$) group. There was no significant change ($P>0.05$) recorded in CAT activity in the VE+CPF relative to that in the S/oil and VE groups, respectively, or between VE and S/oil groups (Fig. 10).

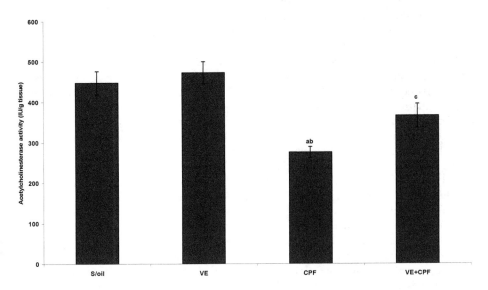

Fig. 10. Effect of chronic administration of soya oil, vitamin E and/or chlorpyrifos on the acetylcholinesterase activity in Wistar rats. [abc]$P<0.01$versus S/oil and VE groups, respectively; [c]$P<0.05$ versus VE group.

4. Discussion

The increase in brain MDA concentration and low SOD and CAT activities in the CPF group is an indication of the ability of this pesticide to elevate lipoperoxidative changes and thereby induce oxidative stress. This was in agreement with the findings from our previous studies (Ambali et al., 2010a; Ambali & Ayo, 2011a, 2011b; Ambali & Aliyu, 2012). The brain due to its biochemical and physiological properties is especially sensitive to free radicals, which destroy its functions and structure (Drewa et al., 1998). The brain is highly vulnerable to oxidative stress because in addition to harboring large amount of oxygen in a relatively small mass, it contains a significant quantity of metals (Fe), and has fewer antioxidant molecules than other organs (Halliwell and Gutteridge, 1999; Naffa-Mazzacoratt et al., 2001). For instance, the CNS is relatively poorly endowed with SOD, CAT, and glutathione peroxidase, and is also relatively lacking in vitamin E (Halliwell & Gutteridge, 1985). CPF is lipophilic and may enhance lipid peroxidation by directly interacting with cellular plasma membrane (Hazarika et al., 2003). The increased MDA concentration which is due to induction of free radical has been shown to alter the composition of membrane lipids, proteins, carbohydrates and DNA. Membrane lipids are vital for the maintenance of cellular integrity and survival (Jain, 1989). Peroxidation of membrane lipids results in the inactivation of enzymes and cross-linking of membrane lipids and proteins and in cell death (Pfafferott et al., 1982; Jain et al., 1983; Jain, 1984). Furthermore, by-products of lipid peroxidation have been shown to cause profound alterations in the structural organization and functions of the cell membrane including decreased membrane fluidity, increased membrane permeability, inactivation of membrane-bound enzymes and loss of essential fatty acids (Van Ginkel & Sevanian, 1994). This lipoperoxidative changes may cause alterations in the structural and functional components of the brain neuronal cells.

The decrease in the SOD and CAT activities in the CPF group has been reported in previous studies (Tuzmen et al., 2007, 2008; Aly et al., 2010; Ambali & Ayo, 2011a) and may reflect the level of oxidative damage caused by the pesticide. SOD is involved in dismutation of the $O_2\bullet-$ to H_2O_2 and oxygen. The significant reduction recorded in the CPF group may be due to either reduction in its synthesis or elevated degradation or inactivation of the enzyme. CAT, on the other hand is known to neutralize H_2O_2 and covert it to H_2O and O_2. The significant decline in the CAT activity observed in group exposed to CPF only may be due to the reduced conversion of $O_2\bullet-$ to H_2O_2 by SOD thereby resulting in the accumulation of $O_2\bullet-$. This accumulated $O_2\bullet-$ inhibits the activity of CAT (Kono & Fridovich, 1982). The decline in the activity of the antioxidant enzymes following chronic CPF exposure in the present study may be due to downregulation in the synthesis of antioxidant enzymes due to persistent toxicant insult (Irshad & Chaudhuri, 2002). Furthermore, $O_2\bullet-$ converts ferroxy state of CAT to ferryl state, which is an inactive form of the enzyme (Freeman & Crapo, 1982), thereby exacerbating the free radical-induced damage to the body tissue.

Pretreatment with vitamin E was shown by the present study to reduce the brain MDA concentration and increase the activities of the antioxidant enzymes, SOD and CAT reflecting its antioxidant properties. α-tocopherol prevents the peroxidation of membrane phospholipids and prevent cell membrane damage through its antioxidant action. The lipophilic character of tocopherol makes it easier to locate the interior of the cell membrane bilayer to exert its antioxidant action. Tocopherol-OH transfers a hydrogen atom with a single electron to a free radical, thus removing the radical before it can interact with the cell membrane (Krishnamoorty et al., 2007). The decreased lipoperoxidation of the membrane due to free radical scavenging effect of vitamin E may have been responsible for the restoration of SOD and CAT activities, since the vitamin may have prevent their full participation in free radical neutralization, hence preserving their activities.

The result also revealed that chronic CPF exposure caused reduction in the brain AChE activity similar to what has been reported in previous studies (Ambali et al., 2010a; Ambali & Ayo, 2011a, 2011b; Ambali & Aliyu, 2012). The ability of CPF to phosphorylate AChE results in impairment of its activity, hence the cholinergic crisis. Apart from this, the induction of lipoperoxidation may have partly contributed to the impaired AChE activity recorded in the CPF group. Oxidative stress affects the activities of various membrane-bound enzymes, including AChE (Mehta et al., 2005) via their direct attack by free radicals or peroxidation of the membrane lipids in which they are embedded (Souza et al., 2010). Besides, OH° has been shown to cause significant reduction in AChE activity in the rat brain (Tsakiris et al., 2000). Vitamin E was shown in the present study to restore the activity of AChE probably due to its antioxidant activity. Vitamin E has been shown in previous studies to restore AChE activity impaired by CPF (Yavuz et al., 2004; Ambali & Aliyu, 2012). The lacrimation and intermittent tremors observed in the CPF group is part of the cholinergic syndrome typical of OP insecticides (Eaton et al., 2008). These cholinergic signs were due to inhibition of AChE by CPF, resulting in accumulation of ACh in the muscarinic

Ameliorative Effect of Vitamin E on Sensorimotor and Cognitive Changes Induced by Chronic Chlorpyrifos
Exposure in Wistar Rats

247

and nicotinic cholinergic receptors. The ability of vitamin E to remedy the CPF-induced cholinergic signs may be attributed its AChE restoration activity. Furthermore, vitamin E has been shown to increase the activity of paraoxonase 1 (Jarvik et al., 2002), an enzyme that increases the detoxification of OP compounds (Shih et al., 1998).

Beam walking across bridges of different cross-sections provides a well-established method of monitoring motor coordination and balance in rodents. The progressive increase in the width at which rats in the CPF group slipped off the beam which indicates impairment of motor coordination has been reported in previous studies (Ambali et al., 2010a; Ambali & Aliyu, 2012). Abou-Donia et al. (2002) observed similar results following repeated exposure of rats to sarin. Beam-walking performance is an integrated form of behavior requiring pertinent level of consciousness, memory, sensorimotor and cortical functions mediated by the cortical area (Abou-Donia et al., 2001). Cortical injury may therefore have been responsible for the deficit in beam-walk performance in the CPF group (Abou-Donia et al., 2001) partly due to oxidative damage. Indeed, CPF and CPF-oxon have been shown to induce apoptosis in rat cortical neuron independent of AChE inhibition (Caughlan et al., 2004). Pretreatment with vitamin E mitigated but did not completely abolish the motor coordination deficits induced by chronic CPF exposure. This is because there was a significant increase in the width at which the VE+CPF group slipped off the beam at week 16 compared to day 0. This shows that oxidative stress may not be the only mechanism involved in motor coordination deficits induced by chronic CPF exposure.

The present study has also shown a significant reduction in forepaw grip time, reflecting deficit in forepaw motor strength following chronic CPF exposure in rats. The result agreed with the finding obtained in an earlier study which showed reduction in hind limb grip strength following repeated CPF administration in rats (Terry et al., 2003). The impairment of motor strength by CPF may have also been due to the decrease in anterograde axonal transport (Terry et al., 2007) or reduced neuronal viability associated with impaired microtubule synthesis and/or function (Prendergast et al., 2007). It has also been postulated that disruption of kinesin-dependent intracellular transport may account for some of the long-term effects of OPs on the peripheral and central nervous system (Gearharta et al., 2007). Reduced hand strength (Miranda et al., 2004) and loss of muscle strength (Steenland et al., 2000) have been observed in humans following prolonged exposure to OPs. Relationship has also been established between higher OP exposure and the development of chronic fatigue syndrome (Tahmaz et al., 2003). Furthermore, the role of muscle (Ambali and Ayo, 2011b) and brain oxidative damage induced by CPF which causes impairment of neuronal viability (Ambali & Ayo, 2011a) hence reduction of motor strength cannot be over emphasized. Although there was a significant deficit in motor strength in the VE+CPF group at weeks 16 and 8 when respectively compared to day 0, the fact that there was no significant change especially at week 16 compared to S/oil and VE groups reflect improvement in motor strength in this group. This may be partly due to reduced brain and perhaps muscle oxidative damage complemented by improvement in AChE activity which improves neuronal transmission.

Chronic CPF exposure has been shown in the present study to interfere with neuromuscular coordination as shown by the decline in the incline plane performance at weeks 8 and 16. The inclined plane test has been used to evaluate integrated muscle function and strength in rodents by evaluating their ability to maintain body position on a board as its angle of inclination is increased. We have earlier demonstrated the ability of acute CPF exposure to impair short-term neuromuscular coordination (Ambali et al., 2010a; Ambali & Aliyu, 2012).

Abou-Donia et al. (2002) similarly showed the ability of the OP warfare agent, sarin to impair incline plane performance in rats. The impairment of neuromuscular coordination may be due to increase in brain oxidative changes induced by CPF, which alters the morphological and functional capacity of the brain region involved in neuromuscular coordination. Oxidative damage to the brain following CPF exposure has been reported in previous studies (Verma, 2001; Ambali et al., 2010a; Ambali & Ayo, 2011a, 2011b; Ambali & Aliyu, 2012). Furthermore, the reduction of AChE activity may have been partly involved in the impaired neuromuscular coordination recorded in the CPF group, since alterations in ACh metabolism may alter neuronal activity.

Although the incline plane performance in the group pretreated with vitamin E at week 16 was significantly lower than that obtained at day 0, the study generally showed that performance in weeks 16 and 8 in the VE+CPF group was not significantly different from that of S/oil or VE group. This shows that the vitamin mitigated the CPF-evoked deficit in neuromuscular coordination. The fact that vitamin E did not completely abolish the CPF-induced impaired incline plane performance shows that oxidative stress and restoration of AChE activity may not be the only factor responsible for the sensorimotor deficit.

The lower ladder score characterized by lower number of missed rungs observed in rats chronically exposed to CPF indicates that the legs of the rats were frequently being held stationary above the rungs for a relatively longer period. This observation demonstrated difficulty in the ability of CPF group to move fast through the obstacles, and hence a deficit in locomotor activity. The deficit in locomotor efficiency observed in the CPF group was dependent on the duration of exposure, with much more impairment recorded at week 16 compared to week 8. The results agreed with the previous findings that slowness of movement is one of the extrapyramidal symptoms (Parkinsonism) observed in humans exposed to non-specific agricultural pesticides, which increased with the duration of exposure (Ritz & Yu, 2000; Alavanja et al., 2004). Thus, the locomotion deficit in the CPF group observed in the present study is part of the sensorimotor deficits occurring in animals chronically exposed to CPF. This impaired mobility may be due to oxidative stress as oxidative damage to the muscle induced by CPF (Ambali & Ayo, 2011b) may have probably caused necrosis thereby impairing locomotion efficiency. Carr et al. (2001) attributed reduced mobility observed in OP poisoning partly to damage in the peripheral musculature, probably due to necrosis of skeletal muscle fibre. Muscle necrosis has been observed following exposure to the OP insecticide, isofenphos and the insecticide metabolite, paraoxon (Dettbarn, 1984; Calore et al., 1999). Similarly, the impaired mobility may be due to inhibition of AChE activity and the subsequent cholinergic paralysis induced by CPF. The severity of the muscle necrosis may be dependent on the level and duration of AChE inhibition (Carr et al., 2001). The amelioration of the locomotor deficits manifested in the improvement of ladder walk and characterized by increase in the number of missed rungs in rats pretreated with vitamin E demonstrated the important role played by oxidative stress and AChE inhibition in the locomotor deficit induced by CPF.

The significant increase in the number of footshocks received by the CPF group relative to the other groups indicates learning impairment. Similarly, the significant reduction in the duration the animal in the CPF group stayed on the platform indicates deficit in memory. This shows that CPF exposure even at low dose is capable of cognitive impairment. CPF-induced cognitive impairment have been reported in several studies in rats (Bushnell et al. 1991; 1994; Prendergast et al., 1997, 1998, 2007; Stone et al., 2000, Moser et al., 2005; Ambali et al., 2010a; Ambali & Aliyu, 2012). In addition, studies in humans have shown persistent cognitive deficits in farmers and pesticide applicators repeatedly exposed to OPs but are symptom-free (Steenland et al., 2000; Dick et al., 2001). The impairment of cognition observed in the CPF group may be due to alteration in ACh metabolism due to reduction of AChE activity. Since ACh has been demonstrated to be involved in cognition, agents such as OPs which alter ACh metabolism may interfere with this role. Many studies have linked central cholinergic system to synaptic plasticity, learning and memory processes (Baskerville et al., 1997; Sachdev et al., 1998). It is believed that OP compounds play a role in memory loss by producing cholinergic dysfunction at the level of the synapse (Carr & Chambers, 1991).

Furthermore, CPF has been shown to induce cytotoxicity directly on the hippocampal cells via the induction of apoptosis, irrespective of its effect on AChE (Terry et al., 2003). Induction of apoptosis has been described as the toxic end-point of CPF neurotoxicity in the brain as it induces structural changes in the brain that may cause functional deficits, including those involved in memory and learning (Caughlan et al., 2004). Apoptosis probably resulting from oxidative damage to cellular macromolecules may have been responsible for the massive degenerative changes in the brain neurons and glial cells of rats chronically exposed to CPF that we reported in an earlier study (Ambali & Ayo, 2011a). CPF-induced oxidative stress may be central to apoptosis, since free radicals have been implicated in apoptotic death of cells (Corcoran et al., 1994; McConkey et al., 1994). Degenerative changes in the neurons leads to functional deficits as it relates to neurotransmission and other brain activities.

Vitamin E has been shown in the present study to improve learning and short-term memory impaired by chronic CPF exposure. We have earlier demonstrated the ability of either vitamin C or E to mitigate short-term cognitive changes induced by acute CPF exposure in rats (Ambali et al., 2010a; Ambali and Aliyu, 2012). The improved learning and short-term memory recorded following pretreatment with vitamin E may be due to its antioxidant and AChE restoration properties. Apart from its antioxidant function, vitamin E influences the cellular response to oxidative stress through modulation of signal-transduction pathways (Azzi et al., 1992), which may have further enhanced the neuronal function. Similarly, neuroprotective effect of vitamin E has been established in several studies (Frantseva et al., 2000a, 2000b; Pace et al., 2003; El-Hossary et al., 2009) and may have contributed in mitigating the behavioural changes induced by CPF in the present study.

5. Conclusion

The present study has shown that the impaired sensorimotor and cognitive changes induced by chronic CPF exposure mitigated by pretreatment with vitamin E are partly due to its antioxidant, neuroprotective and AChE restoration properties.

6. References

Abou-Donia, M.B. (1992). Introduction. *In: Neurotoxicology,* M.B. Abou-Donia (Ed.), 3-24 CRC Press, Boca Raton, FL.

Abou-Donia, M.B.; Dechkovskaia, A.M; Goldstein, L.B.; Bullman S.L. & Khan, W.A. (2002). Sensorimotor deficit and cholinergic changes following coexposure with pyridostigmine bromide and sarin in rats. *Toxicological Sciences*, Vol. 66, pp. 148–158.

Abou-Donia, M.B.; Goldstein, L.B.; Jones, K.H.; Abdel-Rahaman, A.A.; Damodaran, T.; Dechkovskaia, A.M.; Bullman, S.L.; Amir, B.E. & Khan, W.A. (2001). Locomotor and sensorimotor performance deficit in rats following exposure to pyridostigmine bromide, DEET and permethrin alone and in combination. *Toxicological Sciences*, Vol. 60, pp. 305-314.

Alavanja, M.C.; Hoppin, J.A.; & Kamel, F. (2004). Health effects of chronic pesticide exposure: cancer and neurotoxicity. *Annual Review of Public Health*, Vol. 25, pp. 155-197.

Aly, N.; EL-Gendy, K.; Mahmoud F.; & El-Sebae, A.K. (2010). Protective effect of vitamin C against chlorpyrifos oxidative stress in male mice. *Pesticide Biochemistry and Physiology*, Vol. 97, pp. 7–12.

Ambali, S.F. (2009). Ameliorative effect of vitamin C and E on neurotoxicological, hematological and biochemical changes induced by chronic chlorpyrifos administration in Wistar rats. *PhD Dissertation*, Ahmadu Bello University, Zaria, Nigeria, 355pp.

Ambali, S.F. & Aliyu, M.B. (2012). Short-term sensorimotor and cognitive changes induce by acute chlorpyrifos exposure: Ameliorative effect of vitamin E. *Pharmacologia*, Vol 3, No 2, pp. 31-38.

Ambali, S.F.& Ayo, J.O. (2011a) Sensorimotor performance deficits induced by chronic chlorpyrifos exposure in Wistar rats: mitigative effect of vitamin C. *Toxicological and Environmental Chemistry*, Vol. 93, No 6, pp. 1212–1226.

Ambali, S.F. & Ayo, J.O. (2011b). Vitamin C attenuates chronic chlorpyrifos-induced alteration of neurobehavioural parameters in Wistar rats. *Toxicology International* (Accepted manuscript).

Ambali, S.F.; Ayo, J.O.; Ojo, S.A. & Esievo, K.A.N. (2010b). Vitamin E protects rats from chlorpyrifos-induced increased erythrocyte osmotic fragility in Wistar rats. *Food and Chemical Toxicology*, Vol. 48, pp. 3477-3480.

Ambali, S.F.; Idris, S.B.; Onukak, C.; Shittu, M. & Ayo, J.O. (2010a). Ameliorative effects of vitamin C on short-term sensorimotor and cognitive changes induced by acute chlorpyrifos exposure in Wistar rats. *Toxicology and Industrial Health*, Vol. 26, No. 9, pp. 547-558.

Azzi, A.; Boscobonik, D. & Hensey, C. (1992). The protein kinase C family. *European Journal of Biochemistry*, Vol. 208, pp. 547-557.

Bagchi, D.; Bagchi, M.; Hassoun, E.A. & Stohs, S.J. (1995). *In vitro* and *in vivo* generation of reactive oxygen species, DNA damage and lactate dehydrogenase leakage by selected pesticides. *Toxicology*, Vol. 104, pp. 129-140..

Baskerville, K.A.; Schweitzer, J.B. & Herron, P. (1997). Effects of cholinergic depletion on experience dependent plasticity in the cortex of the rat. *Neuroscience* Vol. 80, pp. 1159-1169.

Bazylewicz-Walczak, B.; Majczakowa, W. & Szymczak, M. (1992). Behavioural effects of occupational exposure to organophosphorous pesticides in female greenhouse planting workers. *Neurotoxicology*, Vol. 20, pp. 819-826.

Bushnell, P. J.; Padilla, S. S.; Ward, T.; Pope, C. N. & Olszyk, V. B. (1991). Behavioural and neurochemical changes in rats dosed repeatedly with diisopropylfluorophos-phate. *Journal of Pharmacology and Experimental Therapeutics*, Vol. 256, pp. 741-750.

Bushnell, P.J.; Kelly, K.C. & Ward, T.R. (1994). Repeated inhibition of cholinesterase by chlorpyrifos in rats: behavioural, neurochemical and pharmacological indices of tolerance. *Journal of Pharmacology and Experimental Therapeutics*, Vol. 270, pp. 15-25.

Calore, E.E.; Sesso, A.; Puga, F.R.; Cavaliere, M.J.; Calore, N.M. & Weg, R. (1999). Early expression of ubiquitin in myofibres of rats in organophosphate intoxication. *Ecotoxicology and Environmental Safety*, Vol. 43, pp. 187-194.

Cañadas, F.; Cardona, D.; Dávila, E.; Sánchez-Santed, F. (2005). Long-term neurotoxicity of chlorpyrifos: spatial learning impairment on repeated acquisition in a water maze. *Toxicological Sciences*, Vol. 85, pp.944-951.

Carlock, L.L.; Chen, W.L.; Gordon, E.B.; Killeen, J. C.; Manley, A.; Meyer, L.S.; Mullin, L.S.; Pendino, K.J.; Percy, A.; Sargent, D.E.; & Seaman, L.R. (1999). Regulating and assessing risks of cholinesterase-inhibiting pesticides: Divergent approaches and interpretations. *Journal of Toxicology and Environmental. Health B* 2, pp. 105-160.

Carr, R.L. & Chambers, J.E. (1991). Acute effects of the organophosphate paraoxon on schedule-controlled behaviour and esterase activity in rats: Dose-response relationships. *Pharmacology Biochemistry and Behaviour*, Vol. 40, pp. 929-936.

Carr, R.L.; Chambers, H.W.; Guansco, J.A.; Richardson, J.R.; Tang, J. & Chambers, J.E. (2001). Effect of repeated open-field behaviour in juvenile rats. *Toxicological Sciences*, 59: 260-267.

Casida, J.E. and Quistad, G.B. (2004). Organophosphate toxicology: Safety aspects of non-acetylcholinesterase secondary targets. *Chemical Research in Toxicology*, 17: 983-898.

Caughlan, A.; Newhouse, K.; Namgung, U. & Xia, Z. (2004). Chlorpyrifos induces apoptosis in rat cortical neurons that is regulated by a balance between p38 and ERK/JNK MAP kinases. *Toxicological Sciences*, Vol. 78, pp. 125-134.

Chakraborti, T.K.; Farrar, J.D. & Pope, C.N. (1993). Comparative neurochemical and neurobehavioural effects of repeated chlorpyrifos exposures in young rats. *Pharmacology Biochemistry and Behaviour*, Vol. 46, pp. 219-224.

Clegg, D. J. & van Gemert, M. (1999). Expert panel report of human studies on chlorpyrifos and/or other organophosphate exposures. *Journal of Toxicology and Environmental Health B* 2, pp. 257-279.

Colborn, T. (2006). A case for revisiting the safety of pesticides: A closer look at neurodevelopment. *Environmental Health Perspectives* Vol. 114, pp. 10-17.

Corcoran, G.B.; Fix, L.; Jones, D.P.; Moslen, M.T.; Nicotera, P.; Oberhammer, F.A. & Buttyan, R. (1994). Apoptosis: Molecular control point in toxicity. *Toxicology and Applied Pharmacology*, Vol. 128, pp. 169-181.

Costa, L.G.; Giordano, G.; Guizzetti M. & Vitalone A. (2008). Neurotoxicity of pesticides: a brief review. *Frontiers in Bioscience*, 13:1240-1249.

Dettbarn, W.D. (1984). Pesticide-induced muscle necrosis: mechanisms and prevention. *Fundamental and Applied Toxicology*, Vol. 4, pp. S18-S26.

Dick, R.B.; Steenland, K.; Krieg, E.F. & Hines, C.J. (2001). Evaluation of acute sensory-motor effects and test sensitivity using termiticide workers exposed to chlorpyrifos. *Neurotoxicology and Teratology*, Vol. 23, pp. 381-393.

Dietrich, K.N.; Eskenazi, B.; Schantz, S.; Yolton, K.; Rauh, V. A.; Johnson, C. B.; Alkon, A.; Canfield, R.L.; Pessah, I.N. & Berman, R.F. (2005). Principles and practices of neurodevelopmental assessment in children: Lessons learned from the centers for children's environmental health and disease prevention research. *Environmental Health Perspectives*, Vol. 113; pp. 1437-1446.

Draper, H.H. & Hadley, M. (1990). Malondialdehyde determination as index of lipid peroxidation. *Methods in Enzymology*, Vol. 186, pp. 421-431.

Drewa, G.; Jakbczyk, M. & Araszkiewicz, A. (1998). Role of free 1 radicals in schizophrenia. 2 *Medical Science Monitoring*, Vol. 4, No. 6, pp. 1111-1115..

Eaton, D.L.; Daroff, R.B.; Autrup, H.; Buffler, P.; Costa, L.G.; Coyle, J.; Mckhann, G.; Mobley, W.C.; Nadel, L.; Neubert, D.; Schukte-Hermann, R.; Peter, S. & Spencer, P.S. (2008). Review of the toxicology of chlorpyrifos with an emphasis on human exposure and neurodevelopment. *Critical Reviews in Toxicology* S2, pp. 1-125.

El-Hossary, G.G.; Mansour, S.M. & Mohamed, A.S. (2009). Neurotoxic effects of chlorpyrifos and the possible protective role of antioxidant supplements: an experimental study. *Journal of Applied Science Research*, Vol. 5, No. 9, pp. 1218-1222.

Farag, A.T.; Radwana, A.H.; Sorourb, F.; El Okazyc A.; El-Agamyd, E. & El-Sebae, A. (2010). Chlorpyrifos induced reproductive toxicity in male mice. *Reproductive Toxicology*, Vol. 29, pp. 80-85.

Fiedler, N.; Kipen, H.; Kelly-McNeil, K. & Fenske, R. (1997). Long-term use of organophosphates and neuropsychological performance. *American Journal of Industrial Medicine*, Vol. 32, pp. 487-496.

Frantseva, M.V.; Valazquez, J.L.; Hwang, P.A. & Carlen, P.L. (2000a). Free radicals production correlates with cell death in an *in vitro* model of epilepsy. *European Journal of Neuroscience*, Vol. 12, pp. 1413-1419.

Frantseva, M.V.; Valazquez, J.L.; Tsoraklidis, G.; Mendonca, A.J.; Adamchik, Y.; Mills, L.R.; Carlen, P.L. & Burnham, M.V. (2000b). Oxidative stress in involved in seizure-induced neurodegeneration in the kindling model of epilepsy. *Neuroscience*, Vol. 97, pp. 431-435.

Freeman, B.A. & Crapo, J.D. (1982). Biology of disease: Free radicals and tissue injury. *Laboratory Investigations*, Vol. 47, pp. 412-426.

Freitas, R.M.; Vasconcelos, S.M.M.; de Souza, F.G.F.; Viana, G.S.B. & Fonteles, M.M.F. (2005). Oxidative stress in the hippocampus after pilocarpine induced status epilepticus in Wistar rats. *FEBS Journal*, Vol. 272, pp. 1307-1312.

Gearharta, D.A.; Sicklesb, D.W.; Buccafuscoa, J.J.; Prendergast, M.A. & Terry, Jr, A.V. (2007). Chlorpyrifos, chlorpyrifos-oxon, and diisopropylfluorophosphate inhibit kinesin-dependent microtubule motility. *Toxicology and Applied Pharmacology*, Vol. 218, No.1, pp. 20-29.

Guide for the care and use of laboratory animals, DHEW Publication No. (NIH) 85-23, Revised 1985, Office of Science and Health Reports, DRR/NIH, Bethesda, MD 20892.

Gultekin F.; Ozturk, M. & Akdogan, M. (2000). The effect of organophosphate insecticide chlorpyrifos–ethyl on lipid peroxidation and antioxidant enzymes (in-vitro). *Archives of Toxicology*, Vol. 74, pp. 533- 538.

Gultekin, F.; Delibas, N.; Yasar, S. & Kilinc, I. (2001). *In vivo* changes in antioxidant systems and protective role of melatonin and a combination of vitamin C and vitamin E on oxidative damage in erythrocytes induced by chlorpyrifos-ethyl in rats. *Archives of Toxicology*, Vol. 75, No. 2, pp. 88-96.

Gultekin, F.; Karakoyun, I.; Sutcu, R.; Savik, E.; Cesur, G.; Orhan, H. & Delibas, N. (2007). Chlorpyrifos increases the levels of hippocampal NMDA receptor subunits NR2A and NR2B in juvenile and adult rats. *International Journal of Neuroscience*, Vol. 117, No. 1, pp. 47-62.

Halliwell, B. (1994). Free radicals, antioxidants and human disease: curiosity, cause or consequence? *Lancet*, Vol. 344, pp. 721-724.

Halliwell, B. & Gutteridge, J. C. (1999). *Free Radicals in Biology and Medicine*, 3rd ed., Oxford University Press, London, England.

Halliwell, B. & Gutteridge, J.M.C. (1985). Oxygen radicals and the nervous system. *Trends in Neuroscience*, Vol. 8, pp. 22-26.

Hazarika, A.; Sarkar, S.N.; Hajare, S.; Kataria, M. & Malik, J.K. (2003). Influence of malathion pretreatment on the toxicity of anilofos in male rats: a biochemical interaction study. *Toxicology*, Vol. 185, No. 1-2, pp. 1-8.

He, F. (2000). Neurotoxic effects of insecticides—Current and future research: A review. *Neurotoxicology*, Vol. 21, pp. 829–835.

Hill, R.; Head, S.; Baker, S.; Gregg, M.; Shealy, D.; Bailey, S.; Williams, C.; Sampson, E. & Needham, L. (1995). Pesticide residues in urine of adults living in the United States: reference range concentrations. *Environmental Research*, Vol. 71, pp. 88-108.

Hsu, P. C. & Guo, Y. L. (2002): Antioxidant nutrients and lead toxicity. *Toxicology*, Vol. 180, pp. 33 - 44.

Irshad, M. & Chaudhuri, B.S. (2002). Oxidant-antioxidant system: role and significance in human body. *Indian Journal of Experimental Biology*, Vol. 40, pp. 1233–1239.

Jain, S.K. (1989). Hyperglycaemia can cause membrane lipid peroxidation and osmotic fragility in human red blood cells. *Journal of Biological Chemistry*, Vol. 264, No. 35, pp. 21340-21345.

Jain, S.K. (1984). The accumulation of malonyldialdehyde, a product of fatty acid peroxidation, can disturb aminophospholipid organization in the membrane bilayer of human erythrocytes. *Journal of Biological Chemistry*, Vol. 259, pp. 3391-3394.

Jain, S.K.; Mohandas, N.; Clark, M.R. & Shohet, S.B. (1983). The effect of malonyldialdehyde, a product of lipid peroxidation, on the deformability, dehydration and 51Cr-survival of erythrocytes. *British Journal of Haematology*, Vol. 53, pp. 247-255.

Jarvik, G.P.; Tsai, T.N.; McKinstry, L.A.; Wani, R.; Brophy, V.; Richter, R.J.; Schellenberg, G.D.; Heagerty, P.J.; Hatsukami, T. & Furlong, C.E. (2002). Vitamin C and E intake is associated with increase paraoxonase activity. *Arterioscleriosis, Thrombosis and Vascular Biology,* Vol. 22, pp. 1329 -1333.

Kamel, F.; Engel, L.S.; Gladen, B.C.; Hoppin, J.A.; Alavanja, M.C.R. & Sandler, S.P. (2007). Neurologic symptoms in licensed pesticide applicators in the agricultural health study. *Human and Experimental Toxicology,* Vol. 26, pp. 243-250.

Kamel, F. and Hoppin, J.A. (2004). Association of pesticide exposure with neurologic dysfunction and disease. *Environmental Health Perspectives,* Vol. 112, No. 9, pp. 950-958.

Kamel, F.; Rowland, A.S.; Park, L.P.; Anger, W.K.; Baird, D.D.; Gladen, B.C.; Moreno, T.; Stallone, L. & Sandler, D.P. (2003). Neurobehavioural performance and work experience in Florida farmworkers. *Environmental Health Perspectives,* Vol. 111, pp. 765-772.

Kehrer, J.P. (1993). Free radicals as mediators of tissue injury and disease. *Critical Reviews in Toxicology,* Vol. 23, No 1, pp. 21-48.

Kingston, R.L.; Chen, W.L.; Borron, S.W.; Sioris, L.J.; Harris, C.R. & Engebretsen, K.M. (1999). Chlorpyrifos: a ten-year U.S. poison center exposure experience. *Veterinary and Human Toxicology,* Vol. 41, pp. 87-92.

Kono, Y. & Fridovich I. (1982). Superoxide radical inhibits catalase. *Biological Chemistry,* Vol. 257, pp. 5751-5754.

Krishnamoorthy, G.; Ventaraman, P.; Arunkumar, A.; Vignesh, R. C.; Aruldhas, M. M. & Arunakaran, J. (2007). Ameliorative effect of vitamins (α–tocopherol and ascorbic acid) on PCB (Aroclor 1254)-induced oxidative stress in rat epididymal sperm. *Reproductive Toxicology,* Vol. 23, pp. 239-245.

Iwasaki, M.; Sato, I.; Jin, Y.; Saito, N. & Tsoda, S. (2007). Problems of positive list system revealed by survey of pesticide residue in food. *Journal of Toxicological Sciences,* Vol. 32, No. 2, pp. 179-184.

Lizardi, P.S.; O'Rourke, M.K. & Morris, R.J. (2008). The effects of organophosphate pesticide exposure on hispanic children's cognitive and behavioral functioning. *Journal of Pediatric Psychology,* Vol. 33, No. 1, pp. 91–101.

Lotti M. (2000). *Experimental and Clinical Neurotoxicology.* 2nd Ed., Oxford University Press, New York.

Lowry, H.; Rosebrough, N.J.; Farr, A.L. & Randall, R.J. (1951). Protein measurements with the folin phenol reagent. *Journal of Biological Chemistry,* Vol. 193, pp. 265–275.

Maxwell, S.R. (1995): Prospects for the use of antioxidants therapies. *Drugs,* Vol. 49, pp. 345.

McConkey, D.J., Jondal M.B. and Orrenius, S.G. (1994). Chemical-induced Apoptosis in the Immune System. In: *Immunotoxicology and Immunopharmacology,* (J.H., Dean, M.I., Luster, A.E., Munson & I. Kimber, (Eds.), 473-485, 2nd Edition, Raven Press Ltd. NewYork.

Mehta, A.; Verma, R.S. & Vasthava S. (2005). Chlorpyrifos-induced alterations in rat brain acetylcholine esterase, lipid peroxidation and ATPase. *Indian Journal of Biochemistry and Biophysics,* Vol. 42, pp. 54-58.

Miranda, J.; McConnell, R.; Wesseling, C.; Cuadra, R.; Delgado, E.; Torres, E.; Keifer, M. & Lundberg, I. (2004). Muscular strength and vibration thresholds during two years after acute poisoning with organophosphate insecticides. *Occupational and Environmental Medicine*, Vol. 61, No. 1, pp. e4.

Moser, V.C.; Phillips, P.M.; McDaniel, K.L.; Marshall, R.S.; Hunter, D.L. & Padilla, S. (2005). Neurobehavioural effects of chronic dietary and repeated highlevel spike exposure to chlorpyrifos in rats. *Toxicological Sciences*, Vol. 86, pp. 375-386.

Naffah-Mazzacoratti, M.G.; Cavalheiro, E.A.; Ferreira, E.C.; Abdalla, D.S.P.; Amado, D. & Bellissimo, M.I. (2001). Superoxide dismutase, glutathione peroxidase activities and the hydroperoxide concentration are modified in the hippocampus of epileptic rats. *Epilepsy Research*, Vol. 46, pp. 121-128.

Osfor, M.M.H.; Ibrahim, H.S.; Mohamed, Y.A.; Ahmed, S.M.; Abd El Azeem, A.S. & Hegazy, A.M. (2010). Effect of Alpha Lipoic Acid and Vitamin E on Heavy Metals Intoxication in Male Albino Rats. *Journal of American Science*, Vol. 6, No. 8, pp. 6-63.

Pace, A.; Savarese, A.; Picardo, M.; Maresca V.; Pacetti, U.; Del Monte, G.; Biroccio, A.; Leonetti, C.; Jandolo, B.; Cognetti, F. & Bove, L. (2003). Neuroprotective effect of vitamin e supplementation in patients treated with cisplatin chemotherapy. *Journal of Clinical Oncology*, Vol. 21, pp. 927-931.

Pancetti, F.; Olmos, C.; Dagnino-Subiabre, A.; Rozas, C. & Morales, B. (2007). Noncholinesterase effects induced by organophosphate pesticides and their relationship to cognitive processes: implication for the action of acylpeptide hydrolase. *Journal of Toxicology and Environmental Health, Part B*, Vol. 10, pp. 623–630.

Pfafferott, C.; Meiselman, H.J. and Hochstein, P. (1982). The effect of malonyldialdehyde on erythrocyte deformability. *Blood*, Vol. 59, pp. 12-15.

Pope, C.N. (1999). Organophosphorus pesticides: do they all have the same mechanism of toxicity? *Journal of Toxicology and Environmental Health*, Vol. 2, pp. 161-181.

Pope, C.N.; Chakraborti, T.K.; Chapman, M.L. & Farrar J.D. (1992). Long-term neurochemical and behavioural effects induced by acute chlorpyrifos treatment. *Pharmacology, Biochemistry and Behaviour*, Vol. 42, pp. 251-256.

Prendergast, M.A.; Self, S.L.; Smith, K.J.; Ghayoumi, L.; Mullins, M.M.; Butler, T.R.; Buccafusco, J.J.; Gearhart, D.A. & Terry, A.V. Jr. (2007). Microtubule-associated targets in chlorpyrifos oxon hippocampal neurotoxicity. *Neuroscience*, Vol. 146, No. 1, pp. 330-339.

Prendergast, M.A.; Terry, A.V. Jr. & Buccafusco, J.J. (1997). Chronic, low-level exposure to diisopropyl fluorophosphates causes protracted impairment of spatial navigation learning. *Psychopharmacology* (Berl), Vol. 130, pp. 276-284.

Prendergast, M.A.; Terry, A.V. Jr. & Buccafusco, J.J. (1998). Effects of chronic low-level organophosphate exposure on delayed recall, discrimination and spatial learning in monkeys and rats. *Neurotoxicology and Teratolology*, Vol. 20, pp. 115-122.

Qiao, D.; Seidler, F.J.; Tate, C. A.; Cousins, M. M. & Slotkin, T. A. (2003). Fetal chlorpyrifos exposure: Adverse effects on brain cell development and cholinergic biomarkers emerge postnatally and continue into adolescence and adulthood. *Environmental Health Perspectives*, Vol. 11, pp. 536–544.

Rack, K.D. (1993). Environmental fate of chlorpyrifos. *Review of Environmental Contamination and Toxicology*, Vol. 131, pp. 1–150.

Ritz, B. & Yu, F. (2000). Parkinson's disease mortality and pesticide exposure in California 1984-1994. *International Journal of Epidemiology*, Vol. 29, pp. 323-329.

Sachdev, R.; Lu, S.; Wiley, R. & Ebner, F. (1998). Role of the basal forebrain cholinergic projection in somatosensory cortical plasticity. *Journal of Neurophysiology*, Vol. 79, pp. 3216-3228.

Sally, A.M.; Sharee, A.W., & Janet, D. (2003). What advanced practice Nurses Need to know about free radicals? *International Journal of Advanced Nursing Practice*, Vol. 6, pp. 1.

Sánchez-Santed, F.; Canâdas, F.; Flores, P.; Lo´pez-Grancha, M. & Cardona, D. 2004. Long-term functional neurotoxicity of paraoxon, and chlorpyrifos: Behavioural and pharmacological evidence. *Neurotoxicology and Teratology*, Vol. 26, pp. 305-317.

Saulsbury, M.D.; Heyliger, S.O.; Wang, K. & Johnson, D.J. (2009). Chlorpyrifos induces oxidative stress in oligodendrocyte progenitor cells. *Toxicology*, Vol. 259, pp. 1–9.

Shih, D.M.; Gu, L.; Xia, Y.R.; Navab, M.; Li, W.F.; Hama, S.; Castellani, L.W.; Furlong, C.E.; Costa, L.G.; Fogelman, A.M. & Lusis, A.J. (1998). Mice lacking serum paraoxonase are susceptible to organophosphate toxicity and atherosclerosis. *Nature*, Vol. 394, No. 6690, 284-287.

Sies, H. (1991). Oxidative stress: Introduction. *In* Sies, H. (Ed.), *Oxidative Stress: Oxidants and Antioxidants*, Vol 23., 21-48, Academic Press, San Diego, CA, USA.

Slotkin, T.A. (2004). Cholinergic systems in brain development and disruption by neurotoxicants: nicotine, environmental tobacco smoke, organophosphates. *Toxicology and Applied Pharmacology*, Vol. 198, pp. 132-151.

Slotkin, T.A. (2005). Developmental neurotoxicity of organophosphates: a case study of chlorpyrifos. In: *Toxicity of Organophosphate and Carbamate Pesticides*, (R.C. Gupta, Ed), , 293-314, Elsevier Academic Press, San Diego.

Slotkin, T.A.; Levin, E. D. & Seidler, F. J. (2006). Comparative developmental neurotoxicity of organophosphate insecticides: Effects on brain development are separable from systemic toxicity. *Environmental Health Perspectives*, Vol. 114, pp. 746–751.

Souza, G.F.; Saldanha G.B. & Freitas, R.M. (2010). Lipoic acid increases glutathione peroxidase, Na+, K+-ATPase and acetylcholinesterase activities in rat hippocampus after pilocarpine-induced seizures? *Arquivos de Neuro Psiquiatria*, Vol, 68, pp. 586–591.

Stamper, C.R.; Balduini, W.; Murphy, S.D. & Costa, L.G. (1988). Behavioral and biochemical effects of postnatal parathion exposure in the rat. *Neurotoxicology and Teratology*, Vol. 10, pp. 261-266.

Steenland, K.; Dick, R.B.; Howell, R.J.; Chrislip, D.W.; Hines, C.J.; Reid, T.M.; Lehman, E.; Laber, P.; Krieg, E.F. & Knott, C. (2000). Neurologic function among termiticide applicators exposed to chlorpyrifos. *Environmental Health Perspectives*, Vol. 108, No. 4, pp. 293-300.

Stephens, R.; Spurgeon, A.; Calvert, I.A.; Beach, J.; Levy, L.S.; Berry, H. & Harrington, J.
(1995). Neuropsychological effects of long-term exposure to organophosphates in
sheep dip. *Lancet*, Vol. 315, pp. 1135-1139.

Stone, J.D.; Terry, A.V. Jr.; Pauly, J.R.; Prendergast, M.A. & Buccafusco, J.J. (2000).
Protractive effects of chronic treatment with an acutely sub-toxic regimen of
diisopropylflurophosphate on the expression of cholinergic receptor densities in
rats. *Brain Research*, Vol. 882, pp. 9–18.

Tahmaz, N.; Soutar, A. & Cherrie, J.W. (2003). Chronic fatigue and organophosphate
pesticides in sheep farming: A retrospective study amongst people reporting to a
UK pharmacovigilance scheme. *Annals of Occupational Hygiene*, Vol. 47, No. 4, pp.
261-267.

Terry, A. V. Jr.; Stone, J. D.; Buccafusco, J.J.; Sicles, D. W.; Sood, A. & Prendergast, M.A.
(2003). Repeated exposures to subthreshold doses of chlorpyrifos in rats:
Hippocampal damage, impaired axonal transport, and deficits in spatial learning.
Journal of Pharmacology and Experimental Therapeutics, Vol. 305, pp. 375-384.

Terry, A.V. Jr; Gearhart, D.A.; Beck, W.D. Jr.; Truan, J.N.; Middlemore, M.; Williamson,
L.N.; Bartlett, M.G.; Prendergast, M.A.; Sickles, D.W. & Buccafusco, J.J. (2007).
Chronic intermittent exposure to chlorpyrifos in rats: Protracted effects on axonal
transport, neurotrophin receptors, cholinergic markers, and information
processing. *Journal of Pharmacology and Experimental Therapeutics*, Vol. 322, pp.
1117-1128.

Tsakiris, S.; Angelogianni, P.; Schulpis, K.H. & Stavridis, J.C. (2000). Protective effect of L-
phenylalanine on rat brain acetylcholinesterase inhibition induced by free radicals.
Clinical Biochemistry, Vol. 33, No. 2, pp. 103-106.

Tuzmen, N.; Candan, N. & Kaya, E. (2007). The evaluation of altered antioxidative defense
mechanism and acetylcholinesterase activity in rat brain exposed to chlorpyrifos,
deltamethrin, and their combination. *Toxicology Mechanisms and Methods*, Vol. 17,
No. 535–540.

Tuzmen, N.; Candan, N.; Kaya, E. & Demiryas, N. (2008). Biochemical effects of chlorpyrifos
and deltamethrin on altered antioxidative defense mechanisms and
lipid peroxidation in rat liver. *Cell Biochemistry and Function*, Vol. 26, pp. 119-
124.

Van Ginkel, G. & Sevanian, A., (1994). Lipid peroxidation induced membrane structural
alterations. *Methods in Enzymology*, Vol. 233, pp. 273-288.

Verma, R.S. (2001). Chlorpyrifos-induced alterations in levels of thiobarbitunc acid reactive
substances and glutathione in rat brain. *Indian Journal of Experimental Biology*, Vol.
39, pp. 174-177.

Verma, R.S.; Mnugya, A. & Srivastava, N. (2007). *In vivo* chlorpyrifos induced oxidative
stress: attenuation by antioxidant vitamins. *Pesticides Biochemistry and Physiology*,
Vol. 88 pp. 191–196.

Weiss, B.; Amler, S. & Amler, R.W. (2004). Pesticides. *Pediatrics*, Vol. 113, pp. 1030-
1036.

Yavuz, T.; Delibao, N.; YÂldÂrÂm, B.; Altuntao, I.; CandÂr, O.; Cora, A.; Karahan, N.;
Ãbrioim, E. & Kutsal, A. (2004). Vascular wall damage in rats induced by

methidathion and ameliorating effect of vitamins E and C. *Archives of Toxicology,* Vol. 78, pp. 655-659.

Zhu, H.; Robin, W.; Rockhold, R.W.; Baker, R.C.; Kramer, R.E. & Ho, I.K. (2001). Effects of single or repeated dermal exposure to methyl parathion on behavior and blood cholinesterase activity in rats. *Journal of Biomedical Sciences,* Vol. 8, pp. 467-474.

13

Zero-Inflated Regression Methods for Insecticides

Abdullah Yeşilova[1], M. Salih Özgökçe[2] and Yılmaz Kaya[3]
[1]Yuzuncu Yil University, Faculty of Agriculture, Biometry & Genetic Unit, Van,
[2]Yuzuncu Yil University, Faculty of Agriculture, Plant Protection Department, Van
[3]Siirt University, Faculty of architecture and engineering,
Computer Engineering Department, Siirt,
Turkey

1. Introduction

The numerical abundance of many species sharing the same ecosystem very different levels of the organism and are in constant change, depending on many factors. Due to the heterogeneous strucspeciese of the life cycles of organisms and abiotic resources in the environment based on census population densities derived from overdispersion (variance is higher than means in Poisson distribution) (Cox, 1983; Cameron and Trivedi, 1998) and a large number of zero values (zero-inflated data) is observed (Yeşilova et al, 2011). In such a case, zero-inflated Poisson (ZIP) regression model is a appropriate approach for analyzing a dependent variable having excess zero observations (Lambert, 1992; Böhning, 1998; Böhning et al, 1999; Yau and Lee, 2001; Lee et al, 2001; Khoshgoftaar et al, 2005; Yeşilova et al, 2010). Zero-inflation is also likely in data sets, excess zero observations. In such cases, a zero-inflated negative binominal (ZINB) regression model is an alternative method (Ridout et al, 2001; Yau, 2001; Cheung, 2002; Jansakul, 2005; Long and Frese, 2006; Hilbe, 2007; Yeşilova et al, 2009; Yeşilova et al, 2010). Morever, The Poisson hurdle model and negative binomial hurdle model (Rose and Martin, 2006; Long and Frese, 2006; Hilbe, 2007; Yeşilova et al, 2009; Yeşilova et al, 2010), and zero-inflated generalized Poisson (ZIGP) model (Consul, 1989, Consul and Famoye, 1992; Czado et al., 2007) are widely used in the analysis of zero-inflated data.

In this part, the analysis of data with many zeros for *Notonecta viridis* Delcourt (Heteroptera: Notonectidae) and Chironomidae species (Diptera) were carried out by means of using the models of Poisson Regression (PR), negative binomial (NB) regression, zero-inflated Poisson (ZIP) regression, zero-inflated negative binomial (ZINB) regression and negative binomial hurdle (NBH) model.

Samplings

The study was based on periodical samplings of the coastal band of Van Lake, conducted between July-September 2005 and May-September 2006. Samples were taken at totally twenty sampling points as streams entrance (6 points), settlement coastlines (7 points) and naspeciesal coastlines (7 points). Samples were taken according to Hansen et al. (2000). The invertebrates were collected with a standard sweep net (30 cm width, 1 mm mesh) (Southwood, 1978; Rosenberg, 1997; Hansen et. al, 2000; Yeşilova et al., 2011). Notonectid identification was made by Dmitry A. Gapon (Zoological Institute RAS, Universitetskaya nab., 1, St. Petersburg, Russia).

2. Methods

2.1 Poisson regression

The logarithm of mean of Poisson distribution (μ) is supposed to be a linear function of independent variables (x_i) is,

$$\log(\mu_i) = \left(x_i'\beta\right)$$

Poisson Regression Model can be written as

$$\Pr(y_i / \mu_i, x_i) = \exp(-\mu_i)\mu_i^{y_i}/y_i! \, , \text{yi=0,1,...} \tag{1}$$

In equation 1, y_i denotes dependent variable having Poisson distribution. Likelihood function for PR model is, (Böhning, 1998)

$$LL\left(\beta/y_i, x_i\right) = \sum_{i=1}^{n}\left[y_i x_i'\beta - \exp\left(x_i'\beta\right) - \ln y_i!\right] \tag{2}$$

In equation 2, β are unknown parameters. β can be estimated by maximizing log likelihood function according to ML (Khoshgoftaar et al, 2005; Yau, 2006).

2.2 Negative binomial regression

NB regression model is,

$$\Pr(Y = y_i / x_i) = \frac{\Gamma\left(y_i + \dfrac{1}{\alpha}\right)}{y_i!\Gamma\left(\dfrac{1}{\alpha}\right)}\frac{\left(\alpha\mu_i\right)^{y_i}}{\left(1 + \alpha\mu_i\right)^{y_i + \frac{1}{\alpha}}} \quad \alpha > 0 \tag{3}$$

In equation 3, α is a arbitrary parameter and indicates overdispersion level. Log likelihood function for NB regression model is (Hilbe, 2007; Yau, 2006),

$$LL\left(\beta, \alpha, y\right) = \sum_{i=1}^{n}\begin{bmatrix}\dfrac{1}{\alpha}\log\left(1 + \alpha\mu_i\right) - y_i\log\left(1 + \dfrac{1}{\alpha\mu_i}\right)\\[2mm] + \log\Gamma\left(y_i + \dfrac{1}{\alpha}\right) - \log\Gamma\left(\dfrac{1}{\alpha}\right) - \log y_i!\end{bmatrix}$$

2.3 Zero inflated poisson regression

ZIP regression is [13],

$$\Pr(y_i/x_i) = \begin{cases} \pi_i + (1-\pi_i)\exp(-\mu_i), & y_i = 0 \\ (1-\pi_i)\exp(-\mu_i)\mu_i^{y_i}/y_i!, & y_i > 0 \end{cases} \tag{4}$$

In equation (4), π_i represents the possibility of extra zeros' existence. Log likelihood function for ZIP model is (Yau, 2006),

$$LL = \sum_{i=1}^{n} \begin{pmatrix} Iy_i = 0 \log\left(\pi_i + (1-\pi_i)e^{-\mu_i}\right) \\ + Iy_i > 0 \log\left((1-\pi_i)\dfrac{\mu_i^{y_i} e^{-\mu_i}}{y_i!} \right) \end{pmatrix}$$

$$LL = \sum_{i=1}^{n} \begin{pmatrix} Iy_i = 0 \log\left(\pi_i + (1-\pi_i)e^{-\mu_i}\right) \\ + Iy_i > 0 \begin{pmatrix} \log(1-\pi_i) \\ + y_i \log \mu_i - \mu_i - \log y_i! \end{pmatrix} \end{pmatrix} \tag{5}$$

$I(.)$, given in equation (5) is the indicator function for the specified event. Then μ_i and π_i parameters can be obtained following link functions,

$$\log(\mu) = B\beta \tag{6}$$

and

$$\log\left(\frac{\pi}{1-\pi}\right) = G\gamma \tag{7}$$

In equations 6 and 7, B(nxp) and G(nxq) are covariate matrixes. β and γ are respectively unknown parameter vectors with px1 and qx1 dimension (Yau, 2006).

2.4 Zero inflated negative binomial regression

ZINB regression model is [18],

$$\Pr(y_i/x_i) = \begin{cases} \pi_i + \left(1-\pi_i\right)\left(1+\alpha\mu_i\right)^{-\alpha^{-1}}, & y_i = 0 \\ (1-\pi_i)\dfrac{\Gamma\left(y_i+\dfrac{1}{\alpha}\right)}{y_i!\Gamma\left(\dfrac{1}{\alpha}\right)}\dfrac{(\alpha\mu_i)^{y_i}}{(1+\alpha\mu_i)^{y_i+\frac{1}{\alpha}}}, & y_i > 0 \end{cases} \tag{8}$$

In equation (8), $(\alpha \geq 0)$ indicates an overdispersion parameter. Log likelihood function for ZINB model is (Yau, 2006),

$$
\begin{aligned}
LL(\mu,\alpha,\pi;y) = \sum_i & \left(I_{y_i=0}\log(\pi_i \right. \\
& +\left(1-\pi_i\right)\left(1+\alpha\mu_i\right)^{-\alpha^{-1}} \\
& +I_{y_i>0}\log\left[\left(1-\pi_i\right)\frac{\Gamma\left(y_i+\frac{1}{\alpha}\right)}{y_i!\,\Gamma\left(\frac{1}{\alpha}\right)}\frac{\left(\alpha\mu_i\right)^{y_i}}{\left(1+\alpha\mu_i\right)^{y_i+\frac{1}{\alpha}}}\right]\right)
\end{aligned}
$$

$$
\begin{aligned}
= \sum_i & \left(I_{y_i=0}\log(\pi_i+\left(1-\pi_i\right)\left(1+\alpha\mu_i\right)^{-\alpha^{-1}} \right. \\
& +I_{y_i>0}\left(\log\left(1-\pi_i\right)-\frac{1}{\alpha}\log\left(1+\alpha\mu_i\right) \right. \\
& \quad -y_i\log\left(1+\frac{1}{\alpha\mu_i}\right)+\log\Gamma\left(y_i+\frac{1}{k}\right) \\
& \left.\left. \quad -\log\Gamma\left(\frac{1}{\alpha}\right)-\log y_i! \right) \right)
\end{aligned} \tag{9}
$$

$I(.)$, given in equation 9 is the indicator function for the specified event. The model descripted by Lambert (1992) can be given as,

$$
\log(\mu)=X\beta \quad \text{and} \quad \log\left(\frac{\pi}{1-\pi}\right)=G\gamma
$$

Here, X(nxp) and G(nxq) covariate matrixes, β and γ are respectively unknown parameter vectors with px1 and qx1 dimension. Maximum likelihood estimations for β, α and γ can be obtained by using EM algorithm.

2.5 Negative binomial hurdle model

Log-likelihood for negative binomial hurdle model (Hilbe, 2007),

$$
L=\ln\left(f(0)\right)+\left\{\ln\left[1-f(0)\right]+\ln P(j)\right\} \tag{10}
$$

In equation (10), $f(0)$ indicates the probability of the binary part and $p(j)$ indicates the probability of positive count. The probability of zero for logit model is,

$$
f(0)=P(y=0;x)=1/\left(1+\exp(xb1)\right)
$$

and

$1 - f(0)$ is,

$$\exp(xb1)/(1+\exp(xb1))$$

The log likelihood function for both parts of negative binomial Hurdle Model is,

$$L = \text{cond}\{y == 0, \ln(1/1 - \exp(xb1)),$$

$$\ln(\exp(xb1)/(1+\exp(xb1)))$$

$$+y * \ln(\exp(xb)/(1+\exp(xb)))$$

$$-\ln(1+\exp(xb))/\alpha + \ln\Gamma(y+1/\alpha)$$

$$-\ln\Gamma(y+1) - \ln\Gamma(1/\alpha)$$

$$-\ln(1-(1+\exp(xb))(-1/\alpha))\}$$

2.6 Model selection

Akaiki Information Criteria (AIC) is goodness of criteria used for model selection. AIC,

$$AIC = -2LL + 2r \tag{11}$$

In equations, LL indicates log likelihood, r indicates parameter number and n indicates sample size.

3. Results

In this study, R statistical software program was used. Insect densities were included to the model as dependent variable. Besides years, months, species and station are included as independent variables to the model. The 66 (20.63%) of the 320 dependent variable were zero valued. The distribution of the insect densities was skewed to right because of excess zeros.

Model	AIC
PR	57846.00
ZIP	47791.71
NB	3176.40
ZINB	2819.800
PH	47791.71
NBH	**2803.206**

Table 1. Model selection criteria for PR, NB, ZIP, ZINB, PH and NBH.

In PR analyses, deviance and Pearson Chi-square goodness of statistics higher than one (831.417 and 650.213, respectively). Thus, goodness of statistics represents that there is an overdispersion in insect densities. AIC model selection criteria for the models of PR, NB, ZIP, ZINB, PH, and NBH were given in Table 1. The model with the smallest AIC was NBH regression.

Maximum likelihood (ML) parameter estimations and standard errors for PR were given in Table 2.

	Estimate	Std. Error	z value	$Pr(>|z|)$	e^β
(Intercept)	6.179499	0.054470	113.449	<2e-16 ***	482.992
year	0.118847	0.013069	9.094	<2e-16 ***	1.125244
month	0.175298	0.005066	34.604	<2e-16 ***	1.191246
Station	-0.081353	0.001124	-72.357	<2e-16 ***	0.921917
species	-1.943212	0.018356 -	105.863	<2e-16 ***	0.1432735

*p<0.05, **p<0.01, ***p<0.001

Table 2. Parameter estimations and standard errors for Poisson regression.

ML parameter estimations and standard errors for negative binomial regression were given in Table 2.

	Estimate	Std. Error	z value	$Pr(>|z|)$	e^β
(Intercept)	8.52318	0.99249	8.588	4.16e-16 ***	5029.119
year	-0.15794	0.24824	-0.636	0.525	0.853901
month	-0.08205	0.09168	-0.895	0.372	0.9212259
Station	-0.08031	0.01949	-4.121	4.82e-05 ***	0.9228302
species	-1.92518	0.22452	-8.575	4.56e-16 ***	0.1458495

*p<0.05, **p<0.01, ***p<0.001

Table 3. Parameter estimations and standard errors for negative binomial regression.

ML parameter estimations and standard errors for zero-inflated Poisson regression both count model and logit model were given in Table 4 and Table 5, respectively.

	Estimate	Std. Error	z value	$Pr(>|z|)$	e^β
(Intercept)	6.017745	0.056073	107.32	<2e-16 ***	410.6515
year	0.271101	0.013047	20.78	<2e-16 ***	1.311408
month	0.162333	0.005271	30.80	<2e-16 ***	1.176252
station	-0.046859	0.001122	-41.76	<2e-16 ***	0.954222
species	-2.002676	0.018382	-108.94	<2e-16 ***	0.1349736

*p<0.05, **p<0.01, ***p<0.001

Table 4. Parameter estimations and standard errors for ZIP count model.

	Estimate	Std. Error	z value	Pr(> |z|)	e^β
(Intercept)	-3.92991	1.33906	-2.935	0.00334 **	0.01964544
year	0.50266	0.33705	1.491	0.13587	1.653113
month	0.04405	0.11930	0.369	0.71197	1.045035
station	0.17250	0.02994	5.761	8.36e-09 ***	1.188272
species	-0.44380	0.30013	-1.479	0.13923	0.6415937

*p<0.05, **p<0.01, ***p<0.001

Table 5. Parameter estimations and standard errors for ZIP logit model.

ML parameter estimations and standard errors for zero-inflated negative binomial regression both count model and logit model were given in Table 6 and Table 7, respectively.

	Estimate	Std. Error	z value	Pr(> |z|)	e^β
(Intercept)	9.47226	1.04897	9.030	< 2e-16 ***	12994.22
year	-0.13254	0.20609	-0.643	0.520132	0.8758679
month	-0.16895	0.09356	-1.806	0.070957	0.8445511
station	-0.06233	0.01806	-3.452	0.000557 ***	0.9395728
species	-2.21006	0.20855	-10.597	< 2e-16 ***	0.1096941

*p<0.05, **p<0.01, ***p<0.001

Table 6. Parameter estimations and standard errors for ZINB count model.

	Estimate	Std. Error	z value	Pr(> |z|)	e^β
(Intercept)	-1.21687	4.61145	-0.264	0.791872	0.2961557
year	1.64137	0.88380	1.857	0.063288	5.162237
month	-0.23791	0.22636	-1.051	0.293246	0.7882736
station	0.18398	0.05485	3.354	0.000795 ***	1.201992
species	-3.69139	3.88424	-0.950	0.341934	0.02493732

*p<0.05, **p<0.01, ***p<0.001

Table 7. Parameter estimations and standard errors for ZINB logit model.

ML parameter estimations and standard errors for Poisson hurdle both count model and logit model were given in Table 8 and Table 9, respectively.

	Estimate	Std. Error	z value	Pr(> |z|)	e^β
(Intercept)	6.017745	0.056073	107.32	<2e-16 ***	410.6515
year	0.271101	0.013047	20.78	<2e-16 ***	1.311408
month	0.162333	0.005271	30.80	<2e-16 ***	1.176252
station	-0.046859	0.001122	-41.76	<2e-16 ***	0.954222
species	-2.002676	0.018382	-108.94	<2e-16 ***	0.1349736

*p<0.05, **p<0.01, ***p<0.001

Table 8. Parameter estimations and standard errors for PH count model.

ML parameter estimations and standard errors obtained for the NBH count model was given in Table 8. While stations and species were significant on the insect densities, the effect of years and the effect of months were not significant on the insect densities.

ML parameter estimations and standard errors obtained for the NBH logit model was given in Table 9. The effects months, years and species were not significant on the insect densities. However, the effect of station was significant on the insect densities.

	Estimate	Std. Error	z value	Pr(> |z|)	e^{β}
(Intercept)	3.92991	1.33906	2.935	0.00334 **	50.9024
year	-0.50266	0.33705	-1.491	0.13587	0.6049194
month	-0.04405	0.11930	-0.369	0.71197	0.9569061
station	0.17250	0.02994	-5.761	8.36e-09 ***	1.188272
species	0.44380	0.30013	1.479	0.13923	1.558619

*p<0.05, **p<0.01, ***p<0.001

Table 9. Parameter estimations and standard errors for PH logit model.

ML parameter estimations and standard errors obtained for negative binomial hurdle both count model and logit model were given in Table 10 and Table 11, respectively.

ML parameter estimations and standard errors obtained for the NBH count model was given in Table 10. While stations and species were significant on the insect densities, the effect of years and the effect of months were not significant on the insect densities.

	Estimate	Std. Error	z value	Pr(> |z|)	e^{β}
(Intercept)	9.43372	1.26292	7.470	8.03e-14 ***	12502.95
year	-0.19128	0.24381	-0.785	0.4327	0.8259013
month	-0.17020	0.11124	-1.530	0.1260	0.8434961
station	-0.04587	0.02096	-2.188	0.0287 *	0.9551661
species	-2.33333	0.25071	-9.307	< 2e-16 ***	0.0969723

*p<0.05, **p<0.01, ***p<0.001

Table 10. Parameter estimations and standard errors for NBH count model.

ML parameter estimations and standard errors obtained for the NBH logit model was given in Table 11. The effects months, years and species were not significant on the insect densities. However, the effect of station was significant on the insect densities.

	Estimate	Std. Error	z value	Pr(> |z|)	e^{β}
(Intercept)	3.92991	1.33906	2.935	0.00334 **	50.9024
year	-0.50266	0.33705	-1.491	0.13587	0.6049194
month	-0.04405	0.11930	-0.369	0.71197	0.9569061
station	-0.17250	0.02994	-5.761	8.36e-09 ***	0.8415583
species	0.44380	0.30013	1.479	0.13923	1.558619

*p<0.05, **p<0.01, ***p<0.001

Table 11. Parameter estimations and standard errors for NBH logit model.

Average insect density observed in the year 2005 has shown 17% decrease in reference to the year 2006. Insect densities observed at monthly sampling ranges depending on water temperaspeciese were increased with the rise of temperaspeciese, but specifically after the month of July such intensity was decreased at the rate of 16% ($e^{-0.19128}$ ~ 0.8434961) towards the month of September within the both years. It has been determined that insect intensities observed at different stations have shown differentiation at the rate of 5%. Chironomid larvae which are included in prey of notonectidae fed by different sources of food at aquatic environment have been found at rather lower density in reference to notonectid density. However, it is hard to guess that such decrement has been formed under the impact of notonectidae. Nevertheless notonectidae do not depend on a single host, their sources of food are rather wide range of variety. Small arthropods on the water surface, small crustaceans living in water, larvae of aquatic insects, snails, small fish or larvae of frog are among their preys (Bruce et al., 1990).

4. References

[1] Böhning, D. Zero- Inflated Poisson Models and C. A. MAN. (1998). A Tutorial Collection of Evidence. *Biometrical Journal*, 40(7), 833-843.

[2] Böhning, D., Dietz, E. and Schlattmann, P. (1999). The Zero-Inflated Poisson Model and the Decayed, Missing and Filled Teeth Index in Dental Epidemiology. *Journal of Royal Statistical Society*, A, 162, 195–209.

[3] Bruce, A.M., Pike, E.B. and Fisher, W.J. (1990). A review of treatment processes to meet the EC Sludge Directive. *J. Inst. Wat. Environ. Management*, 4, 1-13.

[4] Cameron, A.C. and Trivedi, P.K. (1998). Regression Analysis of Count Data. New York, Cambridge University Pres.

[5] Cheung, Y.B. (2002). Zero-Inflated Models for Regression Analysis of Count Data. *A Study of Growth and Development. Statistics in Medicine*, 21, 1461-1469.

[6] Consul, P. C. Generalized Poisson distributions, Volume 99 of Statistics: Textbooks and Monographs. New York: Marcel Dekker Inc. Properties and applications 1989.p.1-20.

[7] Consul, P. C. and F. Famoye. (1992). Generalized Poisson regression model. Comm. Statist. Theory Methods. 21(1), 89–109.

[8] Cox, R. (1983). Some Remarks on Overdispersion. *Biometrika*, 70, 269-274.

[9] Czado, C., Erhardt, V., Min, A., Wagner, S. (2007). Dispersion and zero-inflation level applied to patent outsourcing rates Zero-inflated generalized Poisson models with regression effects on the mean, Statistical Modelling. 7(2) : 125-153

[10] Hansen, J., Mki. Sato, R. Ruedy, A. Lacis & V. Oinas. (2000). Global warming in the twenty-first censpeciesy: An alternative scenario,

[11] Hilbe, J.M. (2007). Negative Binomial Regression. Cambridge, UK.

[12] Jansakul, N. (2005). Fitting a Zero-inflated Negative Binomial Model via R. In Proceedings 20th International Workshop on Statistical Modelling. Sidney, Australia, 277-284.

[13] Khoshgoftaar, T.M., Gao, K. and Szabo, R.M. (2005). Comparing Software Fault Predictions of Pure and Zero- inflated Poisson Regression Models. *International Journal of Systems Science*, 36(11), 705-715.

[14] Lambert, D. (1992). Zero-Inflated Poisson Regression, with an Application to Defects in Manaufacspeciesin. *Technometrics*, 34(1), 1-13.

[15] Lee, A.H., Wang, K. and Yau, K. K. W. (2001). Analysis of Zero-Inflated Poisson Data Incorporating Extent of Exposure. *Biometrical Journal*, 43(8), 963-975.

[16] Long, J.S. and Freese, J. (2006). Regression Models for Categorical Depentent Variable Using Stata. A Stata Pres Publication, USA.

[17] McCullagh, P. and Nelder, J. A. (1989). Generalized Linear Models. Second Edition, Chapmann and Hall, London.

[18] Ridout, M., Hinde, J. and Demetrio, C.G.B. (2001). A Score Test for a Zero-Inflated Poisson Regression Model Against Zero-Inflated Negative Binomial Alteratves. *Biometrics*, 57, 219-233.

[19] Rose, C.E, Martin, S.W., Wannemuehler, K.A. and Plikaytis, B.D. (2006). On the of Zero-inflated and Hurdle Models for Medelling Vaccine Adverse event Count Data. *Journal of Biopharmaceutical Statistics*, 16, 463-481.

[20] Rosenberg, D. M., I. J. Davies, D.G. Cobb & A.P. Wiens. (1997). Protocols For Measuring Biodiversity: Benthic Macroinvertebrates in Fresh Waters, http://www.emanrese.ca/eman/ecotools/protocols/freshwater/benthics/intro.ht ml, (Data accessed: 10.10.

[21] Southwood, T.R.E. (1978). Ecological Methods, with Particular Reverence to the Study of Insect Populations, 2nd ed., Chapman and Hall, London and New York, 524 pp.

[22] Yau, K.K.W. and Lee, A.H. (2001). Zero-Inflated Poisson Regression with Random Effects to Evaluate an Occupational Injury Prevention Programme. *Statistics in Medicine*, 20, 2907-2920.

[23] Yau, Z. (2006). Score Tests for Generalization and Zore-Inflation in Count Data Modeling. Unpublished Ph. D. Dissertation, University of South Caroline, Columbia.

[24] Yeşilova, A., Kaydan, B. and Kaya, Y. (2010). Modelling Insect-Egg Data with Excess Zeros using Zero-inflated Regression Models. *Hacettepe Journal of Mathematics and Statistics*. 39(2),273-282.

[25] Yeşilova, A. Y., Kaya, B., Kaki, İ.Kasap (2010)Analysis of Plant Protection Studies with Excess Zeros Using Zero-Inflated and Negative Bi Binomial Hurdle Models *G.U. Journal of Science* .

[26] Yeşilova, A., Özgökçe, M. S., Atlıhan, R., Karaca, İ., Özgökçe, F., Yıldız, Ş. and Kaydan, B. and Kaya, Y. (2011). Investigation of the effects of physico-chemical environmental conditions on population fluctuations of *Notonecta viridis* Delcourt, 1909 (Hemiptera: Notonectidae) in Van Lake by using zero-inflated generalized Poisson regression. *Turkish Journal of Entomology*. 35(2).

Permissions

The contributors of this book come from diverse backgrounds, making this book a truly international effort. This book will bring forth new frontiers with its revolutionizing research information and detailed analysis of the nascent developments around the world.

We would like to thank Sonia Soloneski and Marcelo Larramendy, for lending their expertise to make the book truly unique. They have played a crucial role in the development of this book. Without their invaluable contribution this book wouldn't have been possible. They have made vital efforts to compile up to date information on the varied aspects of this subject to make this book a valuable addition to the collection of many professionals and students.

This book was conceptualized with the vision of imparting up-to-date information and advanced data in this field. To ensure the same, a matchless editorial board was set up. Every individual on the board went through rigorous rounds of assessment to prove their worth. After which they invested a large part of their time researching and compiling the most relevant data for our readers. Conferences and sessions were held from time to time between the editorial board and the contributing authors to present the data in the most comprehensible form. The editorial team has worked tirelessly to provide valuable and valid information to help people across the globe.

Every chapter published in this book has been scrutinized by our experts. Their significance has been extensively debated. The topics covered herein carry significant findings which will fuel the growth of the discipline. They may even be implemented as practical applications or may be referred to as a beginning point for another development. Chapters in this book were first published by InTech; hereby published with permission under the Creative Commons Attribution License or equivalent.

The editorial board has been involved in producing this book since its inception. They have spent rigorous hours researching and exploring the diverse topics which have resulted in the successful publishing of this book. They have passed on their knowledge of decades through this book. To expedite this challenging task, the publisher supported the team at every step. A small team of assistant editors was also appointed to further simplify the editing procedure and attain best results for the readers.

Our editorial team has been hand-picked from every corner of the world. Their multi-ethnicity adds dynamic inputs to the discussions which result in innovative outcomes. These outcomes are then further discussed with the researchers and contributors who give their valuable feedback and opinion regarding the same. The feedback is then collaborated with the researches and they are edited in a comprehensive manner to aid the understanding of the subject.

Apart from the editorial board, the designing team has also invested a significant amount of their time in understanding the subject and creating the most relevant covers. They scrutinized every image to scout for the most suitable representation of the subject and create an appropriate cover for the book.

The publishing team has been involved in this book since its early stages. They were actively engaged in every process, be it collecting the data, connecting with the contributors or procuring relevant information. The team has been an ardent support to the editorial, designing and production team. Their endless efforts to recruit the best for this project, has resulted in the accomplishment of this book. They are a veteran in the field of academics and their pool of knowledge is as vast as their experience in printing. Their expertise and guidance has proved useful at every step. Their uncompromising quality standards have made this book an exceptional effort. Their encouragement from time to time has been an inspiration for everyone.

The publisher and the editorial board hope that this book will prove to be a valuable piece of knowledge for researchers, students, practitioners and scholars across the globe.

List of Contributors

Andréia da Silva Almeida, Francisco Amaral Villela, Maria Ângela André Tillmann and Geri Eduardo Meneghello
Ciência e Tecnologia de Sementes, University Federal de Pelotas, Brazil

Joseph J. Doccola and Peter M. Wild
Arborjet, Inc. Woburn, MA, USA

Ademir Jesus Martins and Denise Valle
Fundação Oswaldo Cruz/ Instituto Oswaldo Cruz/ Laboratório de Fisiologia e Controle de Artrópodes Vetores, Brazil

Anđelka V. Tomašević and Slavica M. Gašić
Institute of Pesticides and Environmental Protection, Belgrade-Zemun, Serbia

Mahdi Banaee
Department of Aquaculture, Natural Resource and Environmental Faculty, Behbahan University of Technology, Iran

Yvonne Rosenberg, Xiaoming Jiang, Lingjun Mao, Segundo Hernandez Abanto and Keunmyoung Lee
PlantVax Inc., USA

Juan D. Claus, Verónica V. Gioria and Gabriela A. Micheloud
Laboratory of Virology, Facultad de Bioquímica y Ciencias Biológicas, Universidad Nacional del Litoral, República Argentina

Gabriel Visnovsky
Chemical and Process Engineering, University of Canterbury, New Zealand

Tom A. Royer and Eric J. Rebek
Oklahoma State University, United States of America

Steven D. Frank
North Carolina State University, United States of America

Carlos E. Bográn
Texas A&M University, United States of America

Beverly A. Wiltz
USDA-ARS, New Orleans, LA, USA

Manuel C. Lagunas-Solar
University of California, Davis, RF Biocidics Inc., Vacaville, California, USA

Amritha G. Kulkarni and B. B. Kaliwal
P.G. Department of Studies in Biotechnology and Microbiology, Karnatak University, India

Suleiman F. Ambali, Joseph O. Ayo, Muftau Shittu and Mohammed U. Kawu
Department of Veterinary Physiology and Pharmacology, Ahmadu Bello University, Zaria, Nigeria

Suleiman O. Salami
Department of Veterinary Anatomy, Ahmadu Bello University, Zaria, Nigeria

Abdullah Yeşilova
Yuzuncu Yil University, Faculty of Agriculture, Biometry & Genetic Unit, Van, Turkey

M. Salih Özgökçe
Yuzuncu Yil University, Faculty of Agriculture, Plant Protection Department, Van, Turkey

Yılmaz Kaya
Siirt University, Faculty of architecture and engineering, Computer Engineering Department, Siirt, Turkey

Printed in the USA
CPSIA information can be obtained
at www.ICGtesting.com
JSHW011450221024
72173JS00005B/1020